Interdisciplinary Concepts in Cardiovascular Health

Ichiro Wakabayashi • Klaus Groschner
Editors

Interdisciplinary Concepts in Cardiovascular Health

Volume I: Primary Risk Factors

Springer

Editors
Ichiro Wakabayashi
Department of Environmental
and Preventive Medicine
Hyogo College of Medicine
Hyogo
Japan

Klaus Groschner
Center for Physiological Medicine
Institute of Biophysics
Medical University of Graz
Graz
Austria

ISBN 978-3-7091-1333-2 ISBN 978-3-7091-1334-9 (eBook)
DOI 10.1007/978-3-7091-1334-9
Springer Wien Heidelberg New York Dordrecht London

Library of Congress Control Number: 2013946028

Printed on acid-free paper

Springer is part of Springer Science+Business Media (www.springer.com)

Preface

According to the recent WHO statistics, cardiovascular diseases are the leading cause of death globally with an estimated 17.3 million people dying from cardiovascular disease in 2008, representing 30 % of all global deaths. Of these deaths, an estimated 7.3 and 6.2 million were attributed to coronary heart disease and stroke, respectively. Undisputedly, cardiovascular diseases are predicted to remain as the leading cause of death. Over the past decades, we have gained substantial knowledge about the molecular and cellular mechanisms underlying disease initiation and progression and in parallel have identified a plethora of potential risk factors. Moreover, epidemiology has recently moved closer to molecular pathology, enabling a first glimpse to be captured of essential molecular mechanisms that determine cardiovascular risk within a certain population. This book is an attempt to provide an overview of the recent development in this field.

There is common consent that pathogenesis of cardiovascular disorders is mainly based on atherosclerosis, which progresses with age. Thus, retardation of atherosclerosis progression is considered the most effective strategy to prevent cardiovascular morbidity and mortality. A wide range of risk factors are involved in the pathogenesis of atherosclerosis. Lifestyle-related factors including diet, nutrition, physical activity, habitual smoking and alcohol consumption, socioeconomic factors, and psychological stress as modifiable factors, in addition to age, gender, race/ethnicity, and genetic polymorphisms as destined factors, are recognized to determine the risk for cardiovascular events. These "primary risk factors" typically promote the genesis of disorders that may be understood as "secondary risk factors" such as obesity, hypertension, diabetes, dyslipidemia, hyperuricemia, and metabolic syndrome. Thromboatherosclerotic alterations in arterial wall structure result from a combination of secondary risk factors and lead to terminal cardiovascular events such as ischemic heart disease and stroke. Our book is accordingly structured into three sections, published as separate volume that give detailed information about our current understanding of primary and secondary risk factors, as well as on terminal cardiovascular events. To sustain cardiovascular health, prevention or compensation of primary risk determinants, as well as early diagnosis and treatment of secondary risk factors, are undoubtedly key strategies.

Accumulated groundbreaking insights into cellular mechanisms involved in the pathogenesis of cardiovascular aging and disease include the identification of vasoactive prostanoids, lipoprotein receptors, and nitric oxide. The general scientific value of these basic findings, as well as their wider impact on our society, is well documented by the Nobel Prizes given for each of these discoveries. This book aims to introduce established principles of both cardiovascular epidemiology and molecular pathophysiology and makes a unique attempt to bridge the gap between epidemiological knowledge and current molecular concepts in cellular pathophysiology. The authors spotlight future avenues for research in basic pathophysiology, as well as in cardiovascular therapy and prevention. The comprehensive overview of cardiovascular pathophysiology provided with this book is expected to help readers to address questions on unresolved pathomechanisms and/or to interpret novel epidemiological findings on cardiovascular disorders.

Finally, the editors would like to express their sincere appreciation to all contributors for their dedicated collaboration in this project. We wish to additionally thank Ms. Karin Osibow for her competent and patient support in editing this book.

We hope our book will enable readers to connect epidemiological knowledge with principles of molecular pathophysiology, thereby promoting the development of new strategies for sustaining cardiovascular health.

Nishinomiya, Japan I. Wakabayashi
Graz, Austria K. Groschner

Contents

Volume I: Primary Risk Factors

Volume II: Secondary Risk Factors

Volume III: Cardiovascular Events

Abbreviations

$^{\bullet}$NO	NO radical
$^{\bullet}$OH	Hydroxyl radical
AA	Arachidonic acid
ACE	Angiotensin-converting enzyme
ACS	Acute coronary syndrome
ADH	Alcohol dehydrogenase
ADL	Activities of daily living
ADP	Adenosine diphosphate
AF	Atrial fibrillation
AFT-1	Activating transcription factor-1
AGE	Advanced glycation end products
AGT	Angiotensinogen
AHA	American Heart Association
Akt/PKB	Akt kinase/protein kinase B
ALA	α-Linolenic acid
ALDH	Aldehyde dehydrogenases
AMI	Acute myocardial infarction
AMPKα-2	Adenosine monophosphate-activated protein kinase alpha-2
APC	Activated protein C
ASO	Arteriosclerosis obliterans
$AT_{1/2}R$	Angiotensin II receptor type 1, 2
ATF-1, ATF-2, ATF-3	Activating transcription factor-1, activating transcription factor-2, activating transcription factor-3
ATP	Adenosine triphosphate
AUC	Area under the curve
BAEC	Bovine aortic endothelial cell
BAG3	BAG family molecular chaperone regulator 3
BH_4	Tetrahydrobiopterin
BMI	Body mass index
BNP	Brain/B-type natriuretic peptide
BP	Blood pressure
CAC	Coronary artery calcium

CAD	Coronary artery disease
CBT	Cognitive behavioral therapy
CCE	Capacitative Ca^{2+} entry
CETP	Cholesteryl ester exchange/transfer protein
cGMP	Cyclic guanosine monophosphate
CHD	Coronary heart disease
CHF	Chronic heart failure
ClO^-	Hypochlorite
COX-1, COX-2	Cyclooxygenase-1, cyclooxygenase-2
CR	Caloric restriction
CRE	C-responsive element
CREB	cAMP-responsive element-binding protein
CRP	C-reactive protein
CS	Citrate synthase
CT	Computed tomography
Cu,Zn-SOD	Cu,Zn-superoxide dismutase
CV	Cardiovascular
CVA	Cerebrovascular attack
CVD	Cardiovascular disease
DBP	Diastolic blood pressure
DCM	Dilative cardiomyopathy
DF	Dietary fibers
DHA	Docosahexaenoic acid
DHEA	Dehydroepiandrosterone
Drp1	Dynamin-related protein
DVT	Deep venous thrombosis
E1	Estrone
E2	Estradiol
E3	Estriol
EDHF	Endothelium-derived hyperpolarizing factor
EGF	Epidermal growth factor
eNOS	Endothelial nitric oxide synthase
EPA	Eicosapentaenoic acid
ER-α, ER-β	Estrogen receptor-alpha, estrogen receptor-beta
ET-1	Endothelin 1
FGF-2	Fibroblast growth factor-2
FH	Familial hypercholesterolemia
FMD	Flow-mediated dilation
FoxO1	Forkhead box O1
FTO	Fat mass and obesity associated
GDH	Glutamate dehydrogenase
GI	Glycemic index
GLP-1	Glucagon-like peptide-1
GLUT4	Glucose transporter type 4

GPCR	G-protein-coupled receptor
GPx	Glutathione peroxidase
GRO	Growth-related oncogene
GRS	Genetic risk score
GWAS	Genome-wide association study
GxE	Gene–environment interactions
GxG	Gene–gene interactions
H_2O_2	Hydrogen peroxide
HCM	Hypertrophic cardiomyopathy
HDL	High-density lipoprotein
HDL-C	High-density lipoprotein cholesterol
HERG	Human ether-a-go-go-related gene
HERS	Heart and Estrogen Replacement Study
HMG-CoA	3-hydroxy-methylglutaryl-coenzyme
HPA	Hypothalamic–pituitary–adrenal
HR	Hazard ratio
HRT	Hormone replacement therapy
HUVEC	Human umbilical vein endothelial cell
ICAM-1	Intercellular adhesion molecule 1
ICE	Ischemic cerebrovascular events
IDL	Intermediate-density lipoprotein
IGT	Impaired glucose tolerance
IHD	Ischemic heart disease
IL-1, IL-4, IL-6, IL-8, -18	Interleukin-1, interleukin-4, interleukin-6, interleukin-8, interleukin-18
iNOS	Inducible nitric oxide synthase
IS	Ischemic stroke
JNK	c-Jun N-terminal kinase
LA	Linoleic acid
LDB3	LIM domain binding 3
LDL	Low-density lipoprotein
LDL-C	Low-density lipoprotein cholesterol
LDL-R	Low-density lipoprotein receptor
L-NAME	Nitro-*L*-arginine methyl ester
LOX-1	Lectin-like oxidized LDL receptor-1
LPL	Lipoprotein lipase
LQT	Long QT
LRP6	Lipoprotein receptor related protein 6
LTA4, LTB4, LTC4, LTD4, LTE4	Leukotriene A4, leukotriene B4, leukotriene C4, leukotriene D4, leukotriene E4
LXRα	Liver X receptor alpha
MAPK	Mitogen-activated protein kinase
MCP-1	Monocyte chemoattractant protein 1
M-CSF	Monocyte colony-stimulating factor
MEF-2A	Myocyte enhancer factor-2A
MET	Metabolic equivalent

MHLW	Japan Ministry of Health, Labour and Welfare
MI	Myocardial infarction
MLC	Myosin light chain
MLP	Muscle LIM protein
mt	Mitochondrial
MUFA	Monosaturated fatty acids
NAD^+	Nicotinamide adenine dinucleotide, oxidized form
NADP	Nicotin amide adenine dinucleotide phosphate
NADPH	Nicotin amide adenine dinucleotide phosphate, reduced form
NCoR	Nuclear receptor corepressor
NEFA	Nonesterified fatty acid
NFκB	Nuclear factor-kappa B
NO	Nitric oxide
NO_2	Nitrogen dioxide radical
NOS	Nitric oxide synthase
NOx	Nicotinamide adenine dinucleotide phosphate oxidase
NRF-1	Nuclear respiratory factor 1
OAG	1-oleoyl-2-acetyl-sn-glycerol
OGA	β-*N*-acetylglucosaminidase
OGT	β-*N*-acetylglucosaminyltransferase
$ONOO^-$	Peroxynitrite
OXPHOS	Oxidative phosphorylation
p53	Tumor protein 53
PAD	Peripheral arterial disease
PAF	Platelet-activating factor
PAI-1	Plasminogen activator inhibitor 1
PARP	Poly (ADP-ribose) polymerase
PDGF	Platelet-derived growth factor
PDGFR	Platelet-derived growth factor receptor
PGC-1α	Peroxisome proliferator-activated receptor coactivator-1 alpha
PGI_2	Prostaglandin I_2, prostacyclin
PI3-K	Phosphoinositide 3-kinase
PITX2	Paired-like homeodomain transcription factor 2
PKC	Protein kinase C
PLC	Phospholipase C
PLN	Phospholamban
PPAR-γ	Peroxisome proliferator-activated receptor gamma
PTP1B	Protein tyrosine phosphatase 1B
PUFA	Polyunsaturated fatty acid
RAGE	Receptors of AGEs
RAP	LDL receptor associated protein

RAS	Renin–angiotensin system
ROCK	Rho-associated coil-forming protein kinase
ROS	Reactive oxygen species
SA-β-gal	Senescence-associated beta-galactosidase
SAPK	Stress-activated protein kinase
SBP	Systolic blood pressure
SCD	Sudden cardiac death
SCFA	Short chain fatty acids
SERM	Selective estrogen receptor modulators
SES	Socioeconomic status
SFA	Saturated fatty acids
SHS	Secondhand smoke
sICAM-1	Soluble ICAM-1
SMRT	Silencing mediator of retinoid and thyroid hormone receptor
SNP	Single nucleotide polymorphism
SOCE	Store-operated Ca^{2+} entry
SOCS3	Suppressor of cytokine signaling 3
SOD	Superoxide dismutase
SORT1	Sortilin 1
sTNFR-2	Soluble tumor necrosis factor receptor 2
sVCAM-1	Soluble VCAM-1
T2DM	Type 2 diabetes mellitus
TC	Total cholesterol
TERT	Telomerase
TFA	Trans-fatty acid
TFAM	Mitochondrial transcription factor A
TG	Triacylglyceride
TGF-β	Transforming growth factor-beta
TIA	Transient ischemic attack
TNF-α	Tumor necrosis factor alpha
tPA	Tissue plasminogen activator
TRPC	Canonical transient receptor potential channel
TSP-1	Thrombospondin-1
TXA_2	Thromboxane A_2
UCP-1, UCP-2, UCP-3	Uncoupling protein-1, uncoupling protein-2, uncoupling protein-3
uPA	Urine plasminogen activator
VCAM-1	Vascular cell adhesion molecule-1
VEGF	Vascular endothelial growth factor
VF	Ventricular fibrillation
VLDL	Very low-density lipoprotein
VTE	Venous thromboembolism
vWF	von Willebrand factor
WHI	Women's Health Initiative
WHO	World Health Organization

Basic Principles of Molecular Pathophysiology and Etiology of Cardiovascular Disorders

1

Michael Poteser, Klaus Groschner, and Ichiro Wakabayashi

Abstract

Molecular etiology of cardiovascular diseases is based on genetic predisposition and environmental factors. While in monogenic forms with Mendelian traits, gene defects are the dominant origin of the disease, the combination and complex interdependence of environmental and genetic factors determine an individual's risk for common forms of cardiovascular disorders. The molecular pathophysiology of Mendelian traits has been well understood, and a wide range of molecular dysfunctions in key proteins have been identified by examining the pathology of rare genetic defects. Pathomechanisms triggered by environmental factors, on the molecular level, are considered to involve crucial posttranslational modification of signaling molecules with a significant dependence on a genetic susceptibility that is determined by common gene variation. Here, we provide a brief introduction to this concept, focusing on selected paradigms of disease-relevant cellular mechanisms such as modification of proteins by reactive oxygen species, phosphorylation, and glycation. Available information bridging molecular knowledge with epidemiological concepts is highlighted.

Keywords

Genetic predisposition • Signaling molecules • Oxidative stress • Protein phosphorylation • Protein glycation

M. Poteser, PhD (✉) • K. Groschner, PhD
Center for Physiological Medicine,
Institute of Biophysics, Medical University of Graz,
Harrachgasse 21/4, Graz A-8010, Austria
e-mail: michael.poteser@medunigraz.at; klaus.groschner@medungraz.at

I. Wakabayashi, MD, PhD
Department of Environmental and Preventive Medicine, Hyogo College of Medicine,
Mukogawa-cho 1-1, Nishinomiya, Hyogo 663-8501, Japan
e-mail: wakabaya@hyo-med.ac.jp

I. Wakabayashi, K. Groschner (eds.), *Interdisciplinary Concepts in Cardiovascular Health*,
DOI 10.1007/978-3-7091-1334-9_1, © Springer-Verlag Wien 2013

1.1 Introduction

Due to the large numbers of different types of organs, tissues, and cells involved, etiologies of cardiovascular diseases (CVDs) cover an accordingly wide spectrum of molecular pathways and mechanisms. A general aspect that applies to all CVDs is provided by the fact that the cause of a disease can be either a genetic or environmental disturbance or, as seen in the vast majority of cases, an interaction of these factors. Consequently, understanding the causal contribution of genetic and environmental risk factors to the pathogenesis is crucial for diagnosis, prognosis, treatment, and public health management. Using the gene-environment context as a basic classification criterion, it appears justified to separate Mendelian/monogenetic diseases, typically showing relatively low prevalence, from multifactorial or polygenic cardiovascular diseases, representing common forms that are often based on balanced interactions between genetic and environmental factors (Fig. 1.1). Nonetheless, secondary modifications in many cases disguise the causal origin of monogenic diseases and generate more complex molecular pathways of pathogenesis.

1.1.1 "Mendelian" and Monogenic Diseases

If the occurrence of a disease in a familiar anamnesis follows a clear Mendelian autosomal-dominant, autosomal-recessive, or gonosomal (X, Y) pattern, it is highly unlikely that multiple independent gene defects cause the disorder in a linked way. Therefore, these diseases are also termed monogenic diseases, as they can be convincingly traced back to single genes. This simple concept of monogenetic diseases has in the last decade been challenged by reports of incomplete penetrance as well as the identification of triallelic forms of inheritance (Eichers et al. 2004), where mutations in a small number of genes interact genetically to manifest a certain phenotype. A monogenic pathophysiology, however, is generally characterized by a relatively clear causal pathway that is a sufficient and, in most cases, inevitable cause of disease. It has been observed in clinical practice that the impact of at least some of these diseases may, nevertheless, be alleviated or worsened by modifications, often age related, of the affected pathway (Miyamoto et al. 2012). Accordingly, environmental factors like smoking or physical activity may have a significant impact on the prognosis of monogenic diseases. Considering the number of possible modifications, the complexity of involved molecular pathways that have been identified in some monogenetic diseases may even approximate those of typical multifactorial disorders.

The prevalence of monogenetic diseases is typically very low and estimated to range from 0.005 % in the rarest forms up to 0.1–0.2 %, as observed in hypertrophic cardiomyopathy (HCM), the most common cardiac disease with Mendelian inheritance. More than 50 % of all cases of HCM occur familially, following an autosomal-dominant pattern, and 22 % of the first-degree relatives in these families were shown to be affected (Greaves et al. 1987). Despite the relatively clear causal origin,

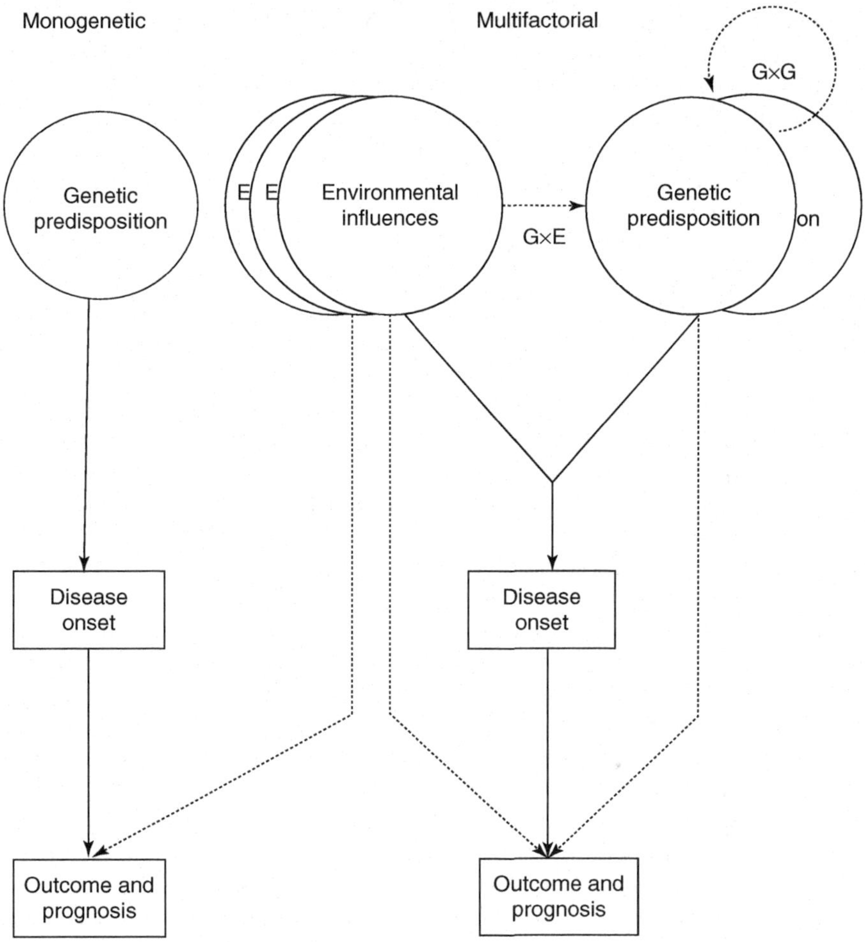

Fig. 1.1 Schematic illustration of monogenetic and multifactorial cardiovascular diseases. In monogenetic diseases (*left*), genetic predisposition is responsible for disease onset induced by a relatively clear molecular pathway. Outcome and prognosis may then be alleviated or worsened by complex modifications of environmental factors. Multifactorial cardiovascular diseases are triggered by an imbalance between environmental (G×E) and genetic factors (G×G). The molecular pathways of the diseases may then again be modulated by heritable or environmental modifications of elements within the underlying molecular pathways

however, the global view on all possible forms of HCM reveals a genetically highly complex situation of pathogenesis, because mutations in several different molecular targets are linked to HCM. While more than 80 % of the mutations involved affect the myosin heavy chain gene (MYH7), troponin T (TNNT2), and myosin-binding protein C (MYBPC3), the HCM phenotype, as characterized by a left ventricular myocardial hypertrophy, may also be caused by mutations affecting a number of other sarcomeric proteins, including troponin I (TNNTI2), α-tropomyosin (TPM1), essential and regulatory myosin light chain (MYL3, MYL2), titin (TTN), β-myosin heavy chain (MYH6), and muscle LIM protein (MLP). These mutations of other sarcomeric proteins have been demonstrated to account for 30–60 % of the sporadic forms of HCM (Richard et al. 2003). In addition to about 300 identified mutations of sarcomeric proteins, mutations of Z-disc proteins have also been found to be involved in HCM. Here LIM domain binding 3 (LDB3) and muscle LIM protein (CSRP3) as well as vinculin and α-actinin 2 were demonstrated to be affected (Gehmlich et al. 2004; Vasile et al. 2006; Wang et al. 2010a). HCM is, therefore, considered a genetically heterogenic disease, as also indicated by the low likeliness of two nonrelative HCM patients to carry the same mutations. An estimated 3 % of affected individuals carry more than one mutation (Van Driest et al. 2004).

Despite the diversity of genetic backgrounds, the resulting molecular pathways leading to HCM involve different elements of a common functional unit in the myocyte. The myosin heavy chain protein forms the thick filament of the sarcomere, while the thin filament is comprised of actin, α-tropomyosin, and the troponins. These structures are stabilized by myosin-binding protein C and titin. A calcium-triggered signal then leads to an ATP-dependent conformational change in the myosin heavy chain, resulting in filament sliding and muscle contraction. Mutations in sarcomeric genes are the sufficient and the primary cause of HCM as demonstrated by studies using transgenic animals in genetically engineered mice and rabbits expressing human HCM mutations that induced the development of histopathological phenotypes comparable to human HCM (Marian et al. 1999; Oberst et al. 1998; Vikstrom et al. 1995). Analysis of protein structure/function relationships of the mutated human β-myosin heavy chain and the broad local distribution of HCM mutations across the protein suggests that the HCM mutations are not necessarily affecting the basic functions of the protein, like ATP hydrolysis or actin binding, but may relate to interaction with other proteins of the sarcomere (Rayment et al. 1995). For some affected sarcomeric proteins, clear mutation-induced functional consequences have been demonstrated. This includes disturbed intrinsic ATPase activity of α-cardiac muscle actin (Muller et al. 2012) or alterations in the concerted conformational change of the α-tropomyosin/actin complex (Rysev et al. 2012). The common result of most sarcomeric mutations involved in HCM appears to be a change in myocyte contractility.

The chain of events leading to left ventricular hypertrophy is still unclear. One proposed mechanism is decreased contractility of cardiac myocytes, leading to cellular stress and production of trophic factors (Marian 2000), which ultimately causes transcriptional reprogramming and hypertrophy. Another hypothesis suggests that abnormal force development of the myocytes triggers a reflectory increase in sympathetic activity, possibly due to changes in diastolic filling (Ostman-Smith and Wettrell 2000). Because both of these hypotheses are supported by several studies (Fineschi et al. 2005; Lim et al. 2001; Pace et al. 2004; Patel et al. 2000), a concerted contribution by these mechanisms might be considered. The resulting left ventricular hypertrophy and fibrosis are typically associated with enhanced systolic function and impaired diastolic function. Depending on the genetic base, the clinical spectrum of HCM is wide ranging, spreading from relatively benign forms to heart failure and sudden cardiac death.

Recent progress in genetics, including the completion of the human genome project, has enabled the identification of a number of cardiovascular target proteins and helped to identify monogenic forms of common disorders. One example is coronary heart disease, a disorder typically displaying multifactorial causality. While common forms are mainly influenced by classical risk factors like dyslipidemia or diabetes mellitus, monogenetic forms have been traced back to LRP6, a protein involved in the Wnt-signaling pathway (Mani et al. 2007), and a less clear causality has been demonstrated for MEF2A, a transcription factor located in the nucleus of proliferating smooth muscle cells (Lieb et al. 2008; Weng et al. 2005). Monogenetic forms were also found in arterial hypertension and atrial fibrillation (see Chap. 4). In arterial hypertension, mutations of the mineralocorticoid receptor cause a partial constitutive activity of the receptor, leading to hypertensive crisis in female members of affected family trees. Inherited mutations in renal ion transporter systems can also cause hypertension, as can be seen in a form of hyperaldosteronism, where a crossing-over induced hybridization of aldosterone synthase and 11β-hydroxylase forms a protein that is no longer under control of by angiotensin II but activated by adrenocorticotropic hormone (ACTH) (Lifton et al. 1992).

1.1.2 Multifactorial Cardiovascular Disorders

Most CVDs can be triggered by genetic as well as environmental factors, which are typically able to induce a disease in patients with genetic predisposition, i.e., who are carrying common variants in susceptibility genes. Environmental factors are characterized, in principle, as avoidable external modulators of cellular pathways and comprise a wide range of physical, chemical, biological, dietary, and psychological factors, including radiation, toxins, pathogens, malnutrition, physical strain, and mental stress.

Most multifactorial diseases are based on complex molecular pathways, making it difficult to distinguish between causation and correlation within the range of disease-promoting factors. The situation is even more complex, because recent scientific advances indicate that genetic and environmental factors are often linked and that different pathological gene expression patterns may interact with other genes as well as environmental factors in a synergistic or antagonistic way. Young scientific disciplines like environmental genomics and epigenetics are positioned to shed light onto these topics using multidisciplinary approaches.

Recently, a database for gene-environment interactions (G×E) of the human genome compiled from literature was presented (Lee et al. 2011), showing 554 significant interactions that were associated with obesity, metabolic disorder, and CVDs. An example of such a gene-environment interaction is provided by a study (Romanoski et al. 2010) demonstrating a significant interaction between genotypes of explanted human endothelial cells and their exposure to oxidized phospholipids, a common scenario found in atherosclerosis. About one third of 59 of the most regulated gene transcripts showed evidence for gene-environment interactions, and a direct local interaction was observed for FGD6, a protein that is hypothesized to represent a Rho family guanine nucleotide exchange factor (RhoGEF) and catalyzing GDP/GTP exchange for small G proteins involved in cytoskeletal pathways.

Another well-established example demonstrating modification of gene expression pattern by environmental factors is found in nonphysiological hypertrophy of the heart. At birth, metabolism of the heart switches from the hypoxic "intrauterine mode" to the oxidation of fatty acids, and a mature set of enzyme proteins is expressed. The fetal metabolic program, however, can become reactivated as soon as the heart is stressed by hypoxia, ischemia, atrophy, or hypertrophy (Kuwahara et al. 2012). This adaptive process of re-expression of fetal genes is seen as a hallmark of pathophysiological hypertrophy and demonstrates the direct effect of environmental conditions on genetic expression (Kehat and Molkentin 2010).

Interactions among genes are defined as phenotypic effects of two or more coincident gene modifications that differ from those observed for independent contributions of each gene. Because of their relatively high abundance within populations, many studies investigating these synergistic effects among genes refer to polymorphisms. Polymorphisms are described as variance among genes with a relatively high frequency of occurrence >1 % that cannot be explained by accident mutations but are rather due to general variability induced by allelic drift, i.e., the frequency of occurrence of an allele, and that are stabilized by evolutionary selection. A large number of polymorphisms have been identified as risk factors in various CVDs (see Chap. 4).

Interactions between different genes (G×G) relevant for cardiovascular pathophysiology have been described in several studies. One example is given by a population-based study, indicating that genes involved in the renin-angiotensin, bradykinin, and fibrinolytic systems synergistically modulate plasma concentrations of plasminogen activator and plasminogen activator inhibitor type 1 (Bentley et al. 2010). Here it is demonstrated that the coincidence of polymorphisms may generate a much higher CVD risk than the simple arithmetic sum of the relative risk factors, as derived from single polymorphisms.

Gene-gene interactions may even link diseases of the cardiovascular system with other disorders (Bondy 2007; Ordovas 2007) and help to explain epidemiological findings of significant comorbidity.

1.2 Common Molecular Pathways in Cardiovascular Disease

In multifactorial CVDs, gene-gene interactions and gene-environment interactions are linked to a plethora of molecular pathways that result in specific cellular dysfunctions, morphological changes, or cell death. The number of pathways that have been identified so far correlates with the diversity of CVDs and, therefore, does not allow for a comprehensive presentation here. However, a selection of molecular mechanisms will be presented within the next paragraphs. Recent scientific advances have helped to identify numerous molecular pathways linked with the pathogenesis of CVDs. It should be kept in mind that most of the results, however, especially those aiming to elucidate molecular pathways, were obtained by using animal or even cellular model systems.

Some identified pathways are involved in several forms of CVDs and may thus be considered as pivotal molecular mechanisms, even if their role may be emphasized in a certain pathophysiological context. The common pathological key events represented by these pathways are, in a broader sense, posttranslational modifications of proteins and other molecules, including oxidation, phosphorylation, and glycation processes.

1.2.1 Oxidative Stress

Reactive oxygen species (ROS) are characterized by unpaired valence electrons in the outer shells of atoms and thereby able to react with a large number of molecular target proteins, lipids, carbohydrates, and nucleic acids. Consequently, ROS have been implicated in many diseases as well as aging-related pathophysiology of the cardiovascular system (Chen et al. 2012).

ROS levels in cells increase after exposure to environmental factors like UV light, heat, or ionizing radiation and can also be produced by metabolism of xenobiotics. However, ROS are also generated by physiological processes, including activity of the mitochondrial respiratory chain and enzymatic reactions as catalyzed by uncoupled eNOS and NADPH oxidases, lipoxygenase, monoamine oxidase, and xanthine oxidase. This general oxidative stress is counterbalanced by the activity of a number of cellular antioxidant enzymes, including glutathione peroxidase (GPx), catalase, and superoxide dismutase (SOD), as well as some exogenously obtained dietary compounds (Shao et al. 2012).

The most common radicals are superoxide anion radical ($^{\bullet}O_2^-$), hydrogen peroxide (H_2O_2), hydroxyl radicals ($^{\bullet}OH$), NO radicals ($^{\bullet}NO$), nitrogen dioxide radicals (NO_2), peroxynitrite ($ONOO^-$), and hypochloride (ClO^-).

The reaction of the superoxide anion, with the main endothelial relaxing factor NO, is extremely efficient. Consequently, one obvious cardiovascular outcome of enhanced superoxide production is decreased NO bioavailability. The reaction product of $^{\bullet}O_2^-$ and NO, peroxynitrite, is known to induce the activity of poly ADP-ribose polymerase (PARP), an enzyme that has been shown to induce cellular dysfunction and metabolic disorders in cells and has been demonstrated to be involved in vascular aging (Burkle et al. 2005; Csiszar et al. 2005).

Production of ROS in the cardiovascular system seems to be tightly linked to the renin-angiotensin system (RAS), because chronic inhibition of the RAS or genetic deletion of the angiotensin II type 1 receptor suppresses peroxynitrite formation during aging. Angiotensin II has been identified as a major stimulus for NAPDH oxidase and, thus, a key element for accumulating oxidative stress during aging (Oudit et al. 2007; Wang et al. 2010b).

Oxidative stress is well known to be involved in initiation and progression of the pathogenesis of atherosclerosis. Atherosclerosis and its clinical representations are directly linked to oxidation of lipids bound to LDLs that accumulate in the network of the extracellular matrix of the subendothelial space. While immobilized, ROS secreted from several pathways of the vessel wall can reach the LDL and induce lipid oxidation (Witztum and Steinberg 1991). LDL oxidation is seen as a two-stage process with an initial, relatively slow oxidation, characterized by only small changes in ApoB and a more rapid second stage initiated by arrival of monocytes, which provides a strong additional source of ROS when converted to macrophages. This oxidative potential of monocytes/macrophages may also contribute to the pathophysiology of hypertension (Lim et al. 2006) and is mainly due to the physiological accumulation of ROS in the phagosome by NOX2 activity (Judkins et al. 2010). This accumulating oxidative stress leads to modification of the LDL receptor and loss of its ability to recognize cholesterol. As a consequence, LDL is transported by the receptors without limiting the cholesterol uptake, resulting in cellular cholesterol accumulation and formation of foam cells (McLaren et al. 2011).

Epidemiological Examples Demonstrating Relevance of Cellular Redox Balance
A lectin-like 52-kD receptor for oxidized LDL (LOX-1) is a scavenger receptor present primarily on vascular endothelial cells and binds and regulates oxidized LDL. Ligand activity of LOX-1 has been found associated with non-HDL cholesterol, triglycerides, history of smoking, and risk of coronary artery disease (Inoue et al. 2010; Uchida et al. 2011).

Oxidation of LDL has been demonstrated to promote atherosclerotic inflammation by affecting monocyte tethering, activation and attachment (McEver 1992). P-selectin, a tethering protein, has been demonstrated to be upregulated and released from endothelial cells in the presence of oxidized LDL (Vora et al. 1997). In addition, oxidized LDL in endothelial cells was shown to induce several pro-inflammatory factors including monocyte chemoattractant protein 1 (MCP-1) (Cushing et al. 1990), monocyte colony-stimulating factor (M-CSF) (Rajavashisth et al. 1990), and human growth-related oncogene (GRO) (Schwartz et al. 1994). Thus, initially limited oxidation of LDL may lead, via a positive feedback mechanism, to accumulated production of ROS. Further development of the atherosclerotic lesion is promoted by activation of transcriptional programs that trigger an inflammatory response and induce arterial calcification that alters the mechanical properties of the vessel wall, enables infiltration by monocytes, and, in advanced stages, triggers plaque rupture.

Recent findings suggest that ROS-generating pathways are modulated by Ca^{2+}, and a crosstalk of signaling between ROS and Ca^{2+} has been demonstrated for several systems (Lee et al. 2010; Sun et al. 2011). While Cav1.2 and other plasmalemmal calcium-conducting channels have been shown to be regulated by ROS (Amberg et al. 2010; Cioffi 2011; Florea and Blatter 2008; Hool et al. 2005; Poteser et al. 2006; Weissmann et al. 2012), intracellular Ca^{2+} stores may be compromised by expressional downregulation of calsequestrin (Hanninen et al. 2010) or ROS-induced modification of the cardiac ryanodine receptor (Donoso et al. 2011). In turn, ROS-generating NADPH oxidase activity (Nox5) was shown to be regulated by intracellular Ca^{2+} levels via CAMKII (El Jamali et al. 2008; Pandey et al. 2011) and $^{\bullet}O_2^-$ production in mitochondria is closely related to Ca^{2+} concentrations (Victor et al. 2009), enabling feed-forward interactions that may finally lead to cellular Ca^{2+} overload (Peng and Jou 2010).

1.2.2 Kinase Pathways

Important cellular functions such as growth and differentiation are governed by environmental and genetic factors that alter signal transduction pathways culminating in altered gene expression. Critical components of many of those pathways are protein kinases, which phosphorylate and regulate cellular substrates including transcription factors. Kinase pathways are characterized by a cascade-like chain of reactions suitable to potentiate even very weak initial events into robust cellular signals.

A specific set of serine/threonine kinases is activated by factors induced via environmental disturbances like heat, inflammatory signals, ischemia, and genotoxic stress. These kinases have therefore been termed stress-activated protein kinases (SAPKs), also known as c-Jun N-terminal kinases (JNK). These SAPKs, together with kinase p38, represent a common element in the pathophysiology of many CVDs (Bogoyevitch et al. 1996; Choukroun et al. 1998; Johnson and Nakamura 2007; Liang and Molkentin 2003; Seko et al. 1997).

A molecular mechanism involving SAPKs is part of the pathophysiology of ischemic-reperfusion injury (Clerk et al. 1998). Ischemic-reperfusion injury is characterized as initial anoxic cell injury, triggering decreased mitochondrial ATP production, activation of hydrolases, and compromised membrane integrity. Resupply of blood in the second phase is then followed by inflammatory cellular responses. Reperfusion of ischemic tissue can activate genetic programs, which induce cells to dedifferentiate, proliferate, or undergo apoptosis. While the exact initial events of the pathway are still unclear, the radical scavenger GSH has been shown to prevent activation of SAPKs, substantiating a key role of oxidative stress. SAPKs, however, are the predominant ATF-2 (activating transcription factor 2) kinases activated by reperfusion (Pombo et al. 1994). After ischemia and reperfusion, SAPKs target transcription factors of multiple promoters, hypothesized to be c-Jun/ATF-2 heterodimers, providing a potent mechanism to induce a large number of genes. Their targets are the ATF/CRE and jun2 TRE motifs of the c-jun promoter (Morooka et al. 1995). The activated genes typically encode for proteins involved in apoptotic processes, inflammatory pathways, and heat shock proteins (Conti et al. 2007).

SAPKs were shown to be involved in inflammatory responses of macrophages and smooth muscle cells of the artherosclerotic plaque (Metzler et al. 2000). Inhibition of p38 prevents the cumulative production of cytokines like IL-1β and TNF-α by interruption of feed-forward amplification on a translational level (Feng et al. 2001).

5′ AMP-activated protein kinase (AMPK) is an enzyme that regulates metabolic functions of heart, as well as liver, brain, and skeletal muscle, and helps to normalize cellular lipid, glucose, and energy imbalances. AMPK is thus seen as an interesting drug target for the treatment of metabolic syndrome and was shown to regulate the function of lipoprotein lipase (LPL) (An et al. 2005), an enzyme that is critical for the fatty acid supply of the cardiomyocyte and the GLUT4 glucose transporter in the heart. AMPK has also been reported to play a role in ion channel activation (Alesutan et al. 2011; Ikematsu et al. 2011), ischemia-reperfusion injury (Kambara et al. 2012; Omar et al. 2012; Paiva et al. 2011), oxidative stress (Colombo and Moncada 2009), hypertrophy (Jiang et al. 2010), and other pathophysiological pathways (Beauloye et al. 2011; Heidrich et al. 2010).

Another group of kinases that has been identified to be critically involved in CVDs including angina, cerebral ischemia, myocardial ischemia, and cardiac hypertrophy is the group of Rho kinases. Rho-associated coil-forming protein kinases (ROCKs) are downstream targets of RhoA, a GTP-triggered molecular switch representing a key element in the regulation of some basic cellular functions like contractility, migration, proliferation, and apoptosis.

Epidemiological Examples Demonstrating the Relevance of Cellular Kinase Pathways

An evidence of involvement of protein phosphorylation in cardiovascular disease, activity of Rho-associated kinases (ROCK) in circulating leukocytes has been proposed to be a marker of cardiovascular risk. ROCK activity reportedly showed significant positive correlations with flow-mediated vasodilation (FMD), reflecting endothelial function, as well as with various cardiovascular risk factors, such as body mass index (BMI), systolic blood pressure, LDL cholesterol, and Framingham risk factor score, a cumulative cardiovascular risk factor (Hidaka et al. 2010; Soga et al. 2011). Moreover, smokers and patients with vasospastic angina have been reported to show increased ROCK activity (Hidaka et al. 2010; Kikuchi et al. 2011).

Typical target proteins of ROCKs are regulatory elements of the cytoskeleton, adducin (Fukata et al. 1999), MARCKS (Mueller et al. 2005), intermediate filaments like desmin and vimentin (Inada et al. 1999), as well as proteins of contractile machinery like calponin (Qu et al. 2008), troponin (Vahebi et al. 2005) and PTEN (Chang et al. 2006). Activation of ROCKs has been demonstrated to regulate the Ca^{2+} sensitivity of the contractile proteins by phosphorylation of MYPT-1, a regulatory subunit of MLC phosphatase, and this modulation is of special importance in contractile cells (Ohtsu et al. 2005).

In addition, gene expression in vascular smooth muscle cells has been demonstrated to depend on this group of kinases. Myocardin expression of vascular smooth muscle cells was shown to be regulated by a signaling cascade that involved

ROCK, p38 MAPK, PP1a, CPI-17, and myocyte enhancer factor-2 (MEF-2) (Iwasaki et al. 2008).

Consequently, ROCKs may contribute to critical cellular mechanisms in many CVDs. Inhibition of ROCKs prevents smooth muscle proliferation induced by platelet-derived growth factor, thrombin, or urotensin-II that has been identified to involve ROCK-induced upregulation of the cyclin-dependent kinase inhibitor p27Kip1 (Liu et al. 2011).

In the endothelium, ROCKs are important regulators of angiogenic migration and permeability and barrier function. ROCKs have also been shown to exert a negative effect on NO production by inhibition of endothelial NO synthase (Bivalacqua et al. 2004), while endogenous NO can inhibit the RhoA/ROCK pathway (Noma et al. 2012).

ROCK inhibitors, such as the pyridine derivative Y-27632 and fasudil, have been demonstrated to normalize arterial pressure in a rat model of arterial hypertension (Tsounapi et al. 2012) as well as human pulmonary hypertension (Li et al. 2009). Upregulation of ROCKs can be observed in arteriosclerotic lesions (Kandabashi et al. 2000, 2002) and contributes to lesion formation in mice (Mallat et al. 2003). ROCK pathways were also shown to be involved in vascular aneurisms (Wang et al. 2005), myocardial ischemia (Bao et al. 2004), and ventricular remodeling after myocardial infarction (Hattori et al. 2004), and the still-growing list of cardiovascular pathophysiological mechanisms involving ROCKs indicates the central role of this protein kinase family in CVD.

1.2.3 Enzymatic and Nonenzymatic Glycation

Glycosylation is a posttranslational protein modification effected by site-specific enzymatic addition of saccharides. Despite the still unclear position between adverse and beneficial effects, glycosylation of proteins has been demonstrated to be an important regulatory mechanism in the cardiovascular system. Under healthy, physiological conditions, glycosylation is important for cell-cell adhesion, protein stability (protection from degradation), and regulation of enzyme activity (Spiro 2002). Carbohydrates may be attached to the nitrogen of asparagine and arginine (N-linked), to the hydroxyl oxygen of serine, threonine, hydroxylysine, or hydroxyproline (O-linked); to the phosphate of phosphoserine (P-linked), at the carbon of tryptophan (C-linked); or via phosphoethanolamine to the carboxy-terminus of proteins (glycation). Thirteen different monosaccharides and 8 amino acid types have been identified to form bonds, enabling 31 different combinations (Spiro 2002). While N-linked glycosylation is the most widely distributed form, the compound formed by attachment of β-N-acetylglucosamine to serine and threonine residues (O-GlcNAcylation) has been recognized as a key regulator of many critical biological processes including translation and transcription, nuclear transport, cytoskeletal reorganization, and signal transduction (Laczy et al. 2009). The reaction is catalyzed by O-GlcNAc-transferase (OGT), reversed by O-GlcNAc-hydrolase (OGA), and depends on the substrate UDP-GlcNAc that is generated by the hexosamine biosynthesis pathway (Marshall et al. 1991). Serine/threonine residues of several proteins including endothelial nitric oxide synthase (eNOS) (Du et al. 2001), estrogen receptor-β (Cheng and Hart 2001), RNA polymerase II (Kelly et al. 1993), and c-Myc (Kamemura et al. 2002) were shown to be phosphorylation sites as well as targets of O-GlcNAcylation. These posttranslational modifications are not mutually exclusive, as some proteins can be phosphorylated and O-GlyNAcylated concomitantly (Wang et al. 2007; Yang et al. 2006). The linkage between ROCK, AMPK, and SAPK pathways (Lima et al. 2011; Cheung and Hart 2008) and O-GlcNAcylation indicates that this modification reflects cellular stress in addition to functioning as a metabolic sensor (Ngoh et al. 2010).

While a number of cardioprotective effects of O-GlcNAcylation were reported (Fulop et al. 2007; Porter et al. 2012), O-GlcNAcylation has also been demonstrated to be responsible for adverse effects of diabetes on the heart (Marsh et al. 2011) and vasculature (Du et al. 2001; Lima et al. 2010). Hyperglycemia represents a major risk factor for macro- and microvascular diseases associated with diabetes mellitus (Nathan 1996), and O-GlcNacylation was shown to contribute to the effects of diabetes on vessels (Buse 2006). The molecular mechanisms by which O-GlcNAcylation contributes to endothelial vascular dysfunction have been demonstrated to be based on modification of several important regulatory proteins that govern blood vessel function, including PGI_2 synthase (Du et al. 2006) and several transcription factors (including specific protein 1, activator protein-1, cAMP response element-binding protein-1, host cell factor-1, nuclear factor of activated T cells, tumor suppressor gene-53) that regulate the expression of proteins like plasminogen activator inhibitor 1 (PAI-1) (Du et al. 2000), thrombospondin-1 (TSP-1) (Raman et al. 2007), and angiopoietin-2 (Ang-2) (Yao et al. 2007). PAI-1 inhibits serine proteases, attenuating the fibrinolytic system in blood and supporting clot formation (Van de Werf et al. 1984). High TSP-1 levels are made responsible for mesangial cell proliferation and increased cellular matrix protein production (Murphy et al. 1999; Yevdokimova et al. 2001), and Ang-2 is an antagonist of Ang-1-promoted Tie2 signaling, leading to an attenuation of blood vessel maturation and stabilization (Chen and Stinnett 2008). The role of PAI-1, TSP-1, and eNOS in diabetes-associated vascular disease is consistently reflected by identification of several polymorphisms in these genes that are linked to vasculopathy (Saely et al. 2008; Santos et al. 2012; Stenina et al. 2004). The O-GlcNAcylation of eNOS at serine 1177, the site for phosphorylation and activation of the enzyme, has been demonstrated to reduce NO production in endothelial cells (Federici et al. 2002). The resulting reduced NO production contributes to impaired relaxation of vessels (Lima et al. 2010). O-GlcNAcylation-mediated decrease in eNOS activity was also shown to be associated with an increase in matrix metalloproteinase activity, an imbalance implicated in diabetes (Galis and Khatri 2002). Matrix metalloproteinases are additionally known to cleave and activate precursors of vasoactive proteins like endothelin 1 (Fernandez-Patron et al. 1999) and cytokines like TNF-α (Gearing et al. 1995) and may thus further impair relaxation and promote inflammatory responses (Fig. 1.2).

Advanced glycation end products (AGE) are products of the reaction of saccharides with lipids, nucleic acids, and proteins. They are generated by a nonenzymatic process, termed "Maillard reaction." Over months and years, the initial products undergo various

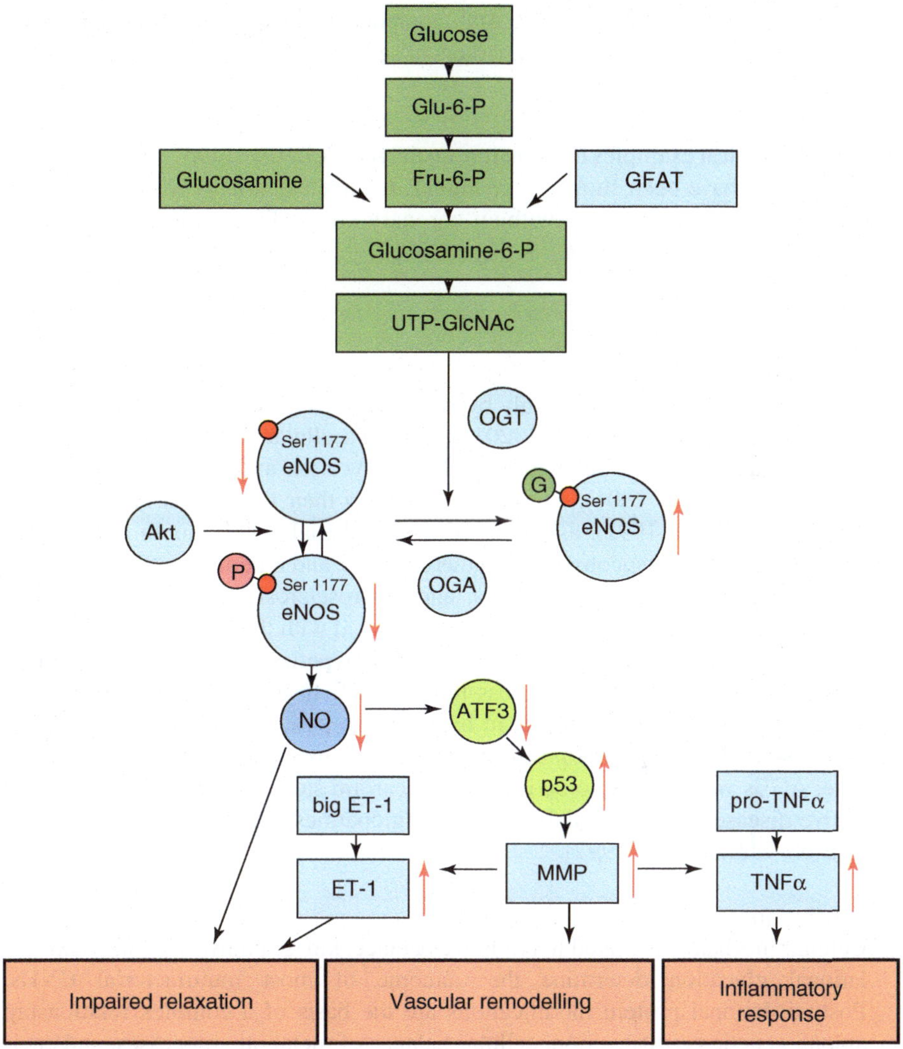

Fig. 1.2 Schematic illustrating molecular mechanisms mediating the adverse effects of O-GlcNAcylation of eNOS on diabetic vasculature. UDP-GlcNAc is produced by the hexosamine biosynthesis pathway (*green*) and transferred to eNOS at serine 1177, an enzyme phosphorylation and activation site, by b-N-acetylglucosaminyltransferase (*OGT*). This modification reduces eNOS activity and NO production in endothelial cells and is contributing to impaired vessel relaxation. In addition, a decrease in eNOS activity leads to activation of the cellular stress-induced transcription factor p53 by reduced activity of its inhibitor ATF3 and enhanced expression and activity of matrix metalloproteinases. These metalloproteinases can directly affect vascular remodeling but may also cleave and activate precursors of vasoactive factors and cytokines that contribute to vascular dysfunction. *Glu-6-P* glucose-6-phosphate, *Fru-6-P* fructose-6-phosphate, *glucosamine-6-P* glucosamine-6-phosphate, *UDP-GlcNAc* UDP-N-acetylglucosamine, *OGA* β-N-acetylglucosaminidase, *Akt* Akt kinase, protein kinase B, *ATF3* activating transcription factor 3, *p53* tumor protein 53, *ET-1* endothelin-1, *MMP* matrix metalloproteinase, *TNF-α* tumor necrosis factor α

changes until finally yielding a stable AGE compound. Consequently, AGEs accumulate by age and are considered as biomarkers of senescence (Simm et al. 2007).

Epidemiological Examples Demonstrating Relevance of Glycation
AGE levels have been shown to be higher in diabetes patients with coronary artery disease than in those without coronary artery disease (Kilhovd et al. 1999) and to be independent predictors of prognosis in heart failure patients (Koyama et al. 2007).

AGEs mediate their effects mainly by cross-linking extracellular matrix (Francis-Sedlak et al. 2010; Kamioka et al. 2011) and intracellular proteins (Bidasee et al. 2003, 2004) and also by affecting intracellular glycation and activation of specific receptors (Singh et al. 2001). Binding of AGEs to their cellular surface receptors (RAGE, Neeper et al. 1992), as expressed in endothelial cells, smooth muscle, and monocytes, promotes vasoconstriction, atherogenesis, and inflammation as associated with coronary dysfunction, thrombosis, and atherosclerosis (Hartog et al. 2007). A polymorphism in RAGE has been found associated with the onset of diabetes, atherosclerosis, and renal dysfunction in patients with hypertension (Kawai et al. 2013).

Plasma levels of pentosidine, a cross-linking AGE, have been demonstrated to be a marker of severity and prognosis in heart failure patients (Koyama et al. 2007; Raposeiras-Roubin et al. 2010). In addition to their involvement in coronary artery disease, AGEs were shown to be involved in carotid stenosis and peripheral artery occlusive disease as well as diabetic microangiopathies (Chen et al. 2004; Gerrits et al. 2008; Murata et al. 1997).

Conclusion
Cellular mechanisms linked to oxidative balance, kinase pathways, and posttranslational glycation determine the outcome of most multifactorial CVDs. Posttranslational protein modifications are the basis of a complex relationship among genetic predisposition, cellular balance of metabolic processes, and environmental factors that interact with each other to contribute to the pathogenesis of CVDs. Consequently, this degree of complexity demands particular attention at the clinical level to identify suitable biomarkers of disease progression and also opens the view on a wide field of potential therapeutic interventions. Many questions in this field are still unresolved, and molecular pathways of CVDs along with epidemiological aspects will remain in the focus of preventive medicine.

References

Alesutan I, Munoz C, Sopjani M, Dermaku-Sopjani M, Michael D, Fraser S, Kemp BE, Seebohm G, Foller M, Lang F (2011) Inhibition of Kir2.1 (KCNJ2) by the AMP-activated protein kinase. Biochem Biophys Res Commun 408:505–510

Amberg GC, Earley S, Glapa SA (2010) Local regulation of arterial L-type calcium channels by reactive oxygen species. Circ Res 107:1002–1010

An D, Pulinilkunnil T, Qi D, Ghosh S, Abrahani A, Rodrigues B (2005) The metabolic "switch" AMPK regulates cardiac heparin-releasable lipoprotein lipase. Am J Physiol Endocrinol Metab 288:E246–E253

Bao W, Hu E, Tao L, Boyce R, Mirabile R, Thudium DT, Ma XL, Willette RN, Yue TL (2004) Inhibition of Rho-kinase protects the heart against ischemia/reperfusion injury. Cardiovasc Res 61:548–558

Beauloye C, Bertrand L, Horman S, Hue L (2011) AMPK activation, a preventive therapeutic target in the transition from cardiac injury to heart failure. Cardiovasc Res 90:224–233

Bentley JP, Asselbergs FW, Coffey CS, Hebert PR, Moore JH, Hillege HL, van Gilst WH (2010) Cardiovascular risk associated with interactions among polymorphisms in genes from the renin-angiotensin, bradykinin, and fibrinolytic systems. PLoS One 5:e12757

Bidasee KR, Nallani K, Yu Y, Cocklin RR, Zhang Y, Wang M, Dincer UD, Besch HR Jr (2003) Chronic diabetes increases advanced glycation end products on cardiac ryanodine receptors/calcium-release channels. Diabetes 52:1825–1836

Bidasee KR, Zhang Y, Shao CH, Wang M, Patel KP, Dincer UD, Besch HR Jr (2004) Diabetes increases formation of advanced glycation end products on sarco(endo)plasmic reticulum Ca^{2+}–ATPase. Diabetes 53:463–473

Bivalacqua TJ, Champion HC, Usta MF, Cellek S, Chitaley K, Webb RC, Lewis RL, Mills TM, Hellstrom WJ, Kadowitz PJ (2004) RhoA/Rho-kinase suppresses endothelial nitric oxide synthase in the penis: a mechanism for diabetes-associated erectile dysfunction. Proc Natl Acad Sci USA 101:9121–9126

Bogoyevitch MA, Gillespie-Brown J, Ketterman AJ, Fuller SJ, Ben-Levy R, Ashworth A, Marshall CJ, Sugden PH (1996) Stimulation of the stress-activated mitogen-activated protein kinase subfamilies in perfused heart. p38/RK mitogen-activated protein kinases and c-Jun N-terminal kinases are activated by ischemia/reperfusion. Circ Res 79:162–173

Bondy B (2007) Common genetic factors for depression and cardiovascular disease. Dialogues Clin Neurosci 9:19–28

Burkle A, Diefenbach J, Brabeck C, Beneke S (2005) Ageing and PARP. Pharmacol Res 52:93–99

Buse MG (2006) Hexosamines, insulin resistance, and the complications of diabetes: current status. Am J Physiol Endocrinol Metab 290:E1–E8

Chang J, Xie M, Shah VR, Schneider MD, Entman ML, Wei L, Schwartz RJ (2006) Activation of Rho-associated coiled-coil protein kinase 1 (ROCK-1) by caspase-3 cleavage plays an essential role in cardiac myocyte apoptosis. Proc Natl Acad Sci USA 103:14495–14500

Chen JX, Stinnett A (2008) Disruption of Ang-1/Tie-2 signaling contributes to the impaired myocardial vascular maturation and angiogenesis in type II diabetic mice. Arterioscler Thromb Vasc Biol 28:1606–1613

Chen AS, Taguchi T, Sugiura M, Wakasugi Y, Kamei A, Wang MW, Miwa I (2004) Pyridoxal-aminoguanidine adduct is more effective than aminoguanidine in preventing neuropathy and cataract in diabetic rats. Horm Metab Res 36:183–187

Chen AF, Chen DD, Daiber A, Faraci FM, Li H, Rembold CM, Laher I (2012) Free radical biology of the cardiovascular system. Clin Sci 123:73–91

Cheng X, Hart GW (2001) Alternative O-glycosylation/O-phosphorylation of serine-16 in murine estrogen receptor beta: post-translational regulation of turnover and transactivation activity. J Biol Chem 276:10570–10575

Cheung WD, Hart GW (2008) AMP-activated protein kinase and p38 MAPK activate O-GlcNAcylation of neuronal proteins during glucose deprivation. J Biol Chem 283:13009–13020

Choukroun G, Hajjar R, Kyriakis JM, Bonventre JV, Rosenzweig A, Force T (1998) Role of the stress-activated protein kinases in endothelin-induced cardiomyocyte hypertrophy. J Clin Invest 102:1311–1320

Cioffi DL (2011) Redox regulation of endothelial canonical transient receptor potential channels. Antioxid Redox Signal 15:1567–1582

Clerk A, Fuller SJ, Michael A, Sugden PH (1998) Stimulation of "stress-regulated" mitogen-activated protein kinases (stress-activated protein kinases/c-Jun N-terminal kinases and p38-mitogen-activated protein kinases) in perfused rat hearts by oxidative and other stresses. J Biol Chem 273:7228–7234

Colombo SL, Moncada S (2009) AMPKalpha1 regulates the antioxidant status of vascular endothelial cells. Biochem J 421:163–169

Conti A, Scala S, D'Agostino P, Alimenti E, Morelli D, Andria B, Tammaro A, Attanasio C, Della Ragione F, Scuderi V, Fabbrini F, D'Esposito M, Di Florio E, Nitsch L, Calise F, Faiella A (2007) Wide gene expression profiling of ischemia-reperfusion injury in human liver transplantation. Liver Transpl 13:99–113

Csiszar A, Pacher P, Kaley G, Ungvari Z (2005) Role of oxidative and nitrosative stress, longevity genes and poly(ADP-ribose) polymerase in cardiovascular dysfunction associated with aging. Curr Vasc Pharmacol 3:285–291

Cushing SD, Berliner JA, Valente AJ, Territo MC, Navab M, Parhami F, Gerrity R, Schwartz CJ, Fogelman AM (1990) Minimally modified low density lipoprotein induces monocyte chemotactic protein 1 in human endothelial cells and smooth muscle cells. Proc Natl Acad Sci USA 87:5134–5138

Donoso P, Sanchez G, Bull R, Hidalgo C (2011) Modulation of cardiac ryanodine receptor activity by ROS and RNS. Front Biosci 16:553–567

Du XL, Edelstein D, Rossetti L, Fantus IG, Goldberg H, Ziyadeh F, Wu J, Brownlee M (2000) Hyperglycemia-induced mitochondrial superoxide overproduction activates the hexosamine pathway and induces plasminogen activator inhibitor-1 expression by increasing Sp1 glycosylation. Proc Natl Acad Sci USA 97:12222–12226

Du XL, Edelstein D, Dimmeler S, Ju Q, Sui C, Brownlee M (2001) Hyperglycemia inhibits endothelial nitric oxide synthase activity by posttranslational modification at the Akt site. J Clin Invest 108:1341–1348

Du X, Edelstein D, Obici S, Higham N, Zou MH, Brownlee M (2006) Insulin resistance reduces arterial prostacyclin synthase and eNOS activities by increasing endothelial fatty acid oxidation. J Clin Invest 116:1071–1080

Eichers ER, Lewis RA, Katsanis N, Lupski JR (2004) Triallelic inheritance: a bridge between Mendelian and multifactorial traits. Ann Med 36:262–272

El Jamali A, Valente AJ, Lechleiter JD, Gamez MJ, Pearson DW, Nauseef WM, Clark RA (2008) Novel redox-dependent regulation of NOX5 by the tyrosine kinase c-Abl. Free Radic Biol Med 44:868–881

Federici M, Menghini R, Mauriello A, Hribal ML, Ferrelli F, Lauro D, Sbraccia P, Spagnoli LG, Sesti G, Lauro R (2002) Insulin-dependent activation of endothelial nitric oxide synthase is impaired by O-linked glycosylation modification of signaling proteins in human coronary endothelial cells. Circulation 106:466–472

Feng Y, Schreiner GF, Chakravarty S, Liu DY, Joly AH (2001) Inhibition of the mitogen activated protein kinase, p38 alpha, prevents proinflammatory cytokine induction by human adherent mononuclear leukocytes in response to lipid loading. Atherosclerosis 158:331–338

Fernandez-Patron C, Radomski MW, Davidge ST (1999) Vascular matrix metalloproteinase-2 cleaves big endothelin-1 yielding a novel vasoconstrictor. Circ Res 85:906–911

Fineschi V, Silver MD, Karch SB, Parolini M, Turillazzi E, Pomara C, Baroldi G (2005) Myocardial disarray: an architectural disorganization linked with adrenergic stress? Int J Cardiol 99:277–282

Florea SM, Blatter LA (2008) The effect of oxidative stress on Ca^{2+} release and capacitative Ca^{2+} entry in vascular endothelial cells. Cell Calcium 43:405–415

Francis-Sedlak ME, Moya ML, Huang JJ, Lucas SA, Chandrasekharan N, Larson JC, Cheng MH, Brey EM (2010) Collagen glycation alters neovascularization in vitro and in vivo. Microvasc Res 80:3–9

Fukata Y, Oshiro N, Kinoshita N, Kawano Y, Matsuoka Y, Bennett V, Matsuura Y, Kaibuchi K (1999) Phosphorylation of adducin by Rho-kinase plays a crucial role in cell motility. J Cell Biol 145:347–361

Fulop N, Zhang Z, Marchase RB, Chatham JC (2007) Glucosamine cardioprotection in perfused rat hearts associated with increased O-linked N-acetylglucosamine protein modification and altered p38 activation. Am J Physiol Heart Circ Physiol 292:H2227–H2236

Galis ZS, Khatri JJ (2002) Matrix metalloproteinases in vascular remodeling and atherogenesis: the good, the bad, and the ugly. Circ Res 90:251–262

Gearing AJ, Beckett P, Christodoulou M, Churchill M, Clements JM, Crimmin M, Davidson AH, Drummond AH, Galloway WA, Gilbert R et al (1995) Matrix metalloproteinases and processing of pro-TNF-alpha. J Leukoc Biol 57:774–777

Gehmlich K, Geier C, Osterziel KJ, Van der Ven PF, Furst DO (2004) Decreased interactions of mutant muscle LIM protein (MLP) with N-RAP and alpha-actinin and their implication for hypertrophic cardiomyopathy. Cell Tissue Res 317:129–136

Gerrits EG, Lutgers HL, Kleefstra N, Graaff R, Groenier KH, Smit AJ, Gans RO, Bilo HJ (2008) Skin autofluorescence: a tool to identify type 2 diabetic patients at risk for developing microvascular complications. Diabetes Care 31:517–521

Greaves SC, Roche AH, Neutze JM, Whitlock RM, Veale AM (1987) Inheritance of hypertrophic cardiomyopathy: a cross sectional and M mode echocardiographic study of 50 families. Br Heart J 58:259–266

Hanninen SL, Ronkainen JJ, Leskinen H, Tavi P (2010) Mitochondrial uncoupling downregulates calsequestrin expression and reduces SR Ca2+ stores in cardiomyocytes. Cardiovasc Res 88:75–82

Hartog JW, Voors AA, Bakker SJ, Smit AJ, van Veldhuisen DJ (2007) Advanced glycation end-products (AGEs) and heart failure: pathophysiology and clinical implications. Eur J Heart Fail 9:1146–1155

Hattori T, Shimokawa H, Higashi M, Hiroki J, Mukai Y, Tsutsui H, Kaibuchi K, Takeshita A (2004) Long-term inhibition of Rho-kinase suppresses left ventricular remodeling after myocardial infarction in mice. Circulation 109:2234–2239

Heidrich F, Schotola H, Popov AF, Sohns C, Chuenemann J, Friedrich M, Coskun KO, von Lewinski D, Hinz J, Bauer M, Mokashi SA, Sossalla S, Schmitto JD (2010) AMPK – activated protein kinase and its role in energy metabolism of the heart. Curr Cardiol Rev 6:337–342

Hidaka T, Hata T, Soga J, Fujii Y, Idei N, Fujimura N, Kihara Y, Noma K, Liao JK, Higashi Y (2010) Increased leukocyte rho kinase (ROCK) activity and endothelial dysfunction in cigarette smokers. Hypertens Res 33:354–359

Hool LC, Di Maria CA, Viola HM, Arthur PG (2005) Role of NAD(P)H oxidase in the regulation of cardiac L-type Ca^{2+} channel function during acute hypoxia. Cardiovasc Res 67:624–635

Ikematsu N, Dallas ML, Ross FA, Lewis RW, Rafferty JN, David JA, Suman R, Peers C, Hardie DG, Evans AM (2011) Phosphorylation of the voltage-gated potassium channel Kv2.1 by AMP-activated protein kinase regulates membrane excitability. Proc Natl Acad Sci USA 108:18132–18137

Inada H, Togashi H, Nakamura Y, Kaibuchi K, Nagata K, Inagaki M (1999) Balance between activities of Rho kinase and type 1 protein phosphatase modulates turnover of phosphorylation and dynamics of desmin/vimentin filaments. J Biol Chem 274:34932–34939

Inoue N, Okamura T, Kokubo Y, Fujita Y, Sato Y, Nakanishi M, Yanagida K, Kakino A, Iwamoto S, Watanabe M, Ogura S, Otsui K, Matsuda H, Uchida K, Yoshimoto R, Sawamura T (2010) LOX index, a novel predictive biochemical marker for coronary heart disease and stroke. Clin Chem 56:550–558

Iwasaki K, Hayashi K, Fujioka T, Sobue K (2008) Rho/Rho-associated kinase signal regulates myogenic differentiation via myocardin-related transcription factor-A/Smad-dependent transcription of the Id3 gene. J Biol Chem 283:21230–21241

Jiang SY, Xu M, Ma XW, Xiao H, Zhang YY (2010) A distinct AMP-activated protein kinase phosphorylation site characterizes cardiac hypertrophy induced by L-thyroxine and angiotensin II. Clin Exp Pharmacol Physiol 37:919–925

Johnson GL, Nakamura K (2007) The c-jun kinase/stress-activated pathway: regulation, function and role in human disease. Biochim Biophys Acta 1773:1341–1348

Judkins CP, Diep H, Broughton BR, Mast AE, Hooker EU, Miller AA, Selemidis S, Dusting GJ, Sobey CG, Drummond GR (2010) Direct evidence of a role for Nox2 in superoxide production, reduced nitric oxide bioavailability, and early atherosclerotic plaque formation in $ApoE^{-/-}$ mice. Am J Physiol Heart Circ Physiol 298:H24–H32

Kambara T, Ohashi K, Shibata R, Ogura Y, Maruyama S, Enomoto T, Uemura Y, Shimizu Y, Yuasa D, Matsuo K, Miyabe M, Kataoka Y, Murohara T, Ouchi N (2012) CTRP9 protein protects against myocardial injury following ischemia-reperfusion through AMP-activated protein kinase (AMPK)-dependent mechanism. J Biol Chem 287:18965–18973

Kamemura K, Hayes BK, Comer FI, Hart GW (2002) Dynamic interplay between O-glycosylation and O-phosphorylation of nucleocytoplasmic proteins: alternative glycosylation/phosphorylation of THR-58, a known mutational hot spot of c-Myc in lymphomas, is regulated by mitogens. J Biol Chem 277:19229–19235

Kamioka M, Ishibashi T, Ohkawara H, Nagai R, Sugimoto K, Uekita H, Matsui T, Yamagishi S, Ando K, Sakamoto T, Sakamoto N, Takuwa Y, Wada I, Shiomi M, Maruyama Y, Takeishi Y (2011) Involvement of membrane type 1-matrix metalloproteinase (MT1-MMP) in RAGE activation signaling pathways. J Cell Physiol 226:1554–1563

Kandabashi T, Shimokawa H, Miyata K, Kunihiro I, Kawano Y, Fukata Y, Higo T, Egashira K, Takahashi S, Kaibuchi K, Takeshita A (2000) Inhibition of myosin phosphatase by upregulated rho-kinase plays a key role for coronary artery spasm in a porcine model with interleukin-1beta. Circulation 101:1319–1323

Kandabashi T, Shimokawa H, Mukai Y, Matoba T, Kunihiro I, Morikawa K, Ito M, Takahashi S, Kaibuchi K, Takeshita A (2002) Involvement of rho-kinase in agonists-induced contractions of arteriosclerotic human arteries. Arterioscler Thromb Vasc Biol 22:243–248

Kawai T, Kamide K, Ito N, Onishi M, Oguro R, Takeya Y, Tatara Y, Maekawa Y, Katsuya T, Ohishi M, Rakugi H (2013) -374 T/A polymorphism in RAGE gene is associated with onset of diabetes mellitus, atherosclerosis, and renal dysfunction in patients with hypertension. Clin Exp Hypertens 35(3):236–241

Kehat I, Molkentin JD (2010) Molecular pathways underlying cardiac remodeling during pathophysiological stimulation. Circulation 122:2727–2735

Kelly WG, Dahmus ME, Hart GW (1993) RNA polymerase II is a glycoprotein. Modification of the COOH-terminal domain by O-GlcNAc. J Biol Chem 268:10416–10424

Kikuchi Y, Yasuda S, Aizawa K, Tsuburaya R, Ito Y, Takeda M, Nakayama M, Ito K, Takahashi J, Shimokawa H (2011) Enhanced Rho-kinase activity in circulating neutrophils of patients with vasospastic angina: a possible biomarker for diagnosis and disease activity assessment. J Am Coll Cardiol 58:1231–1237

Kilhovd BK, Berg TJ, Birkeland KI, Thorsby P, Hanssen KF (1999) Serum levels of advanced glycation end products are increased in patients with type 2 diabetes and coronary heart disease. Diabetes Care 22:1543–1548

Koyama Y, Takeishi Y, Arimoto T, Niizeki T, Shishido T, Takahashi H, Nozaki N, Hirono O, Tsunoda Y, Nitobe J, Watanabe T, Kubota I (2007) High serum level of pentosidine, an advanced glycation end product (AGE), is a risk factor of patients with heart failure. J Card Fail 13:199–206

Kuwahara K, Nishikimi T, Nakao K (2012) Transcriptional regulation of the fetal cardiac gene program. J Pharmacol Sci 119:198–203

Laczy B, Hill BG, Wang K, Paterson AJ, White CR, Xing D, Chen YF, Darley-Usmar V, Oparil S, Chatham JC (2009) Protein O-GlcNAcylation: a new signaling paradigm for the cardiovascular system. Am J Physiol Heart Circ Physiol 296:H13–H28

Lee M, Spokes KC, Aird WC, Abid MR (2010) Intracellular Ca^{2+} can compensate for the lack of NADPH oxidase-derived ROS in endothelial cells. FEBS Lett 584:3131–3136

Lee YC, Lai CQ, Ordovas JM, Parnell LD (2011) A database of gene-environment interactions pertaining to blood lipid traits, cardiovascular disease and type 2 diabetes. J Data Mining Genomics Proteomics 2(1) pii:106

Li F, Xia W, Yuan S, Sun R (2009) Acute inhibition of Rho-kinase attenuates pulmonary hypertension in patients with congenital heart disease. Pediatr Cardiol 30:363–366

Liang Q, Molkentin JD (2003) Redefining the roles of p38 and JNK signaling in cardiac hypertrophy: dichotomy between cultured myocytes and animal models. J Mol Cell Cardiol 35:1385–1394

Lieb W, Mayer B, Konig IR, Borwitzky I, Gotz A, Kain S, Hengstenberg C, Linsel-Nitschke P, Fischer M, Doring A, Wichmann HE, Meitinger T, Kreutz R, Ziegler A, Schunkert H, Erdmann J (2008) Lack of association between the MEF2A gene and myocardial infarction. Circulation 117:185–191
Lifton RP, Dluhy RG, Powers M, Rich GM, Cook S, Ulick S, Lalouel JM (1992) A chimaeric 11 beta-hydroxylase/aldosterone synthase gene causes glucocorticoid-remediable aldosteronism and human hypertension. Nature 355:262–265
Lim DS, Lutucuta S, Bachireddy P, Youker K, Evans A, Entman M, Roberts R, Marian AJ (2001) Angiotensin II blockade reverses myocardial fibrosis in a transgenic mouse model of human hypertrophic cardiomyopathy. Circulation 103:789–791
Lim HS, Patel JV, Lip GY (2006) Reactive oxygen species production by circulating monocytes: insights from pathophysiology to clinical hypertension. J Hum Hypertens 20:307–309
Lima VV, Giachini FR, Carneiro FS, Carneiro ZN, Saleh MA, Pollock DM, Fortes ZB, Carvalho MH, Ergul A, Webb RC, Tostes RC (2010) O-GlcNAcylation contributes to augmented vascular reactivity induced by endothelin 1. Hypertension 55:180–188
Lima VV, Giachini FR, Carneiro FS, Carvalho MH, Fortes ZB, Webb RC, Tostes RC (2011) O-GlcNAcylation contributes to the vascular effects of ET-1 via activation of the RhoA/Rho-kinase pathway. Cardiovasc Res 89:614–622
Liu AJ, Ling F, Wang D, Wang Q, Lu XD, Liu YL (2011) Fasudil inhibits platelet-derived growth factor-induced human pulmonary artery smooth muscle cell proliferation by up-regulation of p27kip(1) via the ERK signal pathway. Chin Med J (Engl) 124:3098–3104
Mallat Z, Gojova A, Sauzeau V, Brun V, Silvestre JS, Esposito B, Merval R, Groux H, Loirand G, Tedgui A (2003) Rho-associated protein kinase contributes to early atherosclerotic lesion formation in mice. Circ Res 93:884–888
Mani A, Radhakrishnan J, Wang H, Mani MA, Nelson-Williams C, Carew KS, Mane S, Najmabadi H, Wu D, Lifton RP (2007) LRP6 mutation in a family with early coronary disease and metabolic risk factors. Science 315:1278–1282
Marian AJ (2000) Pathogenesis of diverse clinical and pathological phenotypes in hypertrophic cardiomyopathy. Lancet 355:58–60
Marian AJ, Wu Y, Lim DS, McCluggage M, Youker K, Yu QT, Brugada R, DeMayo F, Quinones M, Roberts R (1999) A transgenic rabbit model for human hypertrophic cardiomyopathy. J Clin Invest 104:1683–1692
Marsh SA, Dell'Italia LJ, Chatham JC (2011) Activation of the hexosamine biosynthesis pathway and protein O-GlcNAcylation modulate hypertrophic and cell signaling pathways in cardiomyocytes from diabetic mice. Amino Acids 40:819–828
Marshall S, Bacote V, Traxinger RR (1991) Discovery of a metabolic pathway mediating glucose-induced desensitization of the glucose transport system. Role of hexosamine biosynthesis in the induction of insulin resistance. J Biol Chem 266:4706–4712
McEver RP (1992) Leukocyte-endothelial cell interactions. Curr Opin Cell Biol 4:840–849
McLaren JE, Michael DR, Ashlin TG, Ramji DP (2011) Cytokines, macrophage lipid metabolism and foam cells: implications for cardiovascular disease therapy. Prog Lipid Res 50:331–347
Metzler B, Hu Y, Dietrich H, Xu Q (2000) Increased expression and activation of stress-activated protein kinases/c-Jun NH(2)-terminal protein kinases in atherosclerotic lesions coincide with p53. Am J Pathol 156:1875–1886
Miyamoto SD, Stauffer BL, Nakano S, Sobus R, Nunley K, Nelson P, Sucharov CC (2012) Beta-adrenergic adaptation in paediatric idiopathic dilated cardiomyopathy. Eur Heart J [Epub ahead of print]. doi:10.1093/eurheartj/ehs229
Morooka H, Bonventre JV, Pombo CM, Kyriakis JM, Force T (1995) Ischemia and reperfusion enhance ATF-2 and c-Jun binding to cAMP response elements and to an AP-1 binding site from the c-jun promoter. J Biol Chem 270:30084–30092
Mueller BK, Mack H, Teusch N (2005) Rho kinase, a promising drug target for neurological disorders. Nat Rev Drug Discov 4:387–398
Muller M, Mazur AJ, Behrmann E, Diensthuber RP, Radke MB, Qu Z, Littwitz C, Raunser S, Schoenenberger CA, Manstein DJ, Mannherz HG (2012) Functional characterization of the

human alpha-cardiac actin mutations Y166C and M305L involved in hypertrophic cardiomyopathy. Cell Mol Life Sci 69:3457–3479

Murata T, Nagai R, Ishibashi T, Inomuta H, Ikeda K, Horiuchi S (1997) The relationship between accumulation of advanced glycation end products and expression of vascular endothelial growth factor in human diabetic retinas. Diabetologia 40:764–769

Murphy M, Godson C, Cannon S, Kato S, Mackenzie HS, Martin F, Brady HR (1999) Suppression subtractive hybridization identifies high glucose levels as a stimulus for expression of connective tissue growth factor and other genes in human mesangial cells. J Biol Chem 274:5830–5834

Nathan DM (1996) The pathophysiology of diabetic complications: how much does the glucose hypothesis explain? Ann Intern Med 124:86–89

Neeper M, Schmidt AM, Brett J, Yan SD, Wang F, Pan YC, Elliston K, Stern D, Shaw A (1992) Cloning and expression of a cell surface receptor for advanced glycosylation end products of proteins. J Biol Chem 267:14998–15004

Ngoh GA, Facundo HT, Zafir A, Jones SP (2010) O-GlcNAc signaling in the cardiovascular system. Circ Res 107:171–185

Noma K, Kihara Y, Higashi Y (2012) Striking crosstalk of ROCK signaling with endothelial function. J Cardiol 60:1–6

Oberst L, Zhao G, Park JT, Brugada R, Michael LH, Entman ML, Roberts R, Marian AJ (1998) Dominant-negative effect of a mutant cardiac troponin T on cardiac structure and function in transgenic mice. J Clin Invest 102:1498–1505

Ohtsu H, Mifune M, Frank GD, Saito S, Inagami T, Kim-Mitsuyama S, Takuwa Y, Sasaki T, Rothstein JD, Suzuki H, Nakashima H, Woolfolk EA, Motley ED, Eguchi S (2005) Signal-crosstalk between Rho/ROCK and c-Jun NH2-terminal kinase mediates migration of vascular smooth muscle cells stimulated by angiotensin II. Arterioscler Thromb Vasc Biol 25:1831–1836

Omar MA, Verma S, Clanachan AS (2012) Adenosine-mediated inhibition of 5′-AMP-activated protein kinase and p38 mitogen-activated protein kinase during reperfusion enhances recovery of left ventricular mechanical function. J Mol Cell Cardiol 52:1308–1318

Ordovas JM (2007) Genetic links between diabetes mellitus and coronary atherosclerosis. Curr Atheroscler Rep 9:204–210

Ostman-Smith I, Wettrell G (2000) Pathophysiology of hypertrophic cardiomyopathy. Lancet 355:928

Oudit GY, Kassiri Z, Patel MP, Chappell M, Butany J, Backx PH, Tsushima RG, Scholey JW, Khokha R, Penninger JM (2007) Angiotensin II-mediated oxidative stress and inflammation mediate the age-dependent cardiomyopathy in ACE2 null mice. Cardiovasc Res 75:29–39

Pace L, Betocchi S, Losi MA, Della Morte AM, Ciampi Q, Nugnez R, Chiariello M, Salvatore M (2004) Sympathetic nervous function in patients with hypertrophic cardiomyopathy assessed by [123I]-MIBG: relationship with left ventricular perfusion and function. Q J Nucl Med Mol Imaging 48:20–25

Paiva MA, Rutter-Locher Z, Goncalves LM, Providencia LA, Davidson SM, Yellon DM, Mocanu MM (2011) Enhancing AMPK activation during ischemia protects the diabetic heart against reperfusion injury. Am J Physiol Heart Circ Physiol 300:H2123–H2134

Pandey D, Gratton JP, Rafikov R, Black SM, Fulton DJ (2011) Calcium/calmodulin-dependent kinase II mediates the phosphorylation and activation of NADPH oxidase 5. Mol Pharmacol 80:407–415

Patel R, Lim DS, Reddy D, Nagueh SF, Lutucuta S, Sole MJ, Zoghbi WA, Quinones MA, Roberts R, Marian AJ (2000) Variants of trophic factors and expression of cardiac hypertrophy in patients with hypertrophic cardiomyopathy. J Mol Cell Cardiol 32:2369–2377

Peng TI, Jou MJ (2010) Oxidative stress caused by mitochondrial calcium overload. Ann N Y Acad Sci 1201:183–188

Pombo CM, Bonventre JV, Avruch J, Woodgett JR, Kyriakis JM, Force T (1994) The stress-activated protein kinases are major c-Jun amino-terminal kinases activated by ischemia and reperfusion. J Biol Chem 269:26546–26551

Porter K, Medford HM, McIntosh CM, Marsh SA (2012) Cardioprotection requires flipping the 'posttranslational modification' switch. Life Sci 90:89–98

Poteser M, Graziani A, Rosker C, Eder P, Derler I, Kahr H, Zhu MX, Romanin C, Groschner K (2006) TRPC3 and TRPC4 associate to form a redox-sensitive cation channel. Evidence for expression of native TRPC3-TRPC4 heteromeric channels in endothelial cells. J Biol Chem 281:13588–13595

Qu MJ, Liu B, Qi YX, Jiang ZL (2008) Role of Rac and Rho-GDI alpha in the frequency-dependent expression of h1-calponin in vascular smooth muscle cells under cyclic mechanical strain. Ann Biomed Eng 36:1481–1488

Rajavashisth TB, Andalibi A, Territo MC, Berliner JA, Navab M, Fogelman AM, Lusis AJ (1990) Induction of endothelial cell expression of granulocyte and macrophage colony-stimulating factors by modified low-density lipoproteins. Nature 344:254–257

Raman P, Krukovets I, Marinic TE, Bornstein P, Stenina OI (2007) Glycosylation mediates up-regulation of a potent antiangiogenic and proatherogenic protein, thrombospondin-1, by glucose in vascular smooth muscle cells. J Biol Chem 282:5704–5714

Raposeiras-Roubin S, Rodino-Janeiro BK, Grigorian-Shamagian L, Moure-Gonzalez M, Seoane-Blanco A, Varela-Roman A, Alvarez E, Gonzalez-Juanatey JR (2010) Soluble receptor of advanced glycation end products levels are related to ischaemic aetiology and extent of coronary disease in chronic heart failure patients, independent of advanced glycation end products levels: New Roles for Soluble RAGE. Eur J Heart Fail 12:1092–1100

Rayment I, Holden HM, Sellers JR, Fananapazir L, Epstein ND (1995) Structural interpretation of the mutations in the beta-cardiac myosin that have been implicated in familial hypertrophic cardiomyopathy. Proc Natl Acad Sci USA 92:3864–3868

Richard P, Charron P, Carrier L, Ledeuil C, Cheav T, Pichereau C, Benaiche A, Isnard R, Dubourg O, Burban M, Gueffet JP, Millaire A, Desnos M, Schwartz K, Hainque B, Komajda M (2003) Hypertrophic cardiomyopathy: distribution of disease genes, spectrum of mutations, and implications for a molecular diagnosis strategy. Circulation 107:2227–2232

Romanoski CE, Lee S, Kim MJ, Ingram-Drake L, Plaisier CL, Yordanova R, Tilford C, Guan B, He A, Gargalovic PS, Kirchgessner TG, Berliner JA, Lusis AJ (2010) Systems genetics analysis of gene-by-environment interactions in human cells. Am J Hum Genet 86:399–410

Rysev NA, Karpicheva OE, Redwood CS, Borovikov YS (2012) The effect of the Asp175Asn and Glu180Gly TPM1 mutations on actin-myosin interaction during the ATPase cycle. Biochim Biophys Acta 1824:366–373

Saely CH, Muendlein A, Vonbank A, Sonderegger G, Aczel S, Rein P, Risch L, Drexel H (2008) Type 2 diabetes significantly modulates the cardiovascular risk conferred by the PAI-1–675 4G/5G polymorphism in angiographied coronary patients. Clin Chim Acta 396:18–22

Santos KG, Crispim D, Canani LH, Ferrugem PT, Gross JL, Roisenberg I (2012) Relationship of endothelial nitric oxide synthase (eNOS) gene polymorphisms with diabetic retinopathy in Caucasians with type 2 diabetes. Ophthalmic Genet 33:23–27

Schwartz D, Andalibi A, Chaverri-Almada L, Berliner JA, Kirchgessner T, Fang ZT, Tekamp-Olson P, Lusis AJ, Gallegos C, Fogelman AM et al (1994) Role of the GRO family of chemokines in monocyte adhesion to MM-LDL-stimulated endothelium. J Clin Invest 94: 1968–1973

Seko Y, Takahashi N, Tobe K, Kadowaki T, Yazaki Y (1997) Hypoxia and hypoxia/reoxygenation activate p65PAK, p38 mitogen-activated protein kinase (MAPK), and stress-activated protein kinase (SAPK) in cultured rat cardiac myocytes. Biochem Biophys Res Commun 239:840–844

Shao D, Oka S, Brady CD, Haendeler J, Eaton P, Sadoshima J (2012) Redox modification of cell signaling in the cardiovascular system. J Mol Cell Cardiol 52:550–558

Simm A, Wagner J, Gursinsky T, Nass N, Friedrich I, Schinzel R, Czeslik E, Silber RE, Scheubel RJ (2007) Advanced glycation endproducts: a biomarker for age as an outcome predictor after cardiac surgery? Exp Gerontol 42:668–675

Singh R, Barden A, Mori T, Beilin L (2001) Advanced glycation end-products: a review. Diabetologia 44:129–146

Soga J, Noma K, Hata T, Hidaka T, Fujii Y, Idei N, Fujimura N, Mikami S, Maruhashi T, Kihara Y, Chayama K, Kato H, Liao JK, Higashi Y (2011) Rho-associated kinase activity, endothelial function, and cardiovascular risk factors. Arterioscler Thromb Vasc Biol 31:2353–2359

Spiro RG (2002) Protein glycosylation: nature, distribution, enzymatic formation, and disease implications of glycopeptide bonds. Glycobiology 12:43R–56R

Stenina OI, Byzova TV, Adams JC, McCarthy JJ, Topol EJ, Plow EF (2004) Coronary artery disease and the thrombospondin single nucleotide polymorphisms. Int J Biochem Cell Biol 36:1013–1030

Sun L, Yau HY, Lau OC, Huang Y, Yao X (2011) Effect of hydrogen peroxide and superoxide anions on cytosolic Ca2+: comparison of endothelial cells from large-sized and small-sized arteries. PLoS One 6:e25432

Tsounapi P, Saito M, Kitatani K, Dimitriadis F, Ohmasa F, Shimizu S, Kinoshita Y, Takenaka A, Satoh K (2012) Fasudil improves the endothelial dysfunction in the aorta of spontaneously hypertensive rats. Eur J Pharmacol 691:182–189

Uchida K, Suehiro A, Nakanishi M, Sawamura T, Wakabayashi I (2011) Associations of atherosclerotic risk factors with oxidized low-density lipoprotein evaluated by LOX-1 ligand activity in healthy men. Clin Chim Acta 412:1643–1647

Vahebi S, Kobayashi T, Warren CM, de Tombe PP, Solaro RJ (2005) Functional effects of rho-kinase-dependent phosphorylation of specific sites on cardiac troponin. Circ Res 96:740–747

Van de Werf F, Ludbrook PA, Bergmann SR, Tiefenbrunn AJ, Fox KA, de Geest H, Verstraete M, Collen D, Sobel BE (1984) Coronary thrombolysis with tissue-type plasminogen activator in patients with evolving myocardial infarction. N Engl J Med 310:609–613

Van Driest SL, Vasile VC, Ommen SR, Will ML, Tajik AJ, Gersh BJ, Ackerman MJ (2004) Myosin binding protein C mutations and compound heterozygosity in hypertrophic cardiomyopathy. J Am Coll Cardiol 44:1903–1910

Vasile VC, Ommen SR, Edwards WD, Ackerman MJ (2006) A missense mutation in a ubiquitously expressed protein, vinculin, confers susceptibility to hypertrophic cardiomyopathy. Biochem Biophys Res Commun 345:998–1003

Victor VM, Apostolova N, Herance R, Hernandez-Mijares A, Rocha M (2009) Oxidative stress and mitochondrial dysfunction in atherosclerosis: mitochondria-targeted antioxidants as potential therapy. Curr Med Chem 16:4654–4667

Vikstrom KL, Factor SM, Leinwand LA (1995) A murine model for hypertrophic cardiomyopathy. Z Kardiol 84(Suppl 4):49–54

Vora DK, Fang ZT, Liva SM, Tyner TR, Parhami F, Watson AD, Drake TA, Territo MC, Berliner JA (1997) Induction of P-selectin by oxidized lipoproteins. Separate effects on synthesis and surface expression. Circ Res 80:810–818

Wang YX, Martin-McNulty B, da Cunha V, Vincelette J, Lu X, Feng Q, Halks-Miller M, Mahmoudi M, Schroeder M, Subramanyam B, Tseng JL, Deng GD, Schirm S, Johns A, Kauser K, Dole WP, Light DR (2005) Fasudil, a Rho-kinase inhibitor, attenuates angiotensin II-induced abdominal aortic aneurysm in apolipoprotein E-deficient mice by inhibiting apoptosis and proteolysis. Circulation 111:2219–2226

Wang Z, Pandey A, Hart GW (2007) Dynamic interplay between O-linked N-acetylglucosaminylation and glycogen synthase kinase-3-dependent phosphorylation. Mol Cell Proteomics 6:1365–1379

Wang H, Li Z, Wang J, Sun K, Cui Q, Song L, Zou Y, Wang X, Liu X, Hui R, Fan Y (2010a) Mutations in NEXN, a Z-disc gene, are associated with hypertrophic cardiomyopathy. Am J Hum Genet 87:687–693

Wang M, Zhang J, Walker SJ, Dworakowski R, Lakatta EG, Shah AM (2010b) Involvement of NADPH oxidase in age-associated cardiac remodeling. J Mol Cell Cardiol 48:765–772

Weissmann N, Sydykov A, Kalwa H, Storch U, Fuchs B, Mederos y Schnitzler M, Brandes RP, Grimminger F, Meissner M, Freichel M, Offermanns S, Veit F, Pak O, Krause KH, Schermuly RT, Brewer AC, Schmidt HH, Seeger W, Shah AM, Gudermann T, Ghofrani HA, Dietrich A (2012) Activation of TRPC6 channels is essential for lung ischaemia-reperfusion induced oedema in mice. Nat Commun 3:649

Weng L, Kavaslar N, Ustaszewska A, Doelle H, Schackwitz W, Hebert S, Cohen JC, McPherson R, Pennacchio LA (2005) Lack of MEF2A mutations in coronary artery disease. J Clin Invest 115:1016–1020
Witztum JL, Steinberg D (1991) Role of oxidized low density lipoprotein in atherogenesis. J Clin Invest 88:1785–1792
Yang WH, Kim JE, Nam HW, Ju JW, Kim HS, Kim YS, Cho JW (2006) Modification of p53 with O-linked N-acetylglucosamine regulates p53 activity and stability. Nat Cell Biol 8:1074–1083
Yao D, Taguchi T, Matsumura T, Pestell R, Edelstein D, Giardino I, Suske G, Rabbani N, Thornalley PJ, Sarthy VP, Hammes HP, Brownlee M (2007) High glucose increases angiopoietin-2 transcription in microvascular endothelial cells through methylglyoxal modification of mSin3A. J Biol Chem 282:31038–31045
Yevdokimova N, Wahab NA, Mason RM (2001) Thrombospondin-1 is the key activator of TGF-beta1 in human mesangial cells exposed to high glucose. J Am Soc Nephrol 12:703–712

Aging

2

Toshio Hayashi

Abstract

Average life expectancy becomes longer all over the world. Japan is one of the fastest aging nations in the world. With the addition of the low birthrate, the population aging rate (ratio of population 65 years or older) is 23 % and late elderly aged 75 years or older comprise over 10 % of the population. The other character in Japan is the small number of people with independent activities of daily living (ADL). Medical and care expenses for elderly are continuously increasing and the government and general population are interested in gerontology, geriatric care, and sometimes antiaging. Gerontology divides aging into (1) physiological aging, (2) pathological aging/geriatric syndrome, and (3) geriatric disease.

1. Physiological aging is universal, intrinsic, and progressive. It is an inevitable, preprogramed progression and fundamentally difficult to treat, recover from, or prevent. Many longevity genes have been discovered, from those in nematodes to those in mice. Remarkable results have been achieved in the development of longevity in living organisms using molecular biological techniques and disease research from a cellular aging level.
2. Pathological aging occurs in the elderly and has almost the same definition as geriatric syndrome. It has various causes and nursing care is required during treatment. In Western countries, it is called the "Geriatric Giant" due to this care requirement.
3. Geriatric diseases reduce life span and declined functional residual capacity in the elderly. Cerebrovascular attack (CVA) and ischemic heart disease (IHD) both fall into this category. Cellular aging, the basic subunit of physiological aging, undergoes dynamic changes in pathological conditions and may be an important onset factor for geriatric disease. We discuss the effects of aging on cardiovascular disease from these viewpoints.

T. Hayashi, MD, PhD
Department of Geriatrics, Nagoya University Graduate School of Medicine,
Tsuruma-cho 65, Showa-ku, Nagoya, Aichi 466-8550, Japan
e-mail: hayashi@med.nagoya-u.ac.jp

I. Wakabayashi, K. Groschner (eds.), *Interdisciplinary Concepts in Cardiovascular Health*,
DOI 10.1007/978-3-7091-1334-9_2,

Aging is one of independent cardiovascular risks. Recent advances in research are revealing the detailed mechanisms of the effect of aging in the human cardiovascular disease. Some effects of aging are different on people and we are also able to regulate the effects partially. Dividing the status of patients into physiological aging, pathological aging = geriatric syndrome, and disease may help the approach described as above.

Keywords
Late elderly • Physiological aging • Geriatric syndrome • Geriatric disease • Cellular aging

2.1 Background

Average life expectancy for Japanese people is the highest in the world, at 79.64 years for males and 86.39 years for females (Muramatsu and Akiyama 2011). There is also a long life expectancy after 80 years of age – 8 years for males and 11 years for females (Abridged Life Tables for Japan 2010). Japan is one of the fastest aging nations in the world, with the period required for the percentage of population 65 years or older to double from 7 to 14 % less than half that in Western countries (Declining Birth Rate and Aging Population). With the addition of the low birthrate, the population aging rate (ratio of population 65 years or older) is 23 % and late elderly aged 75 years or older comprise over 10 % of the population (Muramatsu and Akiyama 2011). With the national character tending to emphasize prolonging life, the small number of people with independent activities of daily living (ADL) and high proportion of people dependent on care are relative to the high proportion of late elderly living alone or as elderly couples (the utilization rate of nursing care insurance for late elderly is more than 30 %) (Nakamura et al. 2007). Medical expenses are also high and there is intense interest in the government and general population regarding gerontology and geriatric care (Matsuura and Sasaki 2012). Gerontology divides aging into (1) physiological aging, (2) pathological aging/geriatric syndrome, and (3) geriatric disease (Plante 2003).

1. Physiological aging is universal, intrinsic, and progressive. It is defined as being an inevitable, preprogramed progression from birth through development to maturity. To individuals, it means progressive, irreversible damage. It is fundamentally difficult to treat, recover from, or prevent. Although wrinkled and discolored skin is a typical symptom, UV rays are actually heavily involved, making differentiation from pathological aging difficult (Zouboulis and Makrantonaki 2011). The medical field has begun to consider the mechanisms of physiological aging, and many longevity genes have been discovered, from those in nematodes to those in mice (Tomás-Loba et al. 2008). Remarkable results have been achieved in the development of longevity in living organisms using molecular biological techniques and disease research from a cellular aging level.
2. Pathological aging often occurs in the elderly and is a generic term for specific symptoms that has almost the same definition as geriatric syndrome. It has

various causes but refers to a series of symptoms and findings in which nursing care is required during treatment. In Western countries, it is called the "Geriatric Giant" due to this care requirement and acts as an important cue for elderly medical care (Isaacs 1992).

3. Geriatric disease leads to a reduced life span and involves environmental changes in addition to the decline in functional residual capacity in the elderly or a state where overlapping pathological factors and pathological aging have caused the homeostasis of biological functions to breakdown. CVA and IHD both fall into this category. Cellular aging, which is the basic subunit of physiological aging, undergoes dynamic changes in pathological conditions and is now understood as an important onset factor for geriatric disease itself.

Next, we will discuss the effects of aging on cardiovascular disease from the viewpoints of pathological aging and geriatric disease.

2.2 Pathological Aging: Geriatric Syndrome

Pathological aging/geriatric syndrome is located somewhere between physiological aging and geriatric disease. Since the 1982 report on unexplained syncope in elderly patients, patients 50 years or older came to be included and patients have been grouped into three groups depending upon the extent of age-related changes (Steinman et al. 2011; Sonohara et al. 2008). Osteoporosis, urinary incontinence, atherosclerosis, decreased immunity, thrombus formation, and changes in coagulability are also involved. The proportion of arteriosclerotic diseases (IHD, CVA, etc.) that comprise causes of death and nursing dependency increase with age, with their proportion surpassing that of cancer in patients aged 80 years or older (Besdine and Wetle 2010). Senescence or aging is an important risk factor for cardiovascular disease (Strait and Lakatta 2012). In other words, in addition to coronary risk factors such as dyslipidemia and diabetes, the fixed factors of age and sex (menopause) are involved in the progressive atherosclerosis seen in elderly patients. Meanwhile, although diabetes has a multitude of complications, including three major complications, vascular endothelial injury culminating in atherosclerosis is the cause of death in over 50 % of cases (Coccheri 2007). Elderly people exhibit postprandial hyperglycemic impaired glucose tolerance (IGT) with aging and diabetes is highly prevalent in elderly populations. IGT is an independent arteriosclerotic risk factor (Bowman and Armitage 2002). Cellular aging is also related to individual aging, atherosclerosis pathophysiology, severity of diabetes, and cardiovascular disease.

2.3 Cellular Aging: NO and Telomerase

The bioavailability of NO, which is implicated in most vascular endothelial dysfunction, is limited by cellular aging, and scientists are beginning to further understand effects of NO on aging itself and interaction between coronary risk factors, NO, and aging.

2.3.1 Cellular Aging and Telomerase and NO

Cellular aging is often observed in the skin, digestive organs, and vascular endothelial cells, which have fast metabolisms, and is accompanied by reduced NO utility in the endothelium. Conversely, NO action on aging has only been the subject of a short account a number of years ago, and there remain many unclear points (Vasa et al. 2000). Its relationship with elderly atherosclerosis is also intriguing. Telomeres are repetitive DNA sequences found in the termini of chromosomes in eukaryotic organisms that shorten each time cell division occurs. Aging caused by telomere length shortening is defined as replicative senescence and is roughly consistent with physiological aging. Telomere length in lymphocytes as an aging indicator has significant individual variation, and in recent years, significance of telomere length in different cell types such as mononuclear cells and endothelial cells has been debated (Sampson et al. 2006). Telomerase synthesizes with new telomere repeats in RNA protein conjugation and repairs telomere length. Telomerase expression level and function are strongly related to the aging of normal cells and the immortalization of cancer cells. Telomerase activation in vascular endothelial cells occurs in less than 10 % of tumor cells, but measurement has only recently become possible (Hayashi et al. 2006). In elderly aortas during autopsy, excluding the luminal necrotic layer, an aging indicator substance, senescence-associated β-galactosidase (SA-β-gal), existing in lysosomes was observed in sites of arteriosclerotic progression (Fig. 2.1). SA-β-gal is considered as the sign of cellular aging, and it was also observed in the coronary artery in vascular bifurcation of inner membrane surfaces with significant atherosclerosis (Fig. 2.1; Hayashi et al. 2006).

In NO donors in which tolerance is unlikely to occur, DETA/NO lowered SA-β-gal activation in concentration and time dependency in HUVECs and increased telomerase activation (Hayashi et al. 2006). When eNOS was introduced into HEK293 cells that have no intrinsic NOS, both NO production and telomerase activation increased. Similarly, when introduced into HUVEC, NO production increased, SA-β-gal-positive cells decreased, and telomerase was activated.

2.3.2 Dyslipidemia and Cellular Aging

Increased incidence and severity of risk factors including dyslipidemia and hypertension associated with aging and aging itself have contributed to the saying, "a man is as old as his arteries (word of Sir William Osler, 1891, from Lim and Townsend 2009)." Human LDL values increase in both males and females from approximately 1 year after birth to eventually reach a value four times that of other mammals (approximately 25 mg/dl) (Brown et al. 1980). Incidences of cases with TC of 220 mg/dl or higher have increased by approximately twofold in elderly people from 1980 to 1990. In 1990, approximately 25 % of males and 45 % of females showed plasma TC higher than 220 mg/dl (JAPAN Guideline for Diagnosis and

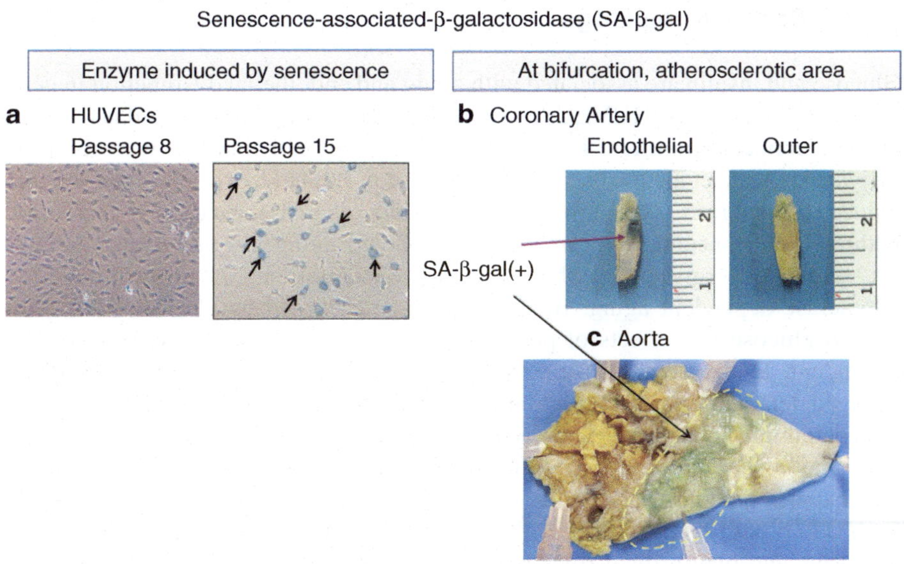

Fig. 2.1 SA-β-gal activity as cellular senescence. (**a**) Representative photographs of SA-β-gal staining in passage 8 and 15 of human umbilical venous endothelial cells (HUVECs). (**b**) SA-β-gal-positive staining was observed in atherosclerotic lesions of the intimal side of human coronary artery, which was obtained by autopsy. No staining was detected in the outer surface and normal surface. (**c**) SA-β-gal-positive staining was observed in atherosclerotic lesions of the intimal side of human thoracic aorta, which was obtained by autopsy. No staining was detected in the non-atherosclerotic area and advanced atherosclerotic area, including the necrotic core and ulcer complicated lesion (Cited from Hayashi et al. 2006)

Prevention of Atherosclerotic Cardiovascular Diseases 2007). For plasma LDL cholesterol levels, almost the same situation occurred. IHD incidence also increases with age and human serum lipid itself is involved in intimal hypertrophy and atherosclerosis formation.

2.3.3 Diabetes Mellitus and Cellular Aging

HUVEC eNOS protein and NO production decreased with 72 h of hyperglycemia administration (Hayashi et al. 2006). These levels recovered by 30 % with the addition of NOS substrates (L-arginine, etc.) or antioxidants (vitamins C and E) (Hayashi et al. 2005). At a normal glucose concentration, NOS substrates and inhibitors do not affect cellular aging. To state this differently, culture stress alone does not cause eNOS substrate deficiency or uncoupling. Effects were additive, with SA-β-gal-positive cells increasing with high glucose treatment and NOS substrate and antioxidants reducing positive cell count (Hayashi et al. 2008a).

2.3.4 Recent Knowledge

Glucose and insulin are associated with aging and endothelial dysfunction in geriatric diabetic patients: HUVECs and HAECs were cultured for 3–28 days at each glucose concentration, and the effect of insulin on cell function was observed (Miyazaki-Akita et al. 2007). Elevated glucose promoted SA-β-gal activation and reduced telomerase activation. NO metabolites decreased and ROS increased. Although all concentrations of insulin promoted aging at normal glucose levels, under hyperglycemia, physiological concentrations of insulin inhibited telomerase-dependent aging. Both cellular aging caused by eNOS siRNA with normal glucose and effects of physiological concentrations of insulin were NO dependent. Hyperglycemia promoted endothelial cell aging and insulin effects changed depending on the conditions, with NO and reactive oxygen also involved. The potential mechanisms underlying the ability of NO to prevent endothelial cell senescence and the possible changes in the NO-mediated anti-senescence effect under pathological conditions are schematically depicted in Fig. 2.2a, b. Forkhead box O (FOXO) transcription factors are involved in multiple signaling pathways and play critical roles in a number of physiological and pathological processes, including differentiation, proliferation, and survival. FOXO is a mammalian homolog of Daf-16, and activation of Akt leads to phosphorylation of FOXO and thereby inhibits the transcription of antioxidant genes such as manganese superoxide dismutase. Finally, FOXO can affect TERT (telomerase reverse transcriptase) activity by regulating levels of ROS under enough amount of BH_4 (tetrahydrobiopterin, Fig. 2.2a). Alternatively, when the signaling pathway such as insulin/IGF-1/PI3-K/Akt is activated, decreased FOXO activity due to phosphorylation leads to decreased FOXO-dependent expression of antioxidant genes (Fig. 2.2b), which is considered to be associated with the potential mechanisms of FOXO-related senescence. It may also contribute to atherosclerosis formation that continues for more than a year with replicative senescence. Intermittent hyperglycemic stimulation also caused cellular aging in a postprandial hyperglycemia model (Hayashi et al. 2010).

Endothelial cell signal transmission including eNOS is performed in caveolae that cover more than a certain percentage of the cell surface area. This percentage changes in tandem with glucose. When one observes THP-1 cells, macrophage system cells with atomic microscopy, localization of NADPH oxidase changes according to glucose concentration (Hayashi et al. 2007a). When prognosis of late-stage aging was looked at with a range of physiologically active substances and nutrition indices, albumin and (not cytokines of BNP) NO metabolite concentrations were significant biomarkers (Hayashi et al. 2007b). Even with the addition of comprehensive functional assessment of the elderly (physical and mental evaluation including cognitive function and ADL), NO metabolites were useful markers comparable to basic ADLs. In our scientific research for the MHLW, we conducted an observational study of 4,014 diabetic patients with a mean age of 67.4 years and reported that HDL was a risk factor for CVA in late-stage elderly (Hayashi et al. 2008b, 2009). Even now, there is no other treatment for maintaining and raising HDL than exercise, and this is related to the fact that ADL and NO have been recognized in elderly prognosis. HDL and eNOS act reciprocally (Mineo and Shaul 2003).

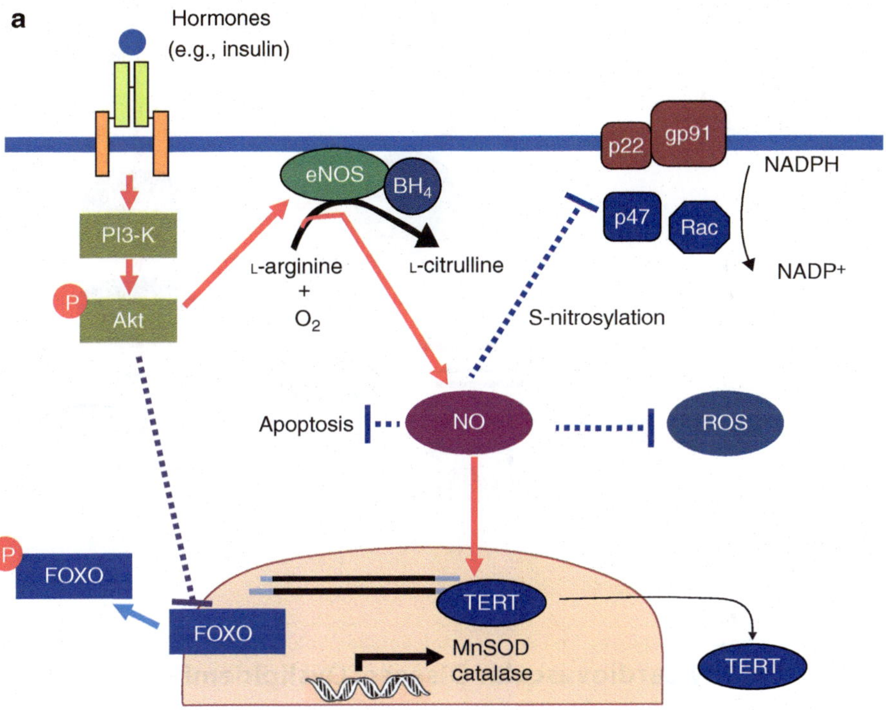

Fig. 2.2 (**a**) Anti-senescence effect of NO in endothelial cells. Under the normal homeostasis condition, endothelial NO is synthesized from L-arginine and oxygen in a reaction catalyzed by eNOS. Hormones such as insulin or estrogen activate eNOS expression through the PI3-K/Akt pathway and/or direct effect via promoter or stabilization of eNOS mRNA. They may also inactivate NADPH oxidase. NO shows antiapoptotic effects and suppresses ROS production by scavenging directly or preventing NADPH, which would result in depletion of TERT export to the cytoplasm from the nucleus. Activation of the PI3-K/Akt pathway can also downregulate FOXO, leading to decreased FOXO-dependent expression of antioxidant genes. However, with the normal condition of endothelium, the balance between the antioxidant effects of NO and ROS production may be kept well. *Solid lines* represent positive regulatory pathways. *Dotted lines* represent negative regulatory pathways. (**b**) Progress of endothelial senescence under pathological conditions. Both the decrease in eNOS expression by the decline in the hormonal signal with age or insulin resistance and the increase in ROS production by endogenous or exogenous sources cause progression of endothelial senescence by unbalance of NO and ROS. Additionally reduction in BH_4 under pathological conditions such as diabetes causes eNOS uncoupling. Under such an uncoupling state, eNOS no longer produces NO and instead generates superoxide. Then, stimulation of the PI3-K/Akt pathway causes not only downregulation of FOXO but also increased eNOS uncoupling, which may result in accelerating endothelial senescence. *Solid lines* represent positive regulatory pathways. *Dotted lines* represent negative regulatory pathways (Cited from Hayashi et al. 2008a)

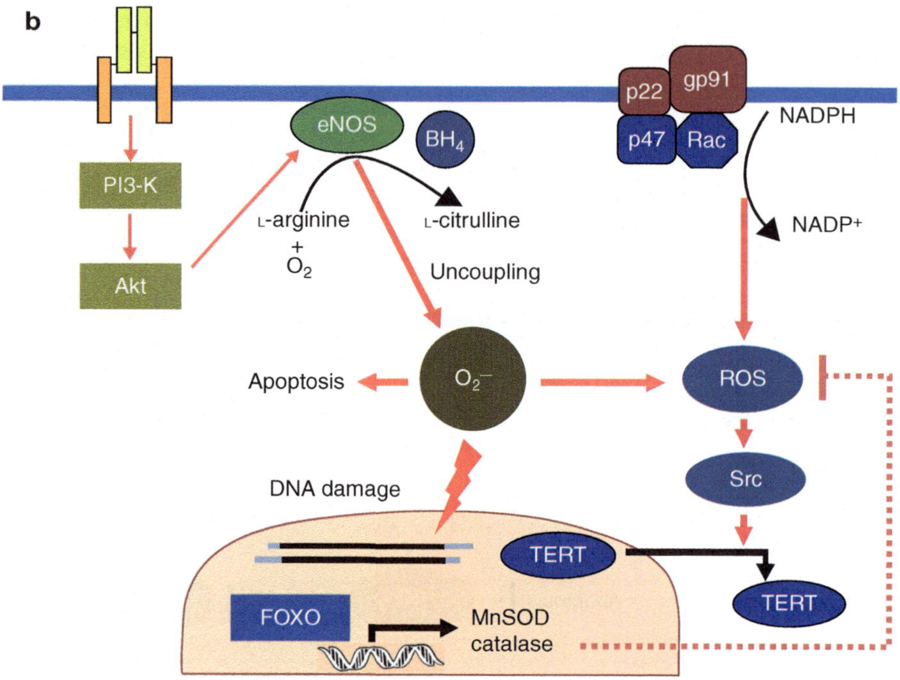

Fig. 2.2 (continued)

2.4 Elderly Cardiovascular Disease: Dyslipidemia

Rankings for causes of death change with aging, and in the late-stage elderly, atherosclerosis holds a higher position than malignancies (Besdine and Wetle 2010). It has been suggested that the increased incidence and severity of risk factors including dyslipidemia and hypertension associated with aging as well as aging itself have contributed to the saying, "a man is as old as his arteries" (word of Sir William Osler, 1891, from Lim and Townsend 2009).

2.4.1 Dyslipidemia

High blood total cholesterol (TC) and low blood HDL cholesterol are also independent risk factors for IHD, CVA (cortical branch), and arteriosclerosis obliterans (ASO) in the elderly. The significance of high blood triglycerides as a risk factor for these atherosclerotic diseases in the elderly is unknown. The secondary preventive effects of anti-dyslipidemic pharmaceuticals on IHD have been proven, even in elderly patients, with large-scale trials. Two large-scale trials conducted in Japan reported primary preventive effects of therapy for dyslipidemia on the risk of cardiovascular disease in subjects that included female patients, but the trials only included some early-stage elderly subjects (Mizuno et al. 2008; Yokoyama et al. 2007). In the West, an investigation including late-elderly subjects (Proper et al.) reported its usefulness, but effects were not as pronounced as expected (Heart Protection Study Collaborative Group 2002; Sever et al. 2003; Shepherd et al. 2002).

Looking at age-related changes in lipid metabolism, the elderly have a lower actual lipid intake than younger people and the rate of absorption from the digestive tract increases slightly with age (Saltzman and Russell 1998). The hepatic ability and rate of apoprotein synthesis do not change with age – production rate of hepatic VLDL increases and LDL receptors decrease. LDL activation decreases with age and lipoproteins rich in small dense cholesterol (chylomicron and VLDL remnants) are stationary for a long time in the circulation. LDL catabolism decreases with age and lipid concentration may rise (Matthan et al. 2005). The above changes in serum lipid levels caused by aging increase the atherosclerotic propensity through increased absorption and production.

Serum TC levels have been previously shown to exhibit minimum values in both Japan and America before 20 years of age and increase along with aging. They peak in the 1960s and subsequently diminish gradually. Although recent studies have shown that TC levels decreased in men and women of almost all ages in the USA, the same trend was seen in each age group (Abbott et al. 1998). Annual examinations involving in-class screening of 20,000 people aged from 20 years to approximately 80 years were conducted for 10 years from 1970 (Sekimoto et al. 1983). It was found that both TC and LDL increased in males of all age groups from their late 20s to their 70s. TC and LDL also increased in females from the late 30s and increased dramatically when women were approximately 50 years of age (influenced by menopause). In women in their 70s, results leveled off. Successive changes over 10 years dictate that if the person is 50 years old and born in the 1920s, both males and females will have approximately 200 mg/dl. If the person is 50 years old and was born in the 1960s, this increases by 20 mg/dl. It has long been reported that TC levels in young Japanese subjects are lower than TC levels in US subjects of the same age range and the same trend can also be seen in other age groups. However, now the situation has changed. Japan Ministry of Health Labour and Welfare (MHLW) statistics have also shown the number of cases with serum TC levels of 220 mg/dl or higher to have increased approximately twofold in the elderly over the 10-year period from 1980 to 1990. In 1990, approximately 25 % of males and 45 % of females showed their plasma TC higher than 220 mg/dl (JAPAN Guideline for Diagnosis and Prevention of Atherosclerotic Cardiovascular Diseases 2007). For plasma LDL cholesterol levels, almost the same situation occurred. The Westernization of the Japanese diet (high-fat diet, increased number of processed foods, increased calorie intake) has also affected this. There was no great difference in statistical levels reported by the MHLW in 2000 when compared with results released in 1990 (Kuzuya et al. 2002). The MHLW survey also includes examinees of medical institutions, and outcomes may differ from those of screened patients. Even in Japan, the public has become more enlightened about dietary therapy, and cholesterol content has come to be displayed on food packages, like in Western nations. Statins have also come to be more widely used.

HDL levels in women fall after menopause due to estrogen deficiency, but it does not change with age in men. HDL is an independent arteriosclerotic risk factor, and the presence or absence of complications is an important consideration for treatment. Recently, the results of a large-scale clinical trial of subjects aged up to their 70s have been released. Our investigation of 4,014 diabetic patients including 1,016 late-elderly diabetic patients older than 75 years (MHLW-funded research group headed by the author) revealed that HDL-C levels and LDL-C/HDL-C ratio are important for male and female in the prevention of ischemic heart disease and cerebrovascular attack (stroke) (Hayashi et al. 2009, 2011, 2013).

There is little evidence regarding the efficacy of treatments for the prevention of cardiovascular disease in patients of 80 years or older.

2.5 Summary

Aging is one of the independent cardiovascular risks and is thought not to be able to regulate its effect on cardiovascular diseases. Recent advances in the research and findings in this field are revealing the detailed mechanisms of the effect of aging in the human cardiovascular disease. It means that some effects of aging are different on people and that we are also able to regulate the effects partially. Dividing the status of patients into physiological aging, pathological aging = geriatric syndrome, and disease may help the approach described as above.

References

Abbott RD, Yano K, Hakim AA, Burchfiel CM, Sharp DS, Rodriguez BL, Curb JD (1998) Changes in total and high-density lipoprotein cholesterol over 10- and 20-year periods (the Honolulu Heart Program). Am J Cardiol 82:172–8

Abridged Life Tables for Japan (2010) By Statistics and Information Department Minister's Secretariat Ministry of Health, Labour and Welfare Japanese Government. Available at http://www.mhlw.go.jp/english/database/db-hw/lifetb10/index.html

Besdine RW, Wetle TF (2010) Improving health for elderly people: an international health promotion and disease prevention agenda. Aging Clin Exp Res 22:219–30

Bowman L, Armitage J (2002) Diabetes and impaired glucose tolerance: a review of the epidemiological and trial evidence for their role in cardiovascular risk. Semin Vasc Med 2:383–9

Brown MS, Kovanen PT, Goldstein JL (1980) Evolution of the LDL receptor concept-from cultured cells to intact animals. Ann N Y Acad Sci 348:48–68

Coccheri S (2007) Approaches to prevention of cardiovascular complications and events in diabetes mellitus. Drugs 67:997–1026

Declining Birth Rate and Aging Population by Statistics Bureau, Director-General for Policy Planning (Statistical Standards and Statistical Research and Training Institute). Available at http://www.stat.go.jp/english/data/handbook/c02cont.htm

Hayashi T, Juliet PA, Matsui-Hirai H, Miyazaki A, Fukatsu A, Funami J, Iguchi A, Ignarro LJ (2005) l-Citrulline and l-arginine supplementation retards the progression of high-cholesterol-diet-induced atherosclerosis in rabbits. Proc Natl Acad Sci USA 102:13681–6

Hayashi T, Matsui-Hirai H, Miyazaki-Akita A, Fukatsu A, Funami J, Ding QF, Kamalanathan S, Hattori Y, Ignarro LJ, Iguchi A (2006) Endothelial cellular senescence is inhibited by nitric oxide: implications in atherosclerosis associated with menopause and diabetes. Proc Natl Acad Sci U S A 103:17018–23

Hayashi T, Juliet PA, Miyazaki A, Ignarro LJ, Iguchi A (2007a) High glucose downregulates the number of caveolae in monocytes through oxidative stress from NADPH oxidase: implications for atherosclerosis. Biochim Biophys Acta 1772:364–72

Hayashi T, Nomura H, Osawa M, Funami J, Miyazaki A, Iguchi A (2007b) Nitric oxide (NO) metabolites are associated with survival in older patients. J Am Geriatr Soc 55:1398–403

Hayashi T, Yano K, Matsui-Hirai H, Yokoo H, Hattori Y, Iguchi A (2008a) Nitric oxide and endothelial cellular senescence. Pharmacol Ther 120:333–9

Hayashi T, Kawashima S, Itoh H, Yamada N, Sone H, Watanabe H, Hattori Y, Ohrui T, Yoshizumi M, Yokote K, Kubota K, Nomura H, Umegaki H, Iguchi A, ; Japan CDM group (2008b) Importance of lipid levels in elderly diabetic individuals: baseline characteristics and 1-year survey of cardiovascular events. Circ J 72:218–25

Hayashi T, Kawashima S, Itoh H, Yamada N, Sone H, Watanabe H, Hattori Y, Ohrui T, Yokote K, Nomura H, Umegaki H, Iguchi A, ; Japan CDM Group (2009) Low HDL cholesterol is associated with the risk of stroke in elderly diabetic individuals: changes in the risk for atherosclerotic diseases at various ages. Diabetes Care 32:1221–3

Hayashi T, Maeda M, Ina K, Matsui-Hirai H, Iguchi A (2010) The diverse role of insulin on endothelial senescence - the difference of continuous and intermittent high glucose. Nitric Oxide Biol Chem 22:S94–S95

Hayashi T, Kawashima S, Nomura H, Itoh H, Watanabe H, Ohrui T, Yokote K, Sone H, Hattori Y, Yoshizumi M, Ina K, Kubota K, ; Japan Cholesterol and Diabetes Mellitus Investigation Group (2011) Age, gender, insulin and blood glucose control status alter the risk of ischemic heart disease and stroke among elderly diabetic patients. Cardiovasc Diabetol 10:86–91

Hayashi T, Araki A, Kawashima S, Sone H, Watanabe H, Ohrui T, Yokote K, Takemoto M, Kubota K, Noda M, Noto H, Ina K, Nomura H (2013) Metabolic predictors of ischemic heart disease and cerebrovascular attack in elderly diabetic individuals: difference in risk by age. Cardiovasc Diabetol 12:10–6

Heart Protection Study Collaborative Group (2002) MRC/BHF Heart Protection Study of cholesterol lowering with simvastatin in 20,536 high-risk individuals: a randomised placebo-controlled trial. Lancet 360(9326):7–22

Isaacs B (1992) The challenge of geriatric medicine. Oxford University Press, London

JAPAN Guideline for Diagnosis and Prevention of Atherosclerotic Cardiovascular Diseases (2007) Edition available at http://www.j-athero.org/en/publications/guideline2007.html

Kuzuya M, Ando F, Iguchi A, Shimokata H (2002) Changes in serum lipid levels during a 10 year period in a large Japanese population. A cross-sectional and longitudinal study. Atherosclerosis 163:313–20

Lim MA, Townsend RR (2009) Arterial compliance in the elderly: its effect on blood pressure measurement and cardiovascular outcomes. Clin Geriatr Med 25:191–205

Matsuura T, Sasaki M (2012) Can the health insurance reforms stop an increase in medical expenditures for middle- and old-aged persons in Japan? Int J Health Care Finance Econ 12:1–25

Matthan NR, Jalbert SM, Lamon-Fava S, Dolnikowski GG, Welty FK, Barrett HR, Schaefer EJ, Lichtenstein AH (2005) TRL, IDL, and LDL apolipoprotein B-100 and HDL apolipoprotein A-I kinetics as a function of age and menopausal status. Arterioscler Thromb Vasc Biol 25:1691–6

Mineo C, Shaul PW (2003) HDL stimulation of endothelial nitric oxide synthase: a novel mechanism of HDL action. Trends Cardiovasc Med 13:226–31

Miyazaki-Akita A, Hayashi T, Ding QF, Shiraishi H, Nomura T, Hattori Y, Iguchi A (2007) 17beta-estradiol antagonizes the down-regulation of endothelial nitric-oxide synthase and GTP cyclohydrolase I by high glucose: relevance to postmenopausal diabetic cardiovascular disease. J Pharmacol Exp Ther 320:591–8

Mizuno K, Nakaya N, Ohashi Y, Tajima N, Kushiro T, Teramoto T, Uchiyama S, Nakamura H, ; MEGA Study Group (2008) Usefulness of pravastatin in primary prevention of cardiovascular events in women: analysis of the Management of Elevated Cholesterol in the Primary Prevention Group of Adult Japanese (MEGA study). Circulation 117:494–502

Muramatsu N, Akiyama H (2011) Japan: super-aging society preparing for the future. Gerontologist 51:425–32

Nakamura A, Ooe Y, Ohno Y, Nakamura T (2007) A three-year-follow-up study on the change in physical and mental functions of the aged by the level of ADL. Evaluation of data from the Japanese elderly care insurance certification of long-term care need. Jpn Hosp 26:41–52

Plante GE (2003) Impact of aging on the body's vascular system. Metabolism 52:31–5

Saltzman JR, Russell RM (1998) The aging gut. Nutritional issues. Gastroenterol Clin North Am 27:309–24

Sampson MJ, Winterbone MS, Hughes JC, Dozio N, Hughes DA (2006) Monocyte telomere shortening and oxidative DNA damage in type 2 diabetes. Diabetes Care 29:283–9

Sekimoto H, Goto Y, Goto Y, Naito C, Yasugi T, Okido M, Kuzuya F, Takeda R, Yamamoto A, Fukuzaki H et al (1983) Changes of serum total cholesterol and triglyceride levels in normal subjects in Japan in the past twenty years. Research committee on familial hyperlipidemia in Japan. Jpn Circ J 47:1351–8

Sever PS, Dahlöf B, Poulter NR, Wedel H, Beevers G, Caulfield M, Collins R, Kjeldsen SE, Kristinsson A, McInnes GT, Mehlsen J, Nieminen M, O''Brien E, Ostergren J, ASCOT investigators (2003) Prevention of coronary and stroke events with atorvastatin in hypertensive patients who have average or lower-than-average cholesterol concentrations, in the Anglo-Scandinavian Cardiac Outcomes Trial-Lipid Lowering Arm (ASCOT-LLA): a multicentre randomised controlled tria. Lancet 361:1149–58

Shepherd J, Blauw GJ, Murphy MB, Bollen EL, Buckley BM, Cobbe SM, Ford I, Gaw A, Hyland M, Jukema JW, Kamper AM, Macfarlane PW, Meinders AE, Norrie J, Packard CJ, Perry IJ, Stott DJ, Sweeney BJ, Twomey C, Westendorp RG, PROSPER Study Group (2002) PROspective Study of Pravastatin in the Elderly at Risk. Pravastatin in elderly individuals at risk of vascular disease (PROSPER). Lancet 360:1623–30

Sonohara K, Kozaki K, Akishita M, Nagai K, Hasegawa H, Kuzuya M, Yokote K, Toba K (2008) White matter lesions as a feature of cognitive impairment, low vitality and other symptoms of geriatric syndrome in the elderly. Geriatr Gerontol Int 8:93–100

Steinman MA, Lund BC, Miao Y, Boscardin WJ, Kaboli PJ (2011) Geriatric conditions, medication use, and risk of adverse drug events in a predominantly male, older veteran population. J Am Ger Soc 59:615–21

Strait JB, Lakatta EG (2012) Aging-associated cardiovascular changes and their relationship to heart failure. Heart Fail Clin 8:143–64

Tomás-Loba A, Flores I, Fernández-Marcos PJ, Cayuela ML, Maraver A, Tejera A, Borrás C, Matheu A, Klatt P, Flores JM, Viña J, Serrano M, Blasco MA (2008) Telomerase reverse transcriptase delays aging in cancer-resistant mice. Cell 135:609–22

Vasa M, Breitschopf K, Zeiher AM, Dimmeler S (2000) Nitric oxide activates telomerase and delays endothelial cell senescence. Circ Res 87:540–2

Yokoyama M, Origasa H, Matsuzaki M, Matsuzawa Y, Saito Y, Ishikawa Y, Oikawa S, Sasaki J, Hishida H, Itakura H, Kita T, Kitabatake A, Nakaya N, Sakata T, Shimada K, Shirato K, ; Japan EPA lipid intervention study (JELIS) Investigators (2007) Effects of eicosapentaenoic acid on major coronary events in hypercholesterolaemic patients (JELIS): a randomised open-label, blinded endpoint analysis. Lancet 369(9567):1090–8

Zouboulis CC, Makrantonaki E (2011) Clinical aspects and molecular diagnostics of skin aging. Clin Dermatol 29:3–14

Gender

3

Toshio Hayashi

Abstract

In all age groups from childhood to old age, women have a longer average life expectancy than men, suggesting that sex (specifically, sex hormones) is closely related to geriatric diseases and aging. Disease structure changes as people age, with cardiovascular disease increasing dramatically in late elderly (older than 75 years of age) females and arteriosclerotic diseases surpassing malignancies as the leading cause of death. After menopause, however, this number increases dramatically, thus reducing the gender gap. The prevalence of cardiovascular disease in women aged 75 years or older is thus equal to that of men of the same age. As this age group has a greater population of females than males, women account for a greater number of cardiovascular disease cases. We will discuss recent knowledge, including the gender differences apparent in risk factors such as dyslipidemia and hypertension. Cardiovascular disease is a typical geriatric disease, and pathological aging contributes to estrogen action and sex differences in atherosclerosis.

Women have a longer average life expectancy than men, and sex hormones, especially estrogen, is closely related to geriatric diseases and aging. Disease structure changes as people age, with cardiovascular disease increasing dramatically in late elderly females and arteriosclerotic diseases surpassing malignancies as the leading cause of death. Although hormone replacement therapy for all postmenopausal women as one same dose of agents failed to prevent cardiovascular disease in elderly female, recent advance of basic and clinical gender-specific medical research made clear the molecular mechanism on the gender differences in cardiovascular diseases and coronary risk factors such as dyslipidemia. Cardiovascular disease is a typical geriatric disease, and pathological aging contributes to estrogen action and sex differences in atherosclerosis, which

T. Hayashi, MD, PhD
Department of Geriatrics, Nagoya University Graduate School of Medicine,
Tsuruma-cho 65, Showa-ku, Nagoya, Aichi 466-8550, Japan
e-mail: hayashi@med.nagoya-u.ac.jp

I. Wakabayashi, K. Groschner (eds.), *Interdisciplinary Concepts in Cardiovascular Health*,
DOI 10.1007/978-3-7091-1334-9_3, © Springer-Verlag Wien 2013

form the main etiology of cardiovascular disease. These gender-dependent differences in risk are important for an individualized strategy to prevent atherosclerotic disease.

Keywords
Estrogen • Nitric oxide • Selective estrogen receptor modulators (SERM) • Gender-based medicine • Hormone replacement therapy

3.1 Introduction

In all age groups from childhood to old age, women have a longer average life expectancy than men, suggesting that sex (specifically, sex hormones) is closely related to geriatric diseases and aging. Disease structure changes as people age, with cardiovascular disease increasing dramatically in late elderly (older than 75 years) females and arteriosclerotic diseases surpassing malignancies as the leading cause of death (Muramatsu and Akiyama 2011). Prior to menopause, fewer women suffer from cardiovascular disease than men. After menopause, however, this number increases dramatically, thus reducing the gender gap (The Lipid Research Clinics Program Epidemiology Committee 1979; Hayashi et al. 2000a). The prevalence of cardiovascular disease in women aged 75 years or older is thus equal to that of men of the same age. As this age group has a greater population of females than males, women account for a greater number of cardiovascular disease cases. This chapter will discuss recent knowledge, including the sex differences apparent in risk factors such as dyslipidemia and hypertension. Aging or senescence is divided into (1) physiological aging, (2) pathological aging/geriatric syndrome, and (3) geriatric disease. Cardiovascular disease is a typical geriatric disease, and pathological aging (The Lipid Research Clinics Program Epidemiology Committee 1979) contributes to estrogen action and sex differences in atherosclerosis, which form the main etiology of cardiovascular disease (Besdine and Wetle 2010).

3.2 Pathophysiology of Sex Differences Apparent in Atherosclerosis: The Anti-atherosclerotic Effects of Estrogen

3.2.1 Introduction

The effect of estrogen on serum lipids has been thoroughly investigated.

After estrogen receptors (ERα) were implicated in the vascular effect mechanism in the 1990s, the presence of a new estrogen receptor (estrogen receptor β,

ERβ) was confirmed in vascular endothelium and smooth muscle (Nakao et al. 1991; Mosselman et al. 1996). The anti-atherosclerotic action exhibited by nitric oxide (NO), which is largely responsible for vascular endothelial function, has also been the subject of research (Hayashi et al. 1989; Flavahan 1992). Shear stress is an important factor in atherosclerosis onset, and we discovered that basal secretion of NO in extracted aortas of domestic rabbits was higher in females (Hayashi et al. 1992).

3.2.2 The Serum Lipid-Mediating Action of Estrogen

Lipid profiles, especially the character of LDL cholesterol, can account for the gender and species differences of ischemic cardiovascular diseases (Fig. 3.1 left and right).

In other words, in animals other than humans, plasma low-density lipoprotein cholesterol (LDL-C) is constant around 25 mg/dl all their life; however, in humans, LDL-C increases up to four times in the first year after birth, and serum lipid itself plays a mandatory role in intimal hypertrophy and atherosclerosis formation (Fig. 3.1, right; Brown et al. 1980). In familial hypercholesterolemia, LDL-C increases two to ten times higher than normal human plasma LDL-C levels. There are no sex differences in LDL levels, and high-density lipoprotein cholesterol (HDL-C) levels are approximately 10 mg/dl higher in females than in males. Estrogen promotes hepatic apolipoprotein A-1 synthesis and inhibits hepatic triacylglycerol (TG) lipase activation. LDL is taken up from the circulation due to accelerated very-low-density lipoprotein (VLDL) production and an increased number of hepatocyte LDL receptors. LDL oxidization is inhibited and coronary risk factor lipoprotein(a) (Lp(a)) is also lowered by estrogen (Brinton 1996; Su et al. 1998). Menopause results in negation of estrogenic effects, leading to lipid changes, and explains approximately 50 % of the postmenopausal cases of rapid atherosclerotic progression (Holm et al. 1999). This drastic change of lipid profile in postmenopausal women reflects the gender differences and change after its 50 years of incidence of ischemic cardiovascular changes (Fig. 3.1, left).

3.2.3 Effects of Estrogen on Nitric Oxide Synthase (NOS)

Most of the remaining anti-atherosclerotic effects of estrogen are believed to arise from direct effects on the vascular wall and, in particular, the endothelium. Estradiol (E2) provides antioxidant action. We discovered that E2 has biphasic action that makes NOS activation and NO production receptor dependent in bovine aortic endothelial cell (BAEC) and human umbilical vein endothelial cell (HUVEC) (Hayashi et al. 1995a). This decreases NOS activation and NO production at pharmacological concentrations (Hayashi et al. 2000a). This suggests that

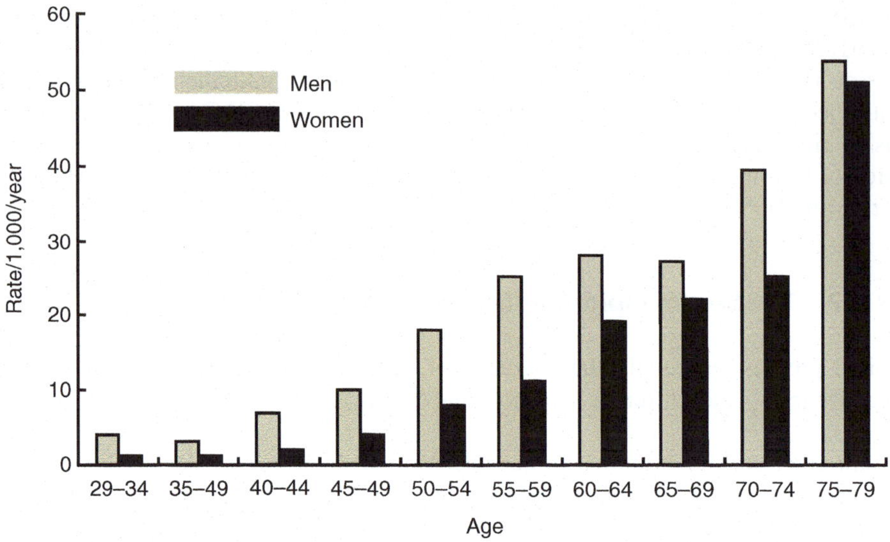

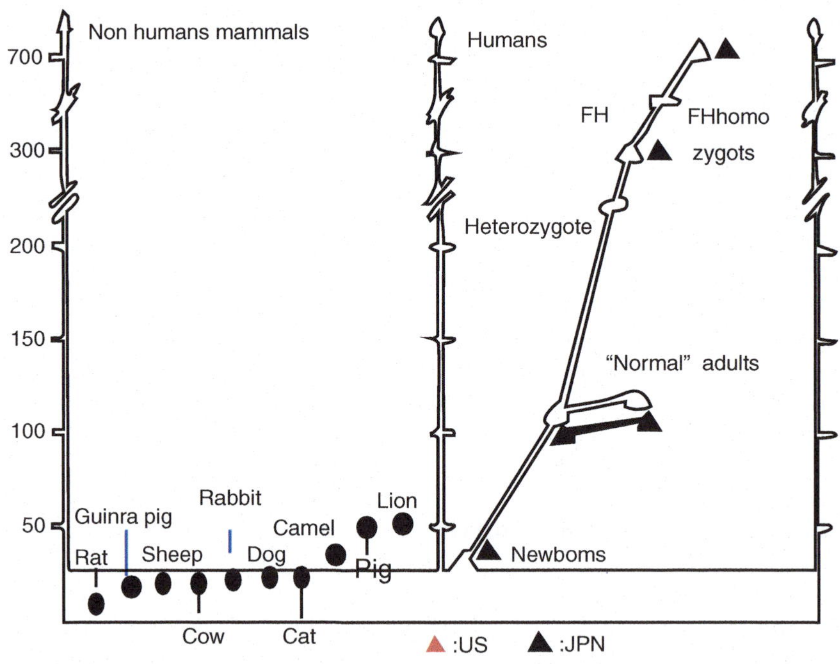

Fig. 3.1 A man is as old as his arteries with gender differences. *Left*: Incidence of ischemic cardiovascular disease: gender differences (From the Framingham Study 1992). *Right*: Plasma LDL concentration of human and nonhuman primates (Cited from Brown et al. (1980) and Goldstein and Brown (1982))

the endothelial nitric oxide synthase (eNOS) sites of E2 action are estrogen response element half-palindromic motifs and AP1 and SP1 regions (Simoncini et al. 2002). E2 inhibits destabilization of endothelial nitric oxide synthase messenger ribonucleic acid (eNOS mRNA) in the presence of tumor necrosis factor-alpha (TNF-α) (Sumi et al. 2001). E2 also increases the number of caveolae, which are eNOS sites of action. Estrogen action diverges at eNOS/NO. T cell- and macrophage-derived inducible nitric oxide synthase (iNOS) was observed around the necrotic foci (soft, friable plaque with abundant inflammatory cells) in the aorta of domestic rabbits and advanced coronary atherosclerotic lesions in humans (Esaki et al. 1997). E2 inhibited macrophage system cell J774 and neutrophil iNOS (Hayashi et al. 1998). Estrogen also has acute effects mediated via eNOS phosphorylation, as well as cell membrane estrogen receptor alpha (ERα) and Akt/PKB, although independent of intranuclear receptors (Hisamoto et al. 2001; Haynes et al. 2000). It also influences atherosclerotic plaque formation (including leukocyte adhesion and migration), angiogenesis/vascularization, and the myocardium. There is a high homology of ERβ with ERα DNA-binding domains, and previous reports have suggested that mediation occurs via ERβ as well as ERα. In dietary cholesterol-loaded rabbits, ERα was observed in the endothelium, while ERβ was observed in the endothelium and around the necrotic foci. When dietary cholesterol loading was conducted on male rabbits and female rabbits which had not undergone ovariectomy, atherosclerotic lesion formation in female rabbits was milder than in male rabbits (Hayashi et al. 2000a, 1995b). With regard to the NO basal secretion observed in extracted blood vessels, reduced NO secretion in females preceded lesion formation. In males, NO basal secretion was originally low. High levels of basal NO secretion in females work defensively against dietary cholesterol-loading stimuli, leading to the hypothesis that E2 has anti-atherosclerotic effects mediated via NO (Hayashi et al. 2000a, 1995b). The effect of estrogen on EDHF was also suggested (Sakuma et al. 2002).

3.2.4 SERM

Estrogen acts on the reproductive and coagulation systems in addition to the cardiovascular system, and hormone replacement therapy (HRT) side effects have obstructed its popularization (Hlatky et al. 2002; Writing Group for the Women's Health Initiative Investigators 2002). Selective estrogen receptor modulators (SERM) were developed taking this into consideration, leading to modulators such as raloxifene being made available. NO-dependent ameliorative effects on endothelial function were also reported, and it is possible that this may solve many issues related to HRT (Ettinger et al. 1999; Wenger et al. 2002).

3.3 Effects of Estrogen on Blood Vessels: Clinical Results

In the arteries of young females, flow-mediated dilation (FMD) fluctuates in tandem with estrus cycles due to estrogen, and FMD decreases after menopause (Hashimoto et al. 1995). When estriol (E3) was administered to elderly females who are more than 25 years postmenopause, however, improved FMD and increased plasma nitrogen oxide (NOx) and cyclic guanosine monophosphate (cGMP) were observed after 30 weeks (Hayashi et al. 2000c).

We created guidelines for HRT for postmenopausal Japanese patients, based on clinical research presided over by the Japanese Ministry of Health, Labour and Welfare as a Longevity Science Research Project (1998–2003) (Watanabe et al. 2004). When 305 women (246 of whom were undergoing HRT, mean age of 63 years) were administered low-dose conjugated estrogen, E3 pharmaceuticals, or E3 patches and observed for an average of 3 years, genital bleeding was observed in 20 % and mastalgia in 16 %. There does, however, appear to be a racial difference with Western patients as no ischemic heart disease, cerebrovascular accidents, or uterine, ovarian, or mammary tumors were observed (data were reported in Japanese only).

In general, HDL concentration is approximately 10 mg/dl higher in females than males, and this difference shrinks following menopause. Within approximately 2 years, HDL decreases by 5–7 %, and TC and LDL increase by approximately 10 %.

Arterial thrombosis is also affected by estrogen in a dimorphic pattern. NO-mediated antithrombotic action occurs at low concentrations, while thrombogenic steroid actions (similar to the contraceptive pill) occur at high concentrations (Battaglioli and Martinelli 2007; Shulman 2011). Deep vein thrombosis is also a potential adverse event. In the Heart and Estrogen Replacement Study (HERS) that targeted patients affected by cardiovascular disease, the most common reason for discontinuation of medication was venous thrombosis in the initial period of estrogen administration (Hlatky et al. 2002; Shulman 2011). The incidence frequency of coagulation factor gene mutations (factor V Leiden mutations) in Asian races is lower than in Western races, and venous thrombosis is uncommon (Ridker et al. 1997). In South Korea, HRT is recommended even now.

3.4 Estrogen and Cellular Aging

3.4.1 Estrogen and Cellular Aging Through NO

Estrogen affects vascular and cellular aging, which is involved in the sex differences apparent in average life expectancy.

Cellular aging markers in cultured endothelium: SA-β-gal positive = senescent cell count was reduced with physiological concentrations of estrogen. This action suggests mediation via NO, as this is inhibited by ER and NOS inhibitor nitro -L-arginine methyl ester (L-NAME) (Fig. 3.2, right; Hayashi et al. 2006a). Even at

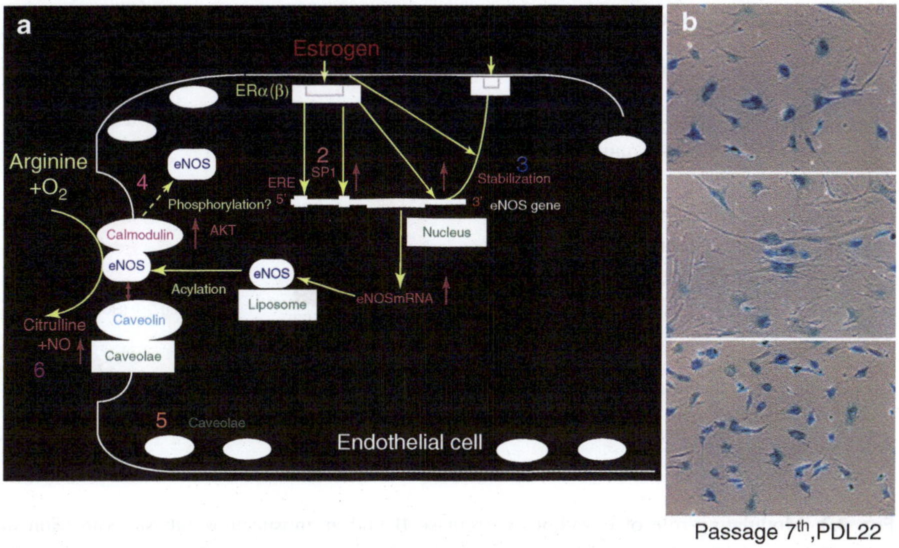

Fig. 3.2 Mechanism of NO release enhancement by estrogen and the effect of aging of estrogen. (**a**) *Left*: Under the normal homeostasis condition, endothelial NO is synthesized from L-arginine and oxygen in a reaction catalyzed by eNOS. Estrogen activates eNOS expression through the PI3-K/Akt pathway and/or direct effect via promoter or stabilization of eNOS mRNA. They may also inactivate NADPH oxidase. NO shows antiapoptotic effects and suppresses ROS production by scavenging directly or preventing NADPH. However, with the normal condition of endothelium, the balance between the antioxidant effects of NO and ROS production may be kept well. *Solid lines* represent positive regulatory pathways. *Dotted lines* represent negative regulatory pathways. *1*. Receptor mediated gender difference (Cited from Hayashi et al. 1992). *2*. Promoter (Cited from Hayashi et al. 1995a). *3*. mRNAstablization (Cited from Sumi et al. 2001). *4*. Calmodulin (Cited from Hisamoto et al. 2001). *5*. Caveolae (Cited from Hayashi et al. 1995a, 2000a). *6*. Superoxide anion (Cited from Hayashi et al. 2000b). *7*. EDHF (Sakuma et al. 2002). *8*, *9* Phosphorylation (Cited from Haynes et al. 2000, Hayashi et al. 2006a). (**b**) *Right*: Effect of estrogen on cellular senescence. Representative photographs of SA-β-gal staining in control, 10-8 M E2-treated, and 10-8M E2- and 10-4M L-NAME-treated cells. Note that treatment with E2 decreased the number of SA-β-gal-positive cells, which was prevented by further treatment with L-NAME. Cells were used in PDL 22 at passage 7 (Cited from Hayashi et al. 2006a). *TNFarufa* tumor necrosing factor arufa, *ERE* etrogen response element (a human transcription factor), *SP1* Specificity Protein 1 (a human transcription factor)

physiological estrogen concentrations, 70 % of atherosclerotic changes in dietary cholesterol-loaded domestic rabbits were inhibited, albeit without changes in serum lipid levels (Hayashi et al. 2000a). In atherosclerosis prevention, estrogen acts via aliquots of NO and exhibits anti-arteriosclerotic effects in proportion to serum estrogen concentration and tissue concentrations of the NO reaction product cGMP (Fig. 3.2, left). The mechanism of maintaining NO effects on atherosclerosis or vascular aging is not due to suppression of the citrulline–arginine system, but rather that of arginase II (Fig. 3.3; Hayashi et al. 2006b). Estrogen thus increases the bioavailability of NO in progressive atherosclerotic vascular disease.

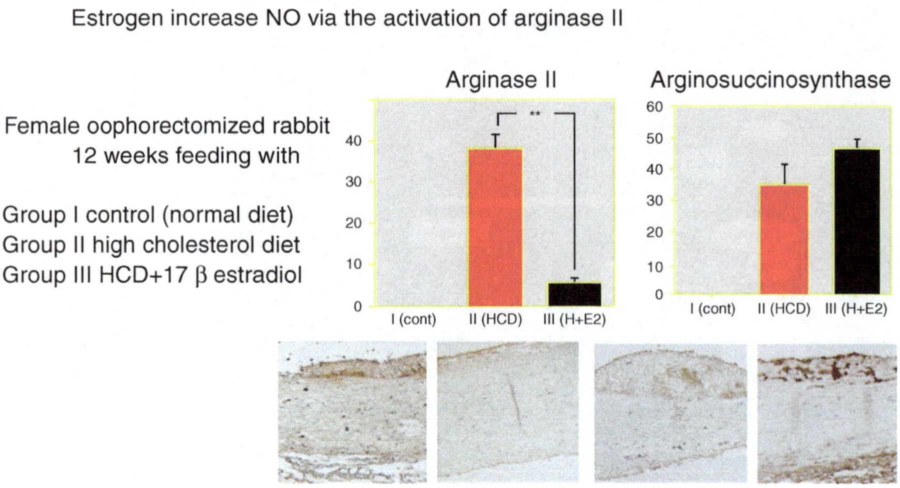

Fig. 3.3 Modulating role of estradiol on arginase II and arginosuccinosynthase expression in hyperlipidemic rabbits as an atheroprotective mechanism. *Upper*: Distribution of arginase II in atherosclerotic aortas. Immunohistochemical analysis with anti-arginase II antibody of thoracic aortas of NZW rabbits from the different groups. **, $P<0.01$. *Lower*: Immunohistochemical analysis with anti-arginase II [*lower left*: group II and *lower middle left*: group III] monoclonal antibodies and the anti-argininosuccinate synthetase monoclonal antibody [*lower middle right*: group II and *lower right*: group III] of the thoracic aortas of NZW rabbits from the atherosclerotic group (original magnification, ×100, scale bars 25 μm; Cited from Hayashi et al. 2006b)

3.4.2 Age-Related Fluctuations in Sex Hormones

The serum concentrations of sex hormones change dramatically over the course of one's lifetime. Adrenal androgens such as dehydroepiandrosterone (DHEA) decrease with age from puberty onward in both males and females (Hinson et al. 2003). Estrogen decreases dramatically in females after menopause and from then on is mainly composed of estrone (E1) from peripheral tissues due to the action of aromatase. Estrogen in males remains at a fixed concentration throughout their entire lifetime and is slightly higher than postmenopausal females. After DHEA was converted into E2 by aromatase, it exerts anti-atherosclerotic effects via NO (Hayashi et al. 2000b).

3.5 Sex Differences in Geriatric Disease and Coronary Risk Factors

Most diseases in elderly patients have atypical symptoms and signs, e.g., asymptomatic angina pectoris in elderly diabetic individuals. Research is also progressing regarding sex differences and their relation to sex hormones for each of these diseases.

3.5.1 Dyslipidemia

Dyslipidemia significantly influences cardiovascular disease sex differences in postmenopausal and elderly patients. High serum LDL cholesterol and low serum HDL cholesterol are also independent risk factors for ischemic heart disease (IHD), cortical branch CVAs, and arteriosclerosis obliterans (ASO) in elderly patients (Aronow 2002). The influence of statins did not live up to the expectations in large-scale clinical trials (Prosper, HPS, ASCOT-LLA) that included Western late elderly subjects (Heart Protection Study Collaborative Group 2002; Sever et al. 2003; Shepherd et al. 2002). Although effects were not seen in women in the Anglo-Scandinavian Cardiac Outcomes Trial – Lipid Lowering Arm (ASCOT-LLA), sex differences are still possible as the study only involved a small number of cases (Shepherd et al. 2002). Meanwhile, our investigation of 4,014 patients (MHLW-funded research group headed by the author) revealed reliable treatment outcomes in women with concomitant diabetes (Hayashi et al. 2008).

In familial hyperlipidemia, there are no sex differences in serum lipid levels, and a sex difference is observed for the age of cardiovascular disease onset (approximately 20 years) (Oosterveer et al. 2009). There is also hardly any fluctuation in serum lipid levels due to E2 administration in dietary cholesterol-loaded animals (Hayashi et al. 2000a). The pleiotropic effect, which is recognized with statins, is also observed with estrogen. More doctors are offering guidance regarding other risk factors appropriate to age and sex.

Human TC levels are lowest in both males and females before 20 years of age and are said to increase with age (Brown et al. 1980; Sekimoto et al. 1983). An annual, in-class cohort study conducted on 20,000 screening examinees revealed trend of increasing TC levels until around 75 years of age (Sekimoto et al. 1983). Cases with TC values of 200 mg/dl or higher were relatively elderly and increased by twofold from 1980 to 1990; approximately 25 % of males and 45 % of females showed plasma TC levels higher than 220 mg/dl (JAPAN Guideline for Diagnosis and Prevention of Atherosclerotic Cardiovascular Diseases 2007). For plasma LDL cholesterol levels, almost the same situation occurred (Kuzuya and Shimokata 2002). Regarding postmenopausal female hyperlipidemia patients, research revealed that coronary heart disease primary preventative effects were observed in Japanese patients with a mean TC of 240 mg/dl, 60 % of whom were females (mega study) (Mizuno et al. 2008). The Japan Lipid Intervention Trial (J-LIT) showed a dramatic increase in cardiovascular disease incidence in females with dyslipidemia due to concomitant IGT abnormalities and smoking (Yokoyama et al. 2007). Use of HRT to prevent ischemic heart disease (IHD) decreased after the results of the Women's Health Initiative (WHI), but HRT has been reported to have an inhibitory effect on calcification that occurs in end-stage atherosclerosis (Writing Group for the Women's Health Initiative Investigators 2002; Manson et al. 2007).

E2 patches are considered very promising for HRT as they bypass hepatic metabolism and do not elevate TG. E3 does not cause serum lipid fluctuations in females directly after menopause, and HDL increases from 6 months after administration in patients with a mean age of 80 years (Hayashi et al. 2000c).

3.5.2 Hypertension

The Framingham study found that the incidence of cardiovascular disease complications increased in both males and females in proportion to hypertension severity (blood pressure) (Engberding and Wenger 2008). Hypertension is a risk factor for renal failure, cerebral infarction, and IHD. Hypertensive cardiovascular disease onset frequency and morbidity rate are both high in females. Senescent elastic aortic atherosclerosis leads to systolic hypertension, and this elastic aortic atherosclerosis leads to decreased diastolic blood pressure, reduced coronary blood flow, and cerebral ischemia (Dao et al. 2005). Systolic hypertension also contributes to myocardial infarction and cerebral infarction onset.

Hypertension onset occurs later in females than in males and often happens after menopause. Hypertension is often complicated by cardiac failure from age 70 onward. There is a debate regarding whether hypertension contributes to estrogen shortage. Animal experiments and cell biological studies have reported that estrogen shortage causes (1) changes related to renal diuresis, (2) expression of angiotensin II receptors and increased secretion of aldosterone, and (3) impaired secretion of arteriolar EDHF and NO (Nawate et al. 2005). On the other hand, it has also been reported that there is no difference in blood pressure fluctuation caused by menopause, HRT, hysterectomy, or before and after menopause (Taddei 2009).

3.5.3 Diabetes Mellitus

As there are multiple branches of the coronary artery, in diabetic angiopathy, diffuse long stenotic lesions are common. There is little morphological change or lipid accumulation. Diabetes is the greatest risk factor for ASO, and elderly females with diabetes are also at high risk for cardiovascular disease. Estrogen prevents vascular endothelial dysfunction caused by hyperglycemia by activating the rate-limiting enzyme GTPCHI of NOS coenzyme BH_4 (Miyazaki-Akita et al. 2007). This is anticipated to aid in the prevention of diabetic cardiovascular diseases. As high-dose estrogen causes hyperglycemia through glucocorticoid action, HRT is contraindicated in severe diabetic cases. In apoE KO mice, estrogen inhibited streptozotocin-induced diabetes and progression of atherosclerosis. Diabetes onset was later in normal female rats than in rats that underwent ovariectomy (Tse et al. 1999). End-stage renal failure is often caused by hypertension and nephritis in males, and there is no sex difference in diabetes origin. Our investigation of 4,014 diabetic patients including 1,016 late elderly diabetic patients older than 75 years (MHLW-funded research group headed by the author) revealed that HDL-C levels are important for the prevention of ischemic heart disease and cerebrovascular attack (stroke) and that insulin treatment might require close attention because of the risk of hypoglycemia in elderly (Hayashi et al. 2009, 2011, 2013).

3.6 Attempt to Implement Gender-Based Medicine in Treatment for Elderly Atherosclerosis

3.6.1 Hormone Replacement Therapy

In a 120,000-person Nurses' Health Study conducted in 1997, postmenopausal HRT caused a decrease of approximately 50 % in cardiovascular disease incidence and mortality, and the American Heart Association (AHA) recommended HRT in their Guidelines for Primary Prevention of Cardiovascular Disease for all postmenopausal women and those with cardiovascular disease risk factors in particular (Grodstein et al. 2001; Grundy et al. 1997). The 1998 HERS study, however, targeting patients with cardiovascular disease, indicated a high rate of cardiovascular disease recurrence 1–3 years after beginning HRT (Sakuma et al. 2002). In 2002, the first Women's Health Initiative (WHI) study aiming to prevent heart attacks found that subjects undergoing HRT had a higher incidence rate of IHD and CVA (Writing Group for the Women's Health Initiative Investigators 2002). A racial difference was also suggested for the side effect of deep vein thrombosis.

3.6.2 Environmental Factors

Finally, we come to the topic of environmental factors. Many environmental factors are acted on by estrogen, but the mechanism of some remains unclear. Firstly, one considers the factors of marriage, childbirth, and child-rearing. Changing attitudes toward females working and social environment deficiencies that make it difficult for working women to raise children have been indicated as having led to the trends toward later marriage and having fewer children in Western countries as well as Japan and Korea. Average life expectancy is growing longer, and modern women need new life values and objectives for the second half of their life after the conclusion of child-rearing following childbirth. The differing status of female employment following childbirth between different countries is one side of the gender difference. An array of issues has been proposed, from mental instability to a health management system including "housewife screening."

There is insufficient societal understanding of menopausal symptoms in women. From the perimenopausal period onward, the incidence of dyslipidemia and diabetes in women increases, and atherosclerotic lesions progress. Modern diet and changing exercise patterns are, however, also to blame. The results of a comparative study of serum lipids targeting subjects from Hiroshima, as well as subjects of Japanese ancestry living in Hawaii and the US mainland, also indicated the importance of environmental factors. Smoking in conjunction with the use of the oral contraceptive pill has been shown to aid in increasing the risk of IHD onset (Imazu et al. 2002). Although the incidence of smoking has decreased in Japanese men, it has risen significantly in women in their 20s from 12.7 % in 1975 to 23.6 % in 1999. The risk for embolic events in women who smoke more than 25 cigarettes per day while using the pill is 39 times greater than in women who neither smoke nor take the pill (Roy 1999; Gregory et al. 2011). There also exist other risk factors for embolic events such as physical factors (noise, cold, heat), hazardous chemical factors, human relationships, and working hours. These factors have been suggested to be more likely to cause the onset and aggravation of hypertension, diabetes, angina, myocardial infarction, and stroke in women than in men.

3.7 Summary

Women have a longer average life expectancy than men, and sex hormones, especially estrogen, are closely related to geriatric diseases and aging. Disease structure changes as people age, with cardiovascular disease increasing dramatically in late elderly (older than 75 years.) females and arteriosclerotic diseases surpassing malignancies as the leading cause of death. The prevalence of cardiovascular disease in late elderly women is thus equal to that of men of the same age. Although hormone replacement therapy for all postmenopausal women as one same dose of agents failed to prevent cardiovascular disease in elderly female, recent advance of basic and clinical gender-specific medical research made clear the molecular mechanism on the gender differences in cardiovascular diseases as well as on the gender differences of coronary risk factors such as dyslipidemia, hypertension, and even environmental factors. Cardiovascular disease is a typical geriatric disease, and pathological aging contributes to estrogen action and sex differences in atherosclerosis, which form the main etiology of cardiovascular disease. These gender-dependent differences in risk are important for an individualized strategy to prevent atherosclerotic disease.

References

Aronow WS (2002) Pharmacologic therapy of lipid disorders in the elderly. Am J Geriatr Cardiol 11:247–256

Battaglioli T, Martinelli I (2007) Hormone therapy and thromboembolic disease. Curr Opin Hematol 14:488–493

Besdine RW, Wetle TF (2010) Improving health for elderly people: an international health promotion and disease prevention agenda. Aging Clin Exp Res 22:219–230

Brinton EA (1996) Oral estrogen replacement therapy in postmenopausal women selectively raises levels and production rates of lipoprotein A-I and lowers hepatic lipase activity without lowering the fractional catabolic rate. Arterioscler Thromb Vasc Biol 16:431–440

Brown MS, Kovanen PT, Goldstein JL (1980) Evolution of the LDL receptor concept-from cultured cells to intact animals. Ann N Y Acad Sci 348:48–68

Castelli WP, Anderson K, Wilson PW, Levy D (1992) Lipids and risk of coronary heart disease. The Framingham Study. Ann Epidemiol 2(1–2):23–8

Dao HH, Essalihi R, Bouvet C, Moreau P (2005) Evolution and modulation of age-related medial elastocalcinosis: impact on large artery stiffness and isolated systolic hypertension. Cardiovasc Res 66:307–317

Engberding N, Wenger NK (2008) Cardiovascular disease prevention tailored for women. Expert Rev Cardiovasc Ther 6:1123–1134
Esaki T, Hayashi T, Muto E, Yamada K, Kuzuya M, Iguchi A (1997) Expression of inducible nitric oxide synthase in T lymphocytes and macrophages of cholesterol-fed rabbits. Atherosclerosis 128:39–46
Ettinger B, Black DM, Mitlak BH, Knickerbocker RK, Nickelsen T, Genant HK, Christiansen C, Delmas PD, Zanchetta JR, Stakkestad J, Glüer CC, Krueger K, Cohen FJ, Eckert S, Ensrud KE, Avioli LV, Lips P, Cummings SR (1999) Reduction of vertebral fracture risk in postmenopausal women with osteoporosis treated with raloxifene: results from a 3-year randomized clinical trial. JAMA 282:637–646
Flavahan NA (1992) Atherosclerosis or lipoprotein-induced endothelial dysfunction. Potential mechanisms underlying reduction in EDRF/nitric oxide activity. Circulation 85:1927–1938
Goldstein JL, Brown MS (1982) Lipoprotein receptors: genetic defense against atherosclerosis. Clin Res 30:417–426
Gregory NC, Ilan B, Osaki Y, Geoffrey FW (2011) The impact of menthol cigarettes on smoking initiation among non-smoking young females in Japan. Int J Environ Res Public Health 8:1–14
Grodstein F, Manson JE, Stampfer MJ (2001) Postmenopausal hormone use and secondary prevention of coronary events in the nurses' health study. A prospective, observational study. Ann Intern Med 135:1–8
Grundy SM, Balady GJ, Criqui MH, Fletcher G, Greenland P, Hiratzka LF, Houston-Miller N, Kris-Etherton P, Krumholz HM, LaRosa J, Ockene IS, Pearson TA, Reed J, Washington R, Smith SC Jr (1997) Guide to primary prevention of cardiovascular diseases. A statement for healthcare professionals from the Task Force on Risk Reduction. American Heart Association Science Advisory and Coordinating Committee. Circulation 95:2329–2331
Guideline for Diagnosis and Prevention of Atherosclerotic Cardiovascular Diseases (2007) Edition. Available at http://www.j-athero.org/en/publications/guideline2007.html
Hashimoto M, Akishita M, Eto M, Ishikawa M, Kozaki K, Toba K, Sagara Y, Taketani Y, Orimo H, Ouchi Y (1995) Modulation of endothelium-dependent flow-mediated dilatation of the brachial artery by sex and menstrual cycle. Circulation 92(12):3431–3435
Hayashi T, Naito M, Ishikawa T, Kuzuya M, Funaki C, Tateishi T, Asai K, Hidaka H, Kuzuya F (1989) Beta-migrating very low density lipoprotein attenuates endothelium-dependent relaxation in rabbit atherosclerotic aortas. Blood Vessels 26:290–299
Hayashi T, Ignarro LJ, Chaudhuri G (1992) Basal release of nitric oxide from aortic rings is greater in female rabbits than in male rabbits: implications for atherosclerosis. Proc Natl Acad Sci USA 89:1125–1163
Hayashi T, Yamada K, Esaki T, Kuzuya M, Satake S, Ishikawa T, Hidaka H, Iguchi A (1995a) Estrogen increases endothelial nitric oxide by a receptor-mediated system. Biochem Biophys Res Commun 213:915–919
Hayashi T, Fukuto JM, Ignarro LJ, Chaudhuri G (1995b) Gender differences in atherosclerosis: possible role of nitric oxide. J Cardiovasc Pharmacol 26:792–802
Hayashi T, Yamada K, Esaki T, Muto E, Chaudhuri G, Iguchi A (1998) Physiological concentrations of 17 beta-estradiol inhibit the synthesis of nitric oxide synthase in macrophages via a receptor-mediated system. J Cardiovasc Pharmacol 31:292–298
Hayashi T, Jayachandran M, Sumi D, Thakur NK, Esaki T, Muto E, Kano H, Asai Y, Iguchi A (2000a) Physiological concentration of 17 b estradiol retards the progression of severe atherosclerosis induced by cholesterol diet plus balloon injury via NO. Arterioscler Thromb Vasc Biol 20:1613–1621
Hayashi T, Esaki T, Muto E, Kano H, Asai Y, Thakur NK, Sumi D, Jayachandran M, Iguchi A (2000b) Dehydroepiandrosterone retards atherosclerosis formation through the conversion to estrogen: the possible role of nitric oxide. Arterioscler Thromb Vasc Biol 20:782–792
Hayashi T, Ito I, Kano H, Endo H, Iguchi A (2000c) Estriol (E3) replacement improves endothelial function and bone mineral density in very elderly women. J Gerontol A Biol Sci Med Sci 55:B183–B190

Hayashi T, Matsui-Hirai H, Miyazaki-Akita A, Fukatsu A, Funami J, Ding QF, Kamalanathan S, Hattori Y, Ignarro LJ, Iguchi A (2006a) Endothelial cellular senescence is inhibited by nitric oxide: implications in atherosclerosis associated with menopause and diabetes. Proc Natl Acad Sci USA 103:17018–17023

Hayashi T, Esaki T, Sumi D, Mukherjee T, Iguchi A, Chaudhuri G (2006b) Modulating role of estradiol on arginase II expression in hyperlipidemic rabbits as an atheroprotective mechanism. Proc Natl Acad Sci USA 103:10485–10490

Hayashi T, Kawashima S, Itoh H, Yamada N, Sone H, Watanabe H, Hattori Y, Ohrui T, Yoshizumi M, Yokote K, Kubota K, Nomura H, Umegaki H, Iguchi A, Japan CDM group (2008) Importance of lipid levels in elderly diabetic individuals: baseline characteristics and 1-year survey of cardiovascular events. Circ J 72:218–25

Hayashi T, Kawashima S, Itoh H, Yamada N, Sone H, Watanabe H, Hattori Y, Ohrui T, Yokote K, Nomura H, Umegaki H, Iguchi A, Japan CDM Group (2009) Low HDL cholesterol is associated with the risk of stroke in elderly diabetic individuals: changes in the risk for atherosclerotic diseases at various ages. Diabetes Care 32:1221–3

Hayashi T, Kawashima S, Nomura H, Itoh H, Watanabe H, Ohrui T, Yokote K, Sone H, Hattori Y, Yoshizumi M, Ina K, Kubota K, Japan Cholesterol and Diabetes Mellitus Investigation Group (2011) Age, gender, insulin and blood glucose control status alter the risk of ischemic heart disease and stroke among elderly diabetic patients. Cardiovasc Diabetol 10:86–91

Hayashi T, Araki A, Kawashima S, Sone H, Watanabe H, Ohrui T, Yokote K, Takemoto M, Kubota K, Noda M, Noto H, Ina K, Nomura H (2013) Metabolic predictors of ischemic heart disease and cerebrovascular attack in elderly diabetic individuals: difference in risk by age. Cardiovasc Diabetol 12:10–16

Haynes MP, Sinha D, Russell KS, Collinge M, Fulton D, Morales-Ruiz M, Sessa WC, Bender JR (2000) Membrane estrogen receptor engagement activates endothelial nitric oxide synthase via the PI3-kinase-Akt pathway in human endothelial cells. Circ Res 87:677–682

Heart Protection Study Collaborative Group (2002) MRC/BHF Heart Protection Study of cholesterol lowering with simvastatin in 20,536 high-risk individuals: a randomised placebo-controlled trial. Lancet 360(9326):7–22

Hinson JP, Brooke A, Raven PW (2003) Therapeutic uses of dehydroepiandrosterone. Curr Opin Investig Drugs 4:1205–1208

Hisamoto K, Ohmichi M, Kurachi H, Hayakawa J, Kanda Y, Nishio Y, Adachi K, Tasaka K, Miyoshi E, Fujiwara N, Taniguchi N, Murata Y (2001) Estrogen induces the Akt-dependent activation of endothelial nitric-oxide synthase in vascular endothelial cells. J Biol Chem 276:3459–3467

Hlatky MA, Boothroyd D, HERS Research Group (2002) Quality of life and depressive symptoms in postmenopausal women after receiving hormone therapy: results from the Heart and Estrogen/Progestin Replacement Study (HERS) trial. JAMA 287:591–597

Holm P, Andersen HL, Andersen MR, Erhardtsen E, Stender S (1999) The direct antiatherogenic effect of estrogen is present, absent, or reversed, depending on the state of the arterial endothelium. A time course study in cholesterol-clamped rabbits. Circulation 100:1727–1733

Imazu M, Sumii K, Yamamoto H, Toyofuku M, Tadehara F, Okubo M, Yamakido M, Kohno N, Onaka AT, Hawaii-Los Angeles-Hiroshima Study (2002) Influence of type 2 diabetes mellitus on cardiovascular disease mortality: findings from the Hawaii-Los Angeles-Hiroshima study. Diabetes Res Clin Pract 57:61–9

Kuzuya M, Shimokata H (2002) Changes in serum lipid levels during a 10 year period in a large Japanese population. A cross-sectional and longitudinal study. Atherosclerosis 163: 313–320

Manson JE, Allison MA, Rossouw JE, Carr JJ, Langer RD, Hsia J, Kuller LH, Cochrane BB, Hunt JR, Ludlam SE, Pettinger MB, Gass M, Margolis KL, Nathan L, Ockene JK, Prentice RL, Robbins J, Stefanick ML, WHI and WHI-CACS Investigators (2007) Estrogen therapy and coronary-artery calcification. N Engl J Med 356:2591–2602

Miyazaki-Akita A, Hayashi T, Ding QF, Shiraishi H, Nomura T, Hattori Y, Iguchi A (2007) 17beta-estradiol antagonizes the down-regulation of endothelial nitric-oxide synthase and GTP cyclohydrolase I by high glucose: relevance to postmenopausal diabetic cardiovascular disease. J Pharmacol Exp Ther 320:591–598

Mizuno K, Nakaya N, Ohashi Y, Tajima N, Kushiro T, Teramoto T, Uchiyama S, Nakamura H, MEGA Study Group (2008) Usefulness of pravastatin in primary prevention of cardiovascular events in women: analysis of the Management of Elevated Cholesterol in the Primary Prevention Group of Adult Japanese (MEGA study). Circulation 117:494–502

Mosselman S, Polman J, Dijkema R (1996) ERβ: identification and characterization of a novel human estrogen receptor. FEBS Lett 392:49–55

Muramatsu N, Akiyama H (2011) Japan: super-aging society preparing for the future. Gerontologist 51:425–432

Nakao J, Chang WC, Murota SI, Orimo H (1991) Estradiol-binding sites in rat aortic smooth muscle cells in culture. Atherosclerosis 38:75–80

Nawate S, Fukao M, Sakuma I, Soma T, Nagai K, Takikawa O, Miwa S, Kitabatake A (2005) Reciprocal changes in endothelium-derived hyperpolarizing factor- and nitric oxide-system in the mesenteric artery of adult female rats following ovariectomy. Br J Pharmacol 144:178–189

Oosterveer DM, Versmissen J, Yazdanpanah M, Hamza TH, Sijbrands EJ (2009) Differences in characteristics and risk of cardiovascular disease in familial hypercholesterolemia patients with and without tendon xanthomas: a systematic review and meta-analysis. Atherosclerosis 207:311–317

Ridker PM, Miletich JP, Hennekens CH, Buring J (1997) Ethnic distribution of factor V Leiden in 4047 men and women. Implications for venous thromboembolism screening. JAMA 277:1305–1307

Roy S (1999) Effects of smoking on prostacyclin formation and platelet aggregation in users of oral contraceptives. Am J Obstet Gynecol 180:S364–S368

Sakuma I, Liu MY, Sato A, Hayashi T, Iguchi A, Kitabatake A, Hattori Y (2002) Endothelium-dependent hyperpolarization and relaxation in mesenteric arteries of middle-aged rats: influence of oestrogen. Br J Pharmacol 135:48–54

Sekimoto H, Goto Y, Goto Y, Naito C, Yasugi T, Okido M, Kuzuya F, Takeda R, Yamamoto A, Fukuzaki H et al (1983) Changes of serum total cholesterol and triglyceride levels in normal subjects in Japan in the past twenty years. Research committee on familial hyperlipidemia in Japan. Jpn Circ J 47:1351–1358

Sever PS, Dahlöf B, Poulter NR, Wedel H, Beevers G, Caulfield M, Collins R, Kjeldsen SE, Kristinsson A, McInnes GT, Mehlsen J, Nieminen M, O'Brien E, Ostergren J, ASCOT Investigators (2003) Prevention of coronary and stroke events with atorvastatin in hypertensive patients who have average or lower-than-average cholesterol concentrations, in the Anglo-Scandinavian Cardiac Outcomes Trial–Lipid Lowering Arm (ASCOT-LLA): a multicentre randomised controlled trial. Lancet 361:1149–1158

Shepherd J, Blauw GJ, Murphy MB, Bollen EL, Buckley BM, Cobbe SM, Ford I, Gaw A, Hyland M, Jukema JW, Kamper AM, Macfarlane PW, Meinders AE, Norrie J, Packard CJ, Perry IJ, Stott DJ, Sweeney BJ, Twomey C, Westendorp RG, PROSPER Study Group (2002) PROspective Study of Pravastatin in the Elderly at Risk. Pravastatin in elderly individuals at risk of vascular disease (PROSPER). Lancet 360:1623–1630

Shulman LP (2011) The state of hormonal contraception today: benefits and risks of hormonal contraceptives: combined estrogen and progestin contraceptives. Am J Obstet Gynecol 205:S9–S13

Simoncini T, Varone G, Fornari L, Mannella P, Luisi M, Labrie F, Genazzani AR (2002) Genomic and nongenomic mechanisms of nitric oxide synthesis induction in human endothelial cells by a fourth-generation selective estrogen receptor modulator. Endocrinology 143:2052–2206

Su W, Campos H, Judge H, Walsh BW, Sacks FM (1998) Metabolism of Apo(a) and ApoB100 of lipoprotein(a) in women: effect of postmenopausal estrogen replacement. J Clin Endocrinol Metab 83:3267–3276

Sumi D, Hayashi T, Jayachandran M, Iguchi A (2001) Estrogen prevents destabilization of endothelial nitric oxide synthase mRNA induced by tumor necrosis factor alpha through estrogen receptor mediated system. Life Sci 69:1651–1660

Taddei S (2009) Blood pressure through aging and menopause. Climacteric 12(Suppl 1):36–40
The Lipid Research Clinics Program Epidemiology Committee (1979) Plasma lipid distributions in selected North American populations: the Lipid Research Clinics Program Prevalence Study. Circulation 60:427–434
Tse J, Martin-McNaulty B, Halks-Miller M, Kauser K, DelVecchio V, Vergona R, Sullivan ME, Rubanyi GM (1999) Accelerated atherosclerosis and premature calcified cartilaginous metaplasia in the aorta of diabetic male Apo E knockout mice can be prevented by chronic treatment with 17 beta-estradiol. Atherosclerosis 144:303–313
Watanabe T, Miyahara Y, Akishita M, Nakaoka T, Yamashita N, Iijima K, Kim H, Kozaki K, Ouchi Y (2004) Inhibitory effect of low-dose estrogen on neointimal formation after balloon injury of rat carotid artery. Eur J Pharmacol 502:265–270
Wenger NK, Barrett-Connor E, Collins P, Grady D, Kornitzer M, Mosca L, Sashegyi A, Baygani SK, Anderson PW, Moscarelli E (2002) Baseline characteristics of participants in the Raloxifene Use for The Heart (RUTH) trial. Am J Cardiol 90:1204–1210
Writing Group for the Women's Health Initiative Investigators (2002) Risks and benefits of estrogen plus progestin in healthy postmenopausal women: principal results from the Women's health Initiative randomized controlled trial. JAMA 288:321–333
Yokoyama M, Origasa H, Matsuzaki M, Matsuzawa Y, Saito Y, Ishikawa Y, Oikawa S, Sasaki J, Hishida H, Itakura H, Kita T, Kitabatake A, Nakaya N, Sakata T, Shimada K, Shirato K, Japan EPA lipid intervention study (JELIS) Investigators (2007) Effects of eicosapentaenoic acid on major coronary events in hypercholesterolaemic patients (JELIS): a randomised open-label, blinded endpoint analysis. Lancet 369(9567):1090–8

Gene Polymorphisms and Signaling Defects

4

Christine Mannhalter, Michael Poteser, and Klaus Groschner

Abstract

Understanding heritability of cardiovascular risk is a key issue in preventive medicine and of particular importance for the development of cost-efficient strategies to preserve cardiovascular health. Recent progress in genotyping technologies enabled detailed insight into human genetic variability. Linkage analysis and genome-wide association screens have complemented candidate gene studies to uncover a large array of genetic variations associated with cardiovascular pathologies. For the majority of identified gene polymorphisms, the clinical impact is as yet unclear and the available information is still of limited value for cardiovascular risk prediction. However, genetic discoveries have greatly improved our understanding of pathophysiological principles. The identification of novel molecular players and target structures contributed significantly to therapeutic progress. In this chapter we aim to present an overview on polymorphisms in genes with proven or at least convincingly suggested causal relation to secondary risk factors and/or cardiovascular phenotypes. Special attention will be paid to functional consequences of genetic variations. Many of the polymorphisms that are causally involved in the pathogenesis of cardiovascular dysfunction impair cellular signaling processes and may in turn affect regulatory genetic networks. Unraveling of molecular mechanisms that link genetic variability to pathological phenotypes is still at a rather early stage. We will present here a selection of these mechanisms and discuss their clinical significance.

C. Mannhalter (✉)
Department of Laboratory Medicine, Medical University of Vienna,
Währinger Gürtel 18-20, Vienna A-1090, Austria
e-mail: christine.mannhalter@meduniwien.ac.at

M. Poteser • K. Groschner
Center for Physiological Medicine,
Institute of Biophysics, Medical University of Graz,
Harrachgasse 21/4, Graz A-8010, Austria
e-mail: klaus.groschner@medunigraz.at

I. Wakabayashi, K. Groschner (eds.), *Interdisciplinary Concepts in Cardiovascular Health*,
DOI 10.1007/978-3-7091-1334-9_4,

Keywords
SNPs • Hypertension • Diabetes • Obesity • Dyslipidemia • Electrical remodeling • Hemostatic disorders

4.1 Introduction

Cardiovascular events represent highly complex pathological phenotypes that are for a large part determined by genetic predisposition. The availability of large-scale genotyping technologies has enabled genome-wide association studies (GWAS) that generated a large body of information on the genomics of cardiovascular disease. This allowed the identification of single-nucleotide polymorphisms (SNPs) in novel gene loci associated with cardiovascular diseases and phenotypes. Thus, GWAS approaches have considerably expanded our knowledge on the molecular players in cardiovascular pathophysiology. Albeit most of the unraveled associations reside outside of coding regions and/or in genomic regions lacking annotated genes, some of the polymorphisms have been demonstrated to cause defined impairment of cellular signaling processes. There is currently a substantial demand in elucidating the mechanisms by which certain genetic variants govern pathogenesis of cardiovascular events. It is expected that the efforts to identify genetic markers for diagnosis, risk prediction, and individual therapy will further improve our understanding of pathomechanisms. Causative SNPs are considered to determine disease susceptibility, onset and progression of the disorder, as well as therapeutic prognosis by modifying specific pathways of cellular communication. Hence, a distorted function or expression of signaling molecules and impairment of signaling pathways may be assigned to causative genetic variants. Molecular (patho)physiology has uncovered an array of disease-relevant signaling pathways and corresponding candidate genes, which have been extensively tested for their role in inheritance of diseases. The candidate genes and gene products that have been identified can be assigned to different functional groups.

- Genetic variants that may promote disease by interfering with certain cell/tissue functions thereby generating a risk situation corresponding to the phenotypes listed as secondary risk factors in Part II of this book.
- Polymorphisms in relevant genes can cause perturbations of cellular homeostasis, of cell-cell communication, as well as of tissue and organ functions. The associated states of cellular dysfunction may represent the key trigger for several cardiovascular risk scenarios and phenotypes.
- A certain set of dysfunctional states may represent joint determinants of the pathogenesis of a risk phenotype (secondary risk factor). As an example, genes that impact on the development of atherosclerotic lesions can be assigned to biochemically and pathophysiologically rather well-defined states such as oxidative stress, endothelial dysfunction, hypercoagulability, dyslipidemia, and inflammation. On the basis of this stratification concept, we discuss here a selection of prominent molecular mechanisms that link gene polymorphisms to cellular dysfunctions. In this chapter we will focus on the genetic basis of defects that promote vascular dysfunction, atherosclerosis, and thrombosis.

4.1.1 Single Nucleotide Polymorphisms (SNPs)

SNPs are frequent variations of single base pairs in the genome occurring with >1 % prevalence in a given population. Several million SNPs are expected to exist in our genome. They are responsible for about 90 % of the genetic heterogeneity. The variations are not evenly distributed over the genome but occur frequently in cytosine-rich regions. SNPs can lead to changes of an amino acid (nonsynonymous SNPs), but often the mutations do not alter the coding sequence and the amino acid (synonymous SNPs). Nonsynonymous SNPs can have functional effects, but frequently these are relatively small. Depending on their location, SNPs can influence the regulation of gene transcription (regulatory SNPs), splicing, or RNA processing and consequently affect the protein concentration. Often, SNPs are considered as successful mutations because they prevailed in a population. The scientific importance of SNPs is based on their high prevalence and their simple detectability.

4.1.2 SNPs and Cardiovascular Diseases

Today, atherosclerosis and arterial thrombosis – with their clinical manifestations myocardial infarction (MI), stroke, and peripheral arterial disease – and venous thromboembolism (VTE), e.g., pulmonary embolism and deep vein thrombosis of the lower limbs, are major causes of morbidity and mortality. Age, sex, systolic blood pressure, hypertension, total and high-density lipoprotein cholesterol levels, smoking, and diabetes as well as newer factors such as N-terminal fragment of prohormone B-type natriuretic peptide levels, von Willebrand factor antigen levels, fibrinogen levels, chronic kidney disease, leukocyte count, C-reactive protein levels, homocysteine levels, uric acid levels, coronary artery calcium (CAC) scores, carotid intima-media thickness, peripheral arterial disease, and pulse wave velocity have been reported as risk predictors for vascular diseases.

A recent meta-analysis interrogated 136 different biomarkers for their relevance in stroke. Three biomarkers (C-reactive protein, P-selectin, and homocysteine) could differentiate between ischemic stroke and healthy control subjects. High levels of admission glucose and high fibrinogen levels were strong predictors of poor prognosis after ischemic stroke and symptomatic intracerebral hemorrhage post-thrombolysis, glutamate was found to be an indicator of progressive stroke, and D-dimer predicted in-hospital death. The study concluded that only few biomarkers that are currently tested have meaningful clinical value (Hasan et al. 2012). The topic "biomarkers" is addressed in more detail in Chap. 15.

Apparently, many established risk factors do not fully capture the overall risk of the diseases. Occlusive coronary thrombus formation superimposed on an atherosclerotic plaque is the ultimate event leading to myocardial infarction (MI). Thus, it was to be expected that the hemostatic system and hemostatic proteins are important players in the pathogenesis of stroke as well as myocardial infarction. The relationship of hemostasis and thrombosis with atherothrombotic cardiovascular disease has been extensively studied in the past decades. Elevated levels of hemostatic factors, such as fibrinogen, plasminogen activator inhibitor (PAI-1), von Willebrand factor (vWF), tissue plasminogen activator (tPA), factor VII (FVII), and D-dimer, were found linked to the development of atherothrombosis and are risk markers for coronary heart disease (CHD), stroke, and other cardiovascular disease (CVD) events (Yang et al. 2007a). Genetic association studies have been able to identify gene variants associated with acute coronary syndrome (ACS), peripheral arterial disease (PAD), ischemic stroke (IS), and VTE.

During the last 20 years, the identification of polymorphisms and the elucidation of their role in arterial and venous thromboses became an important area of research. Recently, next-generation DNA sequencing was introduced for genetic association studies in complex diseases. However, the application of the technique to common diseases is still limited by the high costs and considerable computational burden associated with the analysis of large numbers of samples. Sequencing of a few hundred genes or genomic loci is less expensive, suitable for the analysis of areas of the genome that are deemed to have particular relevance in the pathophysiology of a given disease, and a good alternative to whole-genome (or exome) sequencing. Many SNP variations across an individual's entire genome can be extrapolated from a single microarray experiment.

In a recent study, a weighted genetic risk score (GRS) for MI was calculated for SNPs in genes of the coagulation system. Evidence for association with MI was found for 35 SNPs located in 12 genes: factor V (F5), protein S (PROS1), factor XI (F11), integrin alpha 2 (ITGA2, platelet glycoprotein Ia), factor XII (F12), factor XIIIA (F13A1), plasminogen activator inhibitor-1 (SERPINE1), tissue plasminogen activator (PLAT), von Willebrand factor (VWF), thrombomodulin (THBD), endothelial protein C receptor (PROCR), and factor IX (F9). The GRS differed significantly between cases and controls, and subjects in the highest quintile had a 2.69-fold increased risk for MI compared with those in the lowest quintile (Guella et al. 2011). Already in 2007 Knowles et al. had tested 21 SNPs in coagulation factor and platelet glycoprotein genes and evaluated their association with MI in the ADVANCE study population (Knowles et al. 2007). They had selected SNPs in platelet glycoprotein Ia/IIa (ITGA2), glycoprotein Ibα (GP1BA), glycoprotein IIb (ITGA2B), glycoprotein IIIa (ITGB3), glycoprotein IIIb/platelet glycoprotein IV (CD36), plasminogen activator inhibitor-1 (PAI-1), thrombomodulin (THBD), coagulation factor III/tissue factor (F3), and tissue plasminogen activator (PLAT). In contrast to the recent study, they did not find significant associations of the chosen polymorphisms with acute MI. However, they had evaluated different SNPs and they did not test, e.g., the 4G/5G polymorphism in PAI-1 (rs1799889), which had previously been associated with MI.

In patients with DVT, a molecular-barcode-based technique for sample multiplexing was used on next-generation DNA sequencing platforms to sequence the coding areas of 186 hemostatic/proinflammatory genes (Lotta et al. 2012). Known disease-associated variants as well as novel potentially deleterious variants in disease-associated genes were identified. Interestingly, an increased burden of rare missense mutations in anticoagulant genes was found. Furthermore, the association of DVT with the rs6050 SNP (312Thr>Ala) in the fibrinogen alpha gene (FGA) was observed. The rs6050 SNP had been described before and is now one of the most widely replicated variants in DVT, even though no association of FGA rs6050 with plasmatic fibrinogen activity has been seen. The discovery of rare missense mutations in anticoagulant genes and the association of rs6050 with idiopathic DVT indicates that “unrecognized” anticoagulant deficiencies and mutations to which currently used biochemical assays are not sensitive can represent the cause of DVT.

Regarding the role of SNPs, in stroke the analyses of six cohorts gave inconsistent results. No single locus of significance (approximately $p < 10^7$) could be identified (Lanktree et al. 2010). This is in contrast to a recent meta-analysis which reported that three biomarkers (C-reactive protein, P-selectin, and homocysteine) could differentiate between ischemic stroke and healthy control subjects. Intelligent, appropriately powered, multidisciplinary studies incorporating knowledge from clinical medicine, epidemiology, genetics, and molecular biology will be required to fully characterize the genomic contributors to vascular diseases.

4.2 Hypertension and Vascular Dysfunction

Elevated blood pressure (BP) is an important risk factor for stroke and ischemic heart disease and contributes significantly to the global risk for cardiovascular disease. Hypertension has a substantial heritability. Several genes with large effects have been identified in familial forms of hypertension, including salt sensitivity genes (Vehaskari 2009). However, in spite of these findings, research on the genetics of hypertension in general has been disappointing. The huge efforts made in recruiting and genotyping tens of thousands of individuals and meta-analyzing dozens of cross-sectional, population-based studies on systolic and diastolic BP showed that hypertension is influenced by environmental and lifestyle factors as well as by many genetic loci, each of which has only a small effect on blood pressure regulation. Their interaction indicates a polygenic regulation of blood pressure throughout different periods of life (Kunes and Zicha 2009). In a follow-up investigation in 59,349 individuals of SNPs that had previously been described to be associated with BP, polymorphisms in the following genes could be replicated: LSP1 (lymphocyte-specific protein 1), TNNT3 (troponin type 3), MTHFR (methylene tetrahydrofolate reductase), NPPB (natriuretic peptide B), AGT (angiotensinogen), and ATP2B1 (ATPase). The results provide candidates for further studies to identify mechanisms affecting BP and highlight the utility of studying SNPs. Previously, higher plasma homocysteine levels were observed in individuals with the Val/Val (677TT) genotype in the MTHFR gene, typically found in approximately 10 % of Western populations, and this polymorphism was identified as the main genetic determinant of elevated homocysteine concentration. The association between elevated levels of homocysteine and CAD had been evaluated in many studies. However, the results were conflicting and the question of whether mild hyperhomocysteinemia is a risk factor for coronary artery disease (CAD) is still unclear. Besides the MTHFR 677C>T polymorphism, folate, riboflavin, cobalamin, and vitamin B6 are independent predictors of homocysteine.

Table 4.1 Prominent susceptibility genes in hypertension

Gene	SNP	Variants	Functional consequence	References
MTHFR		677C>T	Homocysteine mediated	Wilson et al. (2011), Johnson et al. (2011b)
ACE	rs4646994		Increased ACE activity	Ned et al. (2012)
AGT	rs11206510	M235T T174M 1166A>C	LDL mediated	Feng et al. (2011)
SCNN1A/ENaC		T493A	Altered Na^+ absorption?	Hsieh et al. (2005)
SCNN1G	rs13331086		Altered Na^+/K^+ plasma levels?	Büsst et al. (2011), Morris (2011)
PAI-1	rs1799768	4G74G	Unknown	Martínez-Calatrava et al. (2007)
LR/lepr		C538T microsatellite polymorphisms in the 3′ UTR	Unknown	Akhter et al. (2012)
NOS3	rs3918226 rs1800779 rs3918226 rs1799983	−690C>T −922A>G −690C>T Glu298Asp G>T	Reduced NO generation?	Zhang et al. (2012)

According to more recent data, MTHFR 677TT appears to be independently associated with hypertension. Riboflavin is required as a cofactor by MTHFR and may have a role in modulating blood pressure, specifically in individuals with the homozygous MTHFR 677TT genotype. However, some observational studies found little or no BP response to riboflavin despite decreases in homocysteine, but some of these studies have not considered the MTHFR 677C>T polymorphism, which could be important. If prospective trials can confirm that the MTHFR mutation conveys a predisposition to hypertension which can be corrected by low-dose riboflavin, the findings could have important implications for the management of hypertension (Wang et al. 2010a, b; Johnson et al. 2011b).

In a Japanese association study, selected candidate gene variants were evaluated for genetic associations with systolic BP (SBP)/diastolic BP (DBP) in 19,426 individuals. Associations with ACE (angiotensin-converting enzyme), ADD1 (alpha-adducin-1), ADRB2 (adrenergic beta-2 receptor), AGT, CYP11B2 (cytochrome P450 11B2), GNB3 (guanine nucleotide-binding protein 3), and NOS3 were tested. BP trait associations at two loci (AGT rs699 and CYP11B2 rs1799998) were replicated. The most significant association was found for CYP11B2 rs1799998. This study provides evidence for two variants in genes with clinical and physiological relevance that are likely to account for a portion of BP variance in the general population and are worth following up via a target gene approach (Takeuchi et al. 2012) (Table 4.1).

4.2.1 Angiotensinogen- and Angiotensin-Converting Enzyme (ACE)

The renin-angiotensin system (RAS) is classically known for its role in the regulation of blood pressure as well as of fluid and electrolyte balance. RAS is involved in the homeostasis of fluid balance and blood pressure via effects on aldosterone secretion by the adrenal cortex and direct vasoconstriction. In the RA system, angiotensinogen (AGT), the precursor of all angiotensin peptides, undergoes two enzymatic cleavages by renin and angiotensin-converting enzyme (ACE) to produce angiotensin I (Ang I) and angiotensin II (Ang II), respectively. Today, there is convincing evidence that angiotensin II and its metabolites play also an important role in the central nervous system and RAS has been implicated in aspects such as cognition, dementia, depression, anxiety, and epilepsy. RAS adversely affects stroke and cardiovascular disease. Polymorphisms in AGT, particularly the M235T and the T174M amino acid exchange, and the 1166A>C polymorphism in angiotensin II type 1 receptor have been reported as candidate risk factors for vascular disease, and the polymorphic variants M235T and T174M of the AGT gene were associated with hypertension (Mohana et al. 2012). Recently, genome-wide association data and resequencing data were reevaluated regarding the effect of AGT variants on BP using sib-pair information. It was shown that the use of sib-pair information can increase the power over using only unrelated samples by more than 40 %. These results indicate that family data are extremely informative when searching for variants underlying complex traits (Feng et al. 2011).

The insertion/deletion (I/D) variant (rs4646994) of the angiotensin I-converting enzyme (ACE) gene is one of the most studied polymorphisms in relation to blood pressure and essential hypertension in humans. However, the results are conflicting. The ACE gene contains a polymorphism characterized by either insertion (I) or deletion (D). The deletion allele occurs in approximately 55 % of the population and is associated with increased activity of the ACE enzyme. Previous research has shown that the insertion/deletion (I/D) polymorphism is a major determinant of plasma ACE activity. The polymorphism itself is thought to be a marker for a closely linked but yet unidentified sequence variant that modulates the expression of the ACE gene. It has been suggested that persons with the DD genotype (those who express, on average, the highest levels of circulating ACE) have an increased risk for myocardial infarction and coronary artery disease, particularly if they are otherwise at low risk. The evaluation of the DD allele in relation to hypertension and type 2 diabetes mellitus (T2DM) gave contradictory results. The variability of the results may be due to ethnic and geographical variations. Apparently, the effect of the polymorphism is stronger in some ethnicities. For instance, in Japan, there is good evidence that the DD genotype of ACE is a genetic risk factor for cerebro- and cardiovascular disease, especially cardiovascular events, in hypertensive patients.

A representative cohort of the United States comprising non-Hispanic white, non-Hispanic black, and Mexican Americans showed that the frequency of the I/D variant differed significantly by race/ethnicity. The D allele was associated with increased SBP but only among non-Hispanic blacks; no other significant associations were observed in any race/ethnic group (Ned et al. 2012). The reasons for these ethnic differences have yet to be elucidated.

4.2.2 Sodium Channels

Epidemiological and animal studies implicate that amiloride-sensitive epithelial sodium channels (ENaC, SCNN1) are responsible for the rate-limiting step of sodium reabsorption in the distal nephron and are important candidates in the development of hypertension. ENaC is a constitutively open channel, and factors controlling the number of active channels at the cell surface are likely to have profound effects on Na^+ absorption and the amount of Na^+ that is excreted in the final urine. The channel is composed of subunits which are encoded by different genes. Hsieh et al. (2005) evaluated the association of two common single-nucleotide polymorphisms (SNP) in the α-subunit of the ENac (αEnaC, SCNN1A) with ischemic cerebrovascular events (ICE). 493Arg carriers exhibited an increased adjusted odds for ICE compared to Trp/Trp carriers. Interaction analysis revealed that the relative risk was highest in women in the lowest age tertile. The effect was independent of traditional vascular risk factors. Several other genetic studies demonstrate the importance of the regulation of Na^+ absorption for the maintenance of the extracellular fluid volume and blood pressure. A number of membrane-bound kinases have been identified that increase ENaC activity at the cell surface. Recently, de-ubiquitylating enzymes have been shown to increase ENaC activity in heterologous expression systems. Currently, *in vivo* physiological studies constitute a major challenge for the understanding of the regulation of ENaC to maintain the Na^+ balance (Schild 2010).

Recently, Harrap and colleagues (Büsst et al. 2011) reported six SNPs in *SCNN1G* associated with systolic BP (SBP). The effect of the minor (*G*) allele of the SNP (*rs13331086*) was most important and remained significant after meta-analysis. The overall effect on SBP was 1.01 mmHg ($P=0.002$) and on DBP was 0.52 mmHg ($P=0.011$), after adjustment for age, sex, and body mass index. Interestingly, the rs13331086 minor allele was associated with urinary potassium excretion but not urinary sodium excretion, urinary sodium/potassium ratio, plasma sodium or potassium, or urinary aldosterone excretion (Büsst et al. 2011; Morris 2011).

4.2.3 Hypertension and the Hemostatic System

Interestingly, an association between a common SNP in the plasminogen activator inhibitor-1 gene and hypertension has been reported. The PAI-1 4G/4G (rs1799768) genotype conferred an elevated relative risk for arterial hypertension, regardless of PAI-1 levels and other hypertension-related factors. The biological and molecular mechanisms are still unresolved (Martínez-Calatrava et al. 2007). Central pressures were higher in women carrying the PAI-1 4G/4G genotype compared to females with the 5G/5G genotype. The association remained after adjustment for potentially confounding factors related to hypertension. No association of the PAI-1 genotype with blood pressure was found in men (Björck et al. 2011).

4.2.4 Hypertension and Leptin

Leptin is a hormone, mainly synthesized in adipocytes, that regulates food intake and energy expenditure of the body (see also Sect. 4.3). Rare mutations in the leptin gene cause obesity. Common polymorphisms of the leptin gene have been associated with obesity; however, recently a strong association of a microsatellite polymorphism in the 3′ untranslated region of the leptin gene with essential hypertension was found in subjects with a high BMI. The association was low in individuals with a normal BMI. Linear regression analysis showed an independent correlation of leptinemia with BMI, and a notable correlation was found between serum leptin concentration and angiotensin II. Currently, the underlying mechanism remains unknown. This association underlines the complex interactions between (patho)physiological systems (Akhter et al. 2012).

4.2.5 Nitric Oxide Synthase 3 (eNOS, NOS3)

Nitric oxide synthase 3 (NOS3) catalyzes production of NO in the endothelium and seems to play a role in cardiovascular disease (CVD). NOS3 variants were significantly associated with CHD and heart failure, and significant pharmacogenetic effects were identified for stroke and all of them cause mortality. The hazard ratio (HR) for CHD was higher in minor allele carriers at NOS3 −690C>T (rs3918226); for NOS3 −922A>G (rs1800779), a higher HR for heart failure was found in minor allele carriers. In contrast, the risk for stroke was lower in −690C>T (rs3918226) and Glu298Asp G>T (rs1799983) minor allele carriers when treated with amlodipine versus lisinopril. Data suggest that NOS3 variants may potentially provide useful clinical information with respect to treatment decisions in the future (Zhang et al. 2012).

4.3 Obesity and Type 2 Diabetes

Abnormalities designated as "secondary risk factors" for cardiovascular events as described in detail in Part II of this book generally show a strong genetic background and typically share susceptibility genes. Consequently, signaling molecules introduced above will be of importance for additional risk situations. One example is NOS3, which is a potential determinant of the severity of diabetic complications. A paradigm for tight overlap in genetic background of secondary risk factors is type 2 diabetes (T2DM) and obesity. The reader will note a significant degree of overlap in genetic risk factors and causative gene variants between the separately discussed disease states. Thus, it is important to emphasize that mutations in susceptibility

Table 4.2 Prominent susceptibility genes in obesity

Gene	SNP	Variants	Functional consequence	References
ADRB3		Trp64Arg		Gjesing et al. (2008)
		Arg64Trp	Increased BMI	Allison et al. (1998)
		Arg64Trp	Increased BMI	Fujisawa et al. (1998)
		Arg64Trp	Increased BMI	Kurokawa et al. (2001)
		Arg64Trp	Increased BMI	Kurokawa et al. (2008)
BDNF		Val66Met	Higher BMI	Shugart et al. (2009)
MC4R		V103I	Reduced risk of obesity	Zobel et al. (2009)
		Val103Ile	Reduced risk of obesity	
		Val103Ile	Reduced risk of obesity	
		Val103Ile	Reduced risk of obesity	
NPY2R	rs12649641		Reduced risk of obesity	Heid et al. (2005), Geller et al. (2004), Young et al. (2007), Stutzmann et al. (2007)
PCSK1		N221D	Increased risk of obesity	Torekov et al. (2006)
PPARGC1B		Ala203Pro	Reduced risk of obesity	Benzinou et al. (2008)
FTO	rs9939609 rs9930506 rs991121980			Andersen et al. (2005)

genes, although causing defined cellular dysfunctions, can give rise to multiple risk phenotypes (secondary risk factors). This overlap in genetic basis of disease can create combined phenotypes representing a malign constellation of multiple "cardiometabolic" risk factors specifically regarding the "metabolic syndrome" (Reaven 1988; Mottillo et al. 2010, see also Chap. 12). Focusing on the genetic background of T2DM and obesity, a list of prominent susceptibility genes that can be considered as causative based on available knowledge of gene product functions is provided in Table 4.2.

Common forms of obesity are polygenic disorders and represent a risk for a majority of chronic diseases. Consequently, obesity is considered as a trait of prominent impact on morbidity and mortality. The tight causative relation between obesity, T2DM, and cardiovascular dysfunctions is well established, and pathogenesis involves a complex interplay of genetic and environmental factors (Temelkova-Kurktschiev and Stefanov 2012). Genetic susceptibility appears to determine the individual risk to develop obesity by more than 40 % and probably even up to 70 % (Maes et al. 1997). The search for genetic variations, which, combined with environmental factors, give rise to pathologically relevant BMI levels, was initially hypothesis driven. Candidate gene studies revealed a small number of relevant genes and causative gene variants, which are robustly associated with obesity and provide the basis of severe but rare, monogenic forms of the disease such as variants of the melanocortin 4 receptor (MC4R), prohormone convertase 1/3 (PCSK1), or brain-derived nerve growth factor (BDNF) as listed in Table 4.2. These studies advanced our understanding of the molecular mechanism involved in obesogenesis but shed little light onto common obesity susceptibility. Nonetheless, with the R64Y mutation of ADRB3, one of the first genetic variation associated with both obesity and T2DM in a most likely causative manner was identified (Eisenach and Wittwer 2010), and for MC4R (Masuzaki et al. 2009), polymorphisms were found that not only give rise to rare monogenic forms of obesity but determine obesity risk also in a polygenic manner as the V103I mutation is negatively associated with BMI (Hinney and Hebebrand 2008).

4.3.1 Fat Mass and Obesity-Associated (FTO)

With the advent of large-scale GWAS, a series of common obesity-associated gene loci were identified including FTO (fat mass and obesity associated) as a repeatedly confirmed susceptibility gene with profound relevance for common obesity (Scuteri et al. 2007; Frayling et al. 2007). FTO was found associated with T2DM and subsequently uncovered as a strong determinant of BMI (Scuteri et al. 2007; Frayling et al. 2007). The FTO gene product is an about 60 kD 2-oxoglutarate-dependent nucleic acid demethylase. Despite the strong impact of FTO polymorphisms on obesity risk, the underlying molecular mechanism is still not understood. Nonetheless, a strong association was found for potentially functional SNPs in FTO and early-onset, severe obesity (Frayling et al. 2007). Recent attempts to unravel the functional coupling of FTO to cellular signaling pathways revealed a potential linkage to neuronal plasticity by interference with the NTRK2/BDNF pathway via interaction with the transcription factor C/EBPb (Rask-Andersen et al. 2011). So far, this neuronal mechanism is rather speculative and needs confirmation by further functional studies. FTO has been suggested a common risk determinant for both obesity and T2DM, and linkage of FTO expression to insulin resistance has been reported (Perry and Frayling 2008). Another interesting aspect is the dependency of the impact of FTO risk alleles on physical activity. Physical activity may specifically attenuate obesity risk in risk A-allele carriers of the FTO rs9939609 genotype (Andreasen et al. 2008).

4.3.2 Leptin and Leptin Receptors

For some prominent candidate genes, GWAS has rather excluded a role in common obesity. One example are the leptin receptors, which play a pivotal role in neuroendocrine signaling for the control of energy balance and adiposity (Münzberg and Myers 2005; Bender et al. 2011). Leptin is formed in adipose tissue and communicates the requirements for food intake and energy expenditure to the central nervous system. Leptin signaling impacts on glucose metabolism by determining insulin sensitivity in a way independent of energy balance control. Thus, leptin is considered a crucial determinant of susceptibility for obesity, T2DM, and cardiovascular risk. Common obesity is associated with disturbances in leptin signaling, most prominently involving impaired leptin sensitivity.

Leptin resistance may originate not only from defects in the leptin receptor but also from impaired downstream signal transduction or availability of the hormone at its target sites within the central nervous system. Importantly, the blood–brain barrier represents the main anatomical structure that controls the traffic of adipose tissue-derived hormone(s) to its target receptors and might therefore be involved in leptin resistance (Münzberg et al. 2005). Moreover, leptin sensitivity may be impaired by defects in signaling elements downstream of the hormone/cytokine receptor. Downregulation of leptin sensitivity is considered as a physiological process in response to altered environmental conditions (Penas-Steinhardt et al. 2011), and this adaptive process involves the protein tyrosine phosphatase 1B (PTP1B) and the suppressor of cytokine signaling (SOCS3). Specifically SOCS appear(s) to play a crucial role in both leptin and insulin resistance. The involvement and relevance of genetic factors in leptin resistance, associated with common obesity, is still unclear. Recently, common variants in the gene encoding for the downstream leptin effector JAK2 (Table 4.2), a janus-type protein kinase, were found associated with susceptibility for metabolic syndrome, suggesting a causative role of this polymorphisms in leptin and/or insulin resistance (Penas-Steinhardt et al. 2011).

4.3.3 Peroxisome Proliferator-Activated Receptor γ (PPARγ)

Another gene of which variants have been found to determine the risk for both obesity and T2DM is PPARG, which encodes a ligand-regulated transcription factor of the nuclear hormone receptor family (PPARγ) that is pivotal for adipogenesis, glucose homeostasis, and insulin resistance (Altshuler et al. 2000). Disturbances in PPARγ signaling and alterations in PPARγ-dependent gene expression have been identified as a hallmark of insulin resistance and T2DM, and a series of SNPs that associate with obesity and/or T2DM have been documented (Jeninga et al. 2009). PPARG harbors not only rare mutations potentially linked to T2DM in a causative manner but also a common variation that is clearly associated with T2DM risk. The common PPARγ P12A polymorphism (rs1801282) has been suggested to exert a profound impact on population risk while modestly affecting individual susceptibility. Interestingly, the less frequent 12A allele was found to confer a decreased risk for the disease although the proline to alanine mutation within the proteins N-terminus reduces transcriptional activity (Altshuler et al. 2000; Gouda et al. 2010). Consistently, another mutation in the N-terminal activation function-1 (AF-1) domain of PPARγ, P113Q, confers enhanced transcriptional activity, which promotes T2DM (Jeninga et al. 2009). This is interesting and unexpected, as impaired PPARγ function has repeatedly been demonstrated to promote insulin resistance due to lack of expression of PPARγ target genes required for insulin signaling. Moreover, pharmacological activation of PPARγ by thiazolidinediones has become an established strategy to increase insulin sensitivity in T2DM therapy. It may be speculated that the 12A variant, which displays modestly reduced transcriptional activity, confers a reduced T2DM risk due to a mildly suppressed function as regulator of adipogenesis. Hence, different functional modifications of PPARγ can result in similar metabolic and cardiovascular outcomes. Recently, this phenomenon was addressed and in part elucidated by resolving the crystal structure of PPARγ associated with its co-receptor RXRα and bound to DNA, ligands, and peptides (Chandra et al. 2008). Several unexpected interactions between missense mutations and specific functional domains were revealed. Furthermore, it is important to note that PPARγ has been recognized to govern also vascular remodeling and inflammatory processes, whereby the transcription factor may impact via this mechanism on atherogenesis and vascular dysfunction (Schiffrin 2005).

4.3.4 Adiponectin (AdipoQ/Acrp30)

An adipose tissue-derived signaling molecule that plays a key role in metabolic and cardiovascular pathophysiology has been identified in the mid-1990s. A 30-kDa adipokine termed adipocyte complement-related protein (Acrp30) or adiponectin (AdiopQ) represents an endogenous regulator of insulin sensitivity and glucose and lipid metabolism. Adiponectin levels are inversely correlated not only with insulin resistance and hyperlipidemia but also with the risk to develop inflammatory diseases and certain types of cancer (Ziemke and Mantzoros 2010). Thus, adiponectin signaling has emerged as a factor involved in obesity-related malignancies. The adiponectin gene is located on chromosome 3q27, within a region that determines susceptibility to T2DM. Production and secretion of the antidiabetic adipokine is governed by PPARγ and is clearly upregulated by thiazolidinediones (Combs et al. 2002). Circulating adiponectin levels are typically low in obese humans (Arita et al. 1999) and are depending on nutrition and exercise. Cellular actions of adiponectin are mediated by two different G-protein-coupled receptors (GPCRs; AdiopR1/R2) and transduced into different signaling cascades including the mTOR, JNK, and STAT3 pathways (Ziemke and Mantzoros 2010). The adipokine was reported to exert cardiovascular protection via enhanced phosphorylation of AMP-activated protein kinase, resulting in phosphorylation and increased activity of NOS3 (eNOS, see Sect. 4.2.5; Morandi et al. 2010). Interestingly, AdipoR1-mediated activation of Ca^{2+} entry and Ca^{2+}/calmodulin-dependent kinase β has recently been suggested as a prominent mechanism by which the adipokine governs mitochondrial functions as well as insulin sensitivity (Iwabu et al. 2010). The adiponectin/AdipoR system may be seen as a critical signaling hub that integrates genetic and environmental/lifestyle factors into cardiovascular risk. Common as well as rare polymorphisms in the adiponectin gene have been found associated with the risk for obesity and T2DM development. Missense mutations as well as a silent mutation that is associated with decreased adiponectin levels were found to increase the risk for obesity and insulin resistance (Stumvoll et al. 2002; Yang et al. 2007b). Interestingly, a SNP (11391 G/A; rs17300539), which is associated with increased levels of the adipokine, apparently not only lacked protection against obesity-associated metabolic disturbances but even increased the risk for childhood obesity and insulin resistance (Morandi et al. 2010). Consequently, enhanced adiponectin levels may be excluded as the basis of the recently documented metabolically healthy state in humans (Karelis et al. 2004).

4.3.5 Calpain 10 and Transcription Factor 7-Like 2 (TCF7L2)

Besides PPARG two other genes were identified as potential basis of polygenic T2DM. These are calpain 10 (CAPN10) encoding for a cysteine protease (Evans et al. 2001) and TCF7L2, which encodes for a high mobility, box-containing transcription factor implicated in the control of glucose homeostasis and vascular functions. TCF7L2 was suggested to affect susceptibility to T2DM (Grant et al. 2006). This is most likely based on its role in regulation of the insulinotropic hormone glucagon like peptide-1 (GLP-1) as one of the TCF7L2 target genes. TCF7L2 functions as a nuclear receptor for β-catenin, which transduces upstream signals from the canonical Wnt pathway to transcriptional control (Prunier et al. 2004). Wnt signaling has been suggested essential for pancreatic β cell proliferation (Papadopoulou and Edlund 2005; Rulifson et al. 2007) and is linked to a variety of other transcriptional regulators (Mulholland et al. 2005). Interestingly, calpain has been proposed as a determinant of β-catenin activity, indicating potential convergence of the mechanisms by which mutations in CAPN10 and TCF7L2 determine insulin secretion and T2DM risk (Benetti et al. 2005). It is important to emphasize that among the currently identified susceptibility genes for T2DM, the majority impacts on β cell function rather than insulin sensitivity.

4.3.6 Inwardly Rectifying Potassium Channel Kir6.2 (KCNJ11) and the Sulfonylurea Receptor 1 SUR1 (ABCC8)

Two functionally well-understood gene products involved in glucose sensing and regulation of pancreatic insulin secretion are the ion channel subunits comprising the β cell ATP-sensitive K^+ channel. KCNJ11 which encodes a 390 aa, two membrane spanning, pore forming channel subunit that combines with the ABC protein SUR1 to form a tetrameric K^+ channel that is inhibited by intracellular ATP. This protein complex represents a paradigm metabolic sensor (Olson and Terzic 2010). Cellular ATP generation reports glucose supply of the pancreatic β cell and is translated by a membrane potential change and modulation of voltage-gated Ca^{2+} entry into appropriate insulin secretion. Gain-of-function mutations in KCNJ11 promote cell hyperpolarization and impair insulin secretion (Flanagan et al. 2009). Such gain-of-function mutations are the basis of rare forms of diabetes in neonates, and a common gain-of-function mutation E23K (rs5219) has been identified, which is associated with T2DM (Gloyn et al. 2003; Nielsen et al. 2003). As expected for a common gene variant, the functional consequence of this mutation on ATP sensitivity is moderate but leads to impaired glucose homeostasis and enhanced T2DM susceptibility.

4.4 Dyslipidemia

Hyper- or dyslipidemia is a trait that represents a traditional (well-recognized) risk factor for atherosclerosis and various forms of vascular disease. The causal relation between low-density lipoprotein cholesterol (LDL-C) levels and progression of atherosclerosis as well as risk for cardiovascular events is well established. It was

Table 4.3 Prominent susceptibility genes in dyslipidemia

Gene	SNP	Variants	Functional consequence	References
CELSR2, PSCR1, SORT1	rs599839		LDL mediated	Schunkert et al. (2010)
	rs646776		LDL mediated	Kathiresan et al. (2009)
PCSK9	rs11206510		LDL mediated	Kathiresan et al. (2009)
MRAS	rs9818870		Adhesion signaling	Erdmann et al. (2009)
LPA, LPAL2	rs2048327		Promotes atherothrombosis	Trégouët et al. (2009)
CXCL12	rs501120		Neointima formation after arterial injury	Schunkert et al. (2010)
CXCL12	rs1746048		Platelet activation in atherosclerotic lesions	Schunkert et al. (2010)
LDL-R	rs1122608		LDL mediated	Kathiresan et al. (2009)
LPA	rs3798220	Ile4399Met	Lp(a) mediated	Luke et al. (2007)

initially substantiated by the pioneering work on the genetic basis of familial hypercholesterinemia with the identification of genetic variations in the low-density lipoprotein receptor (LDL-R) associated with elevated LDL-C and elevated risk for coronary heart disease (Lehmann et al. 1995; Brown and Goldstein 1986). Genes involved in lipid metabolism have long been considered as candidate genes to explain heritability of atherosclerosis. Interestingly, a number of candidate genes, which were identified as the basis of rare, monogenic forms of the disease, were later on found to harbor common mutations representing potential polygenic risk determinants. Recent identification of SNPs of increasing numbers of gene loci associated with blood lipid levels (Kathiresan et al. 2009) has led to the concept of complex cardiovascular disease being based on additive effects of common gene variations. Genes associated with elevated LDL-C were also found associated with cardiovascular risk specifically for myocardial infarction (Willer et al. 2008). The most prominent candidate genes with crucial impact on blood lipid levels encode for apolipoproteins and lipoprotein receptors as well as lipoprotein receptor-associated proteins such as lipoprotein receptor-related protein 6 (LRP6; Tomaszewski et al. 2009). The function and relevance of these signaling molecules are comprehensively discussed in Chap. 13. The GWAS approach for blood lipids revealed a plethora of gene loci for which a causative link to hyperlipidemia is entirely elusive and rediscovered genes with well-known functions in lipid metabolism. Moreover, this hypothesis-generating approach has indeed uncovered a series of novel players in lipid metabolism. We will discuss a few examples of established susceptibility genes and gene polymorphisms that have been suggested as potential basis of heritability of cardiovascular risk (Table 4.3).

4.4.1 Apolipoprotein E (ApoE)

This member of the apolipoprotein family is a 299aa (34 kDa) protein that is for a large part produced in hepatocytes.

The three major human isoforms apoE2/3/4 differ in aa positions 112 and 158 by polymorphisms resulting in C/C (apoE2), C/R (apoE3), or R/R (apoE4) containing proteins encoded by the APOEε2/3/4 allele. This polymorphism of APOE exhibits clear ethnic population dependence (Corbo and Scacchi 1999) with a higher frequency of the ε4 allele in Native Americans, Pygmies, and Papuans, while ε3 is prominent in societies of long agricultural history. Structural apoE diversity can be related to function and pathophysiology as the aa residues in positions 112 and 158 determine the interaction of apoE with lipoprotein particles and LDL receptors, thereby affecting blood cholesterol levels. ApoE2 confers an approximately 50-fold reduction in binding to LDL-R as compared to apoE3 and apoE4 (Raffai et al. 2001). Besides the impact on lipid homeostasis, APOE polymorphisms have been clearly demonstrated to affect signaling pathways involved in immune reactions and inflammatory responses (see Zhang et al. 2011). Polymorphism governs the ability of ApoE to modulate/suppress cytokine secretion such as TNF-α and IL-1β with E2 conferring highest and E4 lowest effectiveness. Consistently, APOE polymorphisms were found to affect mortality infectious diseases (Zhang et al. 2011). Moreover APOE genotype strongly impacts on endothelial function and signaling, and apoE4-VLDL has been shown to counteract the antiapoptotic activity of HDL in the endothelium by suppressing PI3K/Akt pathway via a mechanism involving interaction with LDL-R (DeKroon et al. 2006). Genetic variations of APOE are robustly associated with the risk for a variety of diseases that determine longevity, including not only atherosclerosis and cardiovascular events but also infections as well as neurodegenerative disorders such as Alzheimer's disease (Kim et al. 2009). It is important to note that the role of apoE as a modulator of immune responses is certainly relevant for its significance in cardiovascular disease.

4.4.2 Sortilin 1 (SORT1)

An example for a novel gene and molecular player in lipid homeostasis more recently identified via GWAS is sortilin 1. A genome-wide screen in patients with coronary artery disease identified 1p13.3 as locus associated with both elevated LDL-C levels and cardiovascular events. Among several SNPs within this locus, all residing in a noncoding region, rs12740374, an intronic SNP in CELSR2 (cadherin, EGF LAG-seven-pass G-type receptor), was found to generate a C/EBP transcription factor interaction site and was therefore considered a causative variation (Musunuru et al. 2010). The SNP results in enhanced expression of sortilin 1 in the liver. Sortilin 1 is a multi-ligand receptor and a member of the VSP10P domain receptor family also known as neurotensin receptor 3, which is capable of interaction with apolipoproteins, lipoprotein lipase, and LDL-R-associated protein (RAP). Cellular localization of sortilin 1 includes both the cell surface and intracellular compartments, and a role of the multi-ligand receptor in endocytosis and protein trafficking has been reported (Nielsen et al. 2001). Sortilin 1 is likely to interfere with hepatic secretion of apolipoproteins (Kjolby et al. 2010) and has been demonstrated to interfere with LDL uptake into cells (Linsel-Nitschke et al. 2010).

However, diverging results have been obtained in various experimental models studying the effects of genetic manipulation of sortilin 1 expression on plasma cholesterol levels (Musunuru et al. 2010; Linsel-Nitschke et al. 2010). Thus, the mechanism linking hepatic expression of this novel player in lipid homeostasis and cardiovascular risk needs further investigations.

4.4.3 Proprotein Convertase Subtilisin/Kexin Type 9 (PCSK9)

PCSK9 represents a typical example for a gene harboring both SNPs causing rare, severe mutations resulting in profound changes in LDL-C levels (>100 md/dl) but also more common variants (frequency 19 %) with mild effects (3 mg/dl) (Willer et al. 2008). Proprotein convertase subtilisin/kexin type 9 controls LDL-C plasma levels and influences the cellular fate of LDL-Rs. Representing a secreted proteinase, PCSK9 binds to LDL-Rs and targets the receptors to the lysosome promoting its degradation (Horton et al. 2007, 2009). Consistently, rare gain-of-function mutations are associated with hypercholesterinemia (Abifadel et al. 2003) and clearly increase the risk for cardiovascular events. Protective loss-of-function mutations are more common and reduce LDL-C levels. One example is the nonsynonymous R46L polymorphism (rs11591147, frequency about 3 %) which results in moderate reductions in plasma cholesterol (13 mg/dl) (Sanna et al. 2011) and a strong reduction in cardiovascular risk, which appears not sufficiently explained by the change in LDL-C (Benn et al. 2010). This discrepancy may be taken as an indication for an additional protective mechanism of the R46L mutation besides lowering plasma cholesterol (Benn et al. 2010). However, so far animal models have clearly demonstrated the impact of PCSK9 on lipid metabolism (Frank-Kamenetsky et al. 2008) while lipid-independent effects potentially related to cardioprotection are speculation. Importantly, PCSK9 has recently been demonstrated as a highly attractive target for pharmacological intervention. For example, a monoclonal antibody against PCSK9 has been tested which produces significant reductions of LDL-C in patients with familial hypercholesterinemia (Stein et al. 2012).

4.4.4 Lipoprotein Lp(a) and Apolipoprotein Apo(a)

Epidemiology has identified plasma Lp(a) levels as a predictor of cardiovascular risk (De la Peña-Díaz et al. 2000; Kamstrup 2010). Lp(a) contains the glycoprotein apo(a), which is linked by a disulfide bond to ApoB-100 (Guevara et al. 1993, see also Chap. 13). Apo(a) is structurally related to a family of serine proteases, displaying a high degree of similarity with plasminogen. The apoA gene is localized in close proximity to the plasminogen gene on chromosome 6, and its expression is strongly determined by polymorphisms in the 5′-flanking region of the gene (Suzuki et al. 1997). Moreover, structural polymorphism in apo(a) leading to isoforms with size differences has been recognized as basis of variability in Lp(a) plasma levels. The apo(a) size variation arises from the presence of a variable number of repetitive domains, which are homologous to the plasminogen kringle-4 (K4). The structural similarity to plasminogen combined with a lack of fibrinolytic protease function results in a substantial antifibrinolytic potential based on competitive interference with the plasminogen-fibrin interaction (Loscalzo et al. 1990). Apo(a) polymorphism affects both Lp(a) plasma levels and antifibrinolytic activity and thereby cardiovascular risk. Size polymorphism based on a copy number variant has been reported as a basis of variation of plasma Lp(a) levels among ethnic groups (Kraft et al. 1996a, b), and a common nonsense mutation in the K4-2 domain was identified, which results in low Lp(a) levels (Parson et al. 2004). More recently, a nonsynonymous SNP (rs3798220) encoding an Ile>Met exchange in the pseudoprotease domain was found associated with elevated circulating Lp(a) and cardiovascular risk (Luke et al. 2007). Notably, apo(a) polymorphism represents an established paradigm for overlap in the genetic basis of disturbances in lipid metabolism and hemostasis, specifically affecting the fibrinolytic system.

4.5 Arrhythmias: Structural and Electrical Remodeling

With respect to their genetic background, arrhythmias and structural diseases of the heart are positioned at some point between rare genetic variants, typically associated with high severity levels, and common forms, characterized by a rather small increment of risk. As several examples indicate, a single pathological phenotype may be linked to different genetic determinants of predisposition and may be either based on rare genetic modifications and Mendelian traits or associated with common polymorphisms combined with a significant dependency on environmental factors.

A classical structural disease of the heart is represented by primary dilated cardiomyopathy (DCM) that affects either the left or both heart ventricles showing reduced contractility and dilatation without changes in cardiac wall thickness. While secondary dilated cardiomyopathy is induced by infection, toxins, autoimmune diseases, or disorders of the endocrine system, most cases of primary dilated cardiomyopathy are due to the idiopathic form of the disease that has a clear monogenetic origin. In addition to many HCM-related gene products, dystrophin and desmin (Arbustini et al. 2000; Li et al. 1999; Taylor et al. 2007) have been identified as affected proteins, as well as actin, δ-sarcoglycan, titin, laminin, and phospholamban (Herman et al. 2012; Takeda 2003; van der Zwaag et al. 2012). In addition to rare monogenetic forms, several common gene variants were shown to affect the risk for DCM. A GWAS study genotyping 664 DCM cases in Germany identified heat shock protein B7 (HSPB7, SNP rs1739843), a protein involved in cardiac stress response, to be associated with DCM risk (Stark et al. 2010). Variations in this gene have also been found associated with sporadic and advanced heart failure (Cappola et al. 2010; Matkovich et al. 2010). Polymorphisms of presenilin-1 (SNP rs177415), a protein involved in cell apoptosis and cardiac development, were found associated with the susceptibility for DCM (Li et al. 2011). A recent GWAS study (Villard et al. 2011) identified SNP rs2234962 located within the sequence encoding for

BAG family molecular chaperone regulator 3 (BAG3), a finding that is supported by a study indicating that this gene may also be affected in rare forms of DCM (Norton et al. 2011).

Cardiomyopathies as well as heart failure are typically associated with arrhythmogenesis and a certain level of electrical remodeling. Nonetheless, genetic risk for arrhythmias has largely been identified in terms of polymorphisms in genes encoding for ion channels that determine cardiac excitability and electrical properties, thus representing classical "channelopathies." Sudden cardiac death (SCD) is a pathological phenotype responsible for 50 % of cardiovascular mortality (Kolder et al. 2012). The primary functional cause, a ventricular fibrillation (VF), may be induced by rare genetic variations associated with a clear familial heritability pattern, based on channelopathies. These forms of SCD typically occur in younger individuals. However, the majority of cases occurs in individuals above the age of 40 years and is based on VF as a consequence of coronary artery disease (Zipes and Wellens 1998). Genome-wide association studies helped to identify a number of frequent gene variants linked to these common forms of SCD. However, for many genes found in these studies, a direct causal contribution of their encoded proteins in pathophysiological pathways is not immediately evident. These genes and gene products may rather affect expression of other genes. To determine common causal molecular targets of genetic variability linked to SCD, recent studies have been investigating intermediate phenotypes of SCD risk, i.e., the effect of frequent genetic variants on specific mechanisms and processes that have a clear causal connection to SCD (Kolder et al. 2012). Some loci identified by these studies affect proteins already known to be associated with Mendelian disorders like long QT syndrome and short QT syndrome, including potassium channels. It is of note that in addition to mutations in genes encoding for ion channels, polymorphisms of nitric synthase 1 neuronal adaptor protein (NOS1AP) (Arking et al. 2006), Na/Ka ATPase β1 subunit (ATP1B1) (Lingrel 2010), and phospholamban (PLN) (Periasamy et al. 2008), key elements of myocardial function, were also found associated with VF risk.

A cardiac disorder that receives increasing attention due to its high prevalence and association with substantial health-care costs is atrial fibrillation (AF). Representing the most common form of arrhythmia with a clear familial clustering and a wide array of established, general risk factors for cardiovascular diseases such as age, gender, diabetes, and hypertension, AF has been intensively investigated in terms of its genetic basis (Rienstra et al. 2012). AF represents a paradigm for a vicious feed-forward pathogenetic process that involves structural and electrical remodeling of the atrium. Recent studies have revealed a number of common genetic variations with examples that give a hint to the molecular pathophysiology of this disorder.

4.5.1 Human Ether-a-gogo Related Gene (HERG)

VF is closely related to changes in the structure and function of HERG (KCNQ1, KCNH2, KCNE1) and/or its regulatory subunits (Amin et al. 2012; Nishio et al. 2009; Sinner et al. 2008). A polymorphism of KCNH2 (SNP rs1805123) associated with VF leads to the substitution of lysine 897 by threonine in the delayed rectifier potassium channel. The exact effect of the genetic variant on channel function is still a matter of controversy (Bezzina et al. 2003; Männikkö et al. 2010; Paavonen et al. 2003). However, a rather rare SNP (rs1805128) affecting KCNE1 (D85N) has been demonstrated to be associated with functional defects in delayed rectifiier potassium channel (IKr, Nishio et al. 2009) and a large gene survey identified the KCNE1-D85N as a possible modulator of drug-induced torsades de pointes (Kääb et al. 2012).

4.5.2 Voltage-Gated Sodium Channels (NaV)

Another intermediate phenotype involved in VF and SCD is represented by the mechanisms responsible for PR interval and QRS duration. Here, voltage-gated sodium channels (Nav1.5, SCN5A) (Shinlapawittayatorn et al. 2011) have been demonstrated to be affected by polymorphisms that are also targets of Mendelian traits leading to cardiac conduction disease (Schott et al. 1999). SCN10A, encoding Nav1.8, is the gene affected by SNP rs6795970, a single-nucleotide polymorphism responsible for changes in PR interval and QRS duration (Holm et al. 2010).

4.5.3 Paired-Like Homeodomain Transcription Factor 2 (PITX2)

GWAS studies have identified a series of genetic loci that confer susceptibility to AF. Among these loci in particular, 4q25 near the developmental transcription factor PITX2 has received particular attention (Liu et al. 2012). PITX2 is essential in embryonic development and governs development of cardiac left-right asymmetry, pulmonary myocardium, and the sleeve of cardiomyocytes that reaches into the pulmonary vein, which represents a critical origin of triggered activities leading to AF. PITX2 is a downstream target of Wnt/β-catenin signaling and, in the heart, a pivotal player in developmental signaling pathways involving factors such as Nodal, Shox2, and Tbx3 (Poelmann et al. 2008; Tessari et al. 2008; Christoffels et al. 2010).

PITX2 is the closest gene to common risk variants on chromosome 4q25 with the first identified by Gudbjartsson et al. (2007) (rs2200733 and rs2220464; see Table 4.4).

The gene is responsible for development and identity of atrial and ventricular excitability properties as well as pulmonary myocardium. This key role has been established by genetic mouse models demonstrating a potential role in AF (Wang et al. 2010a). PITX2 is considered as an inhibitor of genetic programs essential for pacemaker functions and to govern the potential for ectopic activity. Human atrium displays expression of the PITX2c isoform which was found downregulated in AF patients (Kirchhof et al. 2011). Moreover, PITX2c-deficient mice were shown to display increased sensitivity to programmed stimulation-induced AF. Importantly, lack of PITX2c was suggested to affect ion channel expression in the atrium (Chinchilla et al. 2011), including KCNQ1 and may be also other K^+ channels essential for atrial repolarization.

Table 4.4 Prominent susceptibility genes linked to arrhythmias: structural and electrical remodeling

Gene	SNP[a]	Variants	Functional consequence	References
Dystrophin	rs128627257		Defective myocardial force transmission	Arbustini et al. (2000), Li et al. (1999)
Desmin	rs121913005		Defective myocardial force transmission	Taylor et al. (2007)
Actin	rs104894546		Defective myocardial force transmission	Takeda (2003)
δ-Sarcoglycan	rs104893869		Defective myocardial force transmission	Takeda (2003)
Titin	rs199469665 rs199469666		Defective myocardial force transmission	Herman et al. (2012)
Laminin	rs80356682 rs121913570		Defective myocardial force transmission	Takeda (2003)
HSPB7	rs10927875		Susceptibility for DCM	Stark et al. (2010), Cappola et al. (2010), Matkovich et al. (2010)
Presenilin-1	rs177415		Susceptibility for DCM	Li et al. (2011)
BAG3	rs2234962		Susceptibility for DCM	Villard et al. (2011), Norton et al. (2011)
NOS1AP	rs12143842 rs2880058 rs10494366 rs12029454 rs4657178 rs16857031		VF risk	Arking et al. (2006)
ATP1B1	rs10919071		VF risk	Lingrel (2010)
Phospholamban	rs12210810		VF risk	Periasamy et al. (2008)
PLN			Susceptibility for DCM	van der Zwaag et al. (2012)
KCNQ1	rs12296050			Amin et al. (2012)
KCNH2	rs2968863			Nishio et al. (2009)
KCNE1	rs1805128			Sinner et al. (2008)
KCNH2	rs1805123	Lys897Thre	Associated with VF	Bezzina et al. (2003), Männikkö et al. (2010), Paavonen et al. (2003)
KCNE1	rs1805128	D85N	Functional defects in I_{Kr}	Nishio et al. (2009)
			Drug-induced torsades des pointes	Kääb et al. (2012)
SCN5A	rs12053903		Susceptibility for DCM	Shinlapawittayatorn et al. (2011)
	rs2051211		AF risk	Schott et al. (1999)
	rs41312391		SCD risk	
	rs2200733			

(continued)

Table 4.4 (continued)

Gene	SNP[a]	Variants	Functional consequence	References
SCN10A	rs6795970		Changes in PR interval and QRS duration	Holm et al. (2010)
PITX2	rs2200733 rs2220464		Common risk variants on chromosome 4q25	Gudbjartsson et al. (2007)
KCNN3	rs133763339		Associated with lone AF	Ellinor et al. (2012)

[a]SNP–numbers not necessarily mentioned in given references

4.5.4 Small Conductance Ca^{2+}-Regulated Potassium Channel (KCNN3)

Cardiac electrophysiologists have for a long time ignored the presence and/or significance of small-/intermediate-type Ca^{2+}-activated K^{+} channels. Only recently these channels got into the spotlight of cardiology since expression in human hearts (Xu et al. 2003) specifically in atrium has been demonstrated. Ellinor et al. (2012) recently identified a polymorphism (rs133763339) on chromosome 1q21 that is associated with lone AF. This common variation affects a member of the intermediate/small conductance Ca^{2+}-activated K^{+} channel family that appears of importance for atrial action potential morphology (KCNN3). In fact, pharmacological experiments suggest that inhibition of KCNN3 prolongs atrial action potential duration and is able to terminate AF (Diness et al. 2010). Thus, KCNN3 has emerged as an attractive target for AF therapy. It is important to note that AF is of pivotal importance as a cardiovascular risk factor promoting heart failure and stroke due to disturbances in cardiac mechanics, hemodynamics, and hemostasis.

4.6 Hemostatic Dysfunctions

In the following chapter, proteins and SNPs important for hemostatic balance will be discussed regarding their contribution to the risk of vascular diseases.

The coagulation system is essential to prevent excessive bleeding and to enable wound healing after trauma. Thus, an active coagulation system is needed. However, coagulation has to be carefully balanced to avoid bleeding as well as thrombosis. This balance is guaranteed by the well-tuned interaction of the pro- and the anticoagulatory system and fibrinolysis. In case of increased or decreased concentrations, hyperactivity, or subnormal function of any one of the participating components, the fine-tuning is disturbed resulting in clinical phenotypes. During the last two decades, great effort has been made to understand the molecular genetic mechanism of the hemostatic system. The DNA sequences of all known coagulation factors have been elucidated, and many genetic alterations affecting the function of coagulation factors have been identified (Table 4.5).

Table 4.5 Genes and polymorphisms relevant for atherothrombosis and vascular diseases

Gene	SNP	Variants	Functional consequence	References
Fibrinogen	rs1800790	–455G>A	Increased fibrinogen plasma level	Carty et al. (2010), Dehghan et al. (2009), Albert et al. (2009), Lovely et al. (2011)
	rs6050	Thr331Ala	Modestly lower plasma level	
Prothrombin	rs1799963	20210G>A	Increased prothrombin plasma level	Zee et al. (2009), Satra et al. (2011), Delluc et al. (2010)
FV Leiden	rs 6025	506Arg>Gln	Impacts on inactivation of FV leads to APC resistance	Tomaiuolo et al. (2012), Satra et al. (2011), Delluc et al. (2010), Maitland-van der Zee et al. (2009)
FXII	rs1801020	–4C>T	Association with lower FXII plasma level	Johnson et al. (2011a), Calafell et al. (2010)
FXIII	rs5985	Val35Leu	Increased FXIII activity, decreased clot stability, and resistance to fibrinolysis	Debette and Seshadri (2009), Slowik et al. (2005)
FVII	rs 6046	Arg413Gln	Associated with lower FVII activity	Ken-Dror et al. (2010), Kudaravalli et al. (2002)
	rs510335	–401G/T	Associated with lower FVII activity	
	rs5742910	323 0/10 bp insertion	Associated with lower FVII activity	
	rs510317	–402G>A	Associated with higher FVIIag and FVIIc levels	
FVIII	rs1800291	Asp1260Glu	Associated with FVIII levels	Viel et al. (2007), Reiner et al. (2009)
PAI-1	rs34857375	–675(4G/5G)	SNP modulates PAI-1 levels, influences inhibition of fibrinolysis	Satra et al. (2011), Zorio et al. (2008), Tsantes et al. (2008), Ye et al. (2006), Hoekstra et al. (2004)
GPIa (ITGA2)	rs 1801106	Glu534Lys	Influences GPIa activity and expression density	Kunicki et al. (2012), Wang et al. (2010b), Kunicki et al. (2009)
GPIIIa (ITGB3)	rs5918	Leu59Pro	Higher number of activated platelets	Kucharska-Newton et al. (2011), Galasso et al. (2010), Bray (2000)

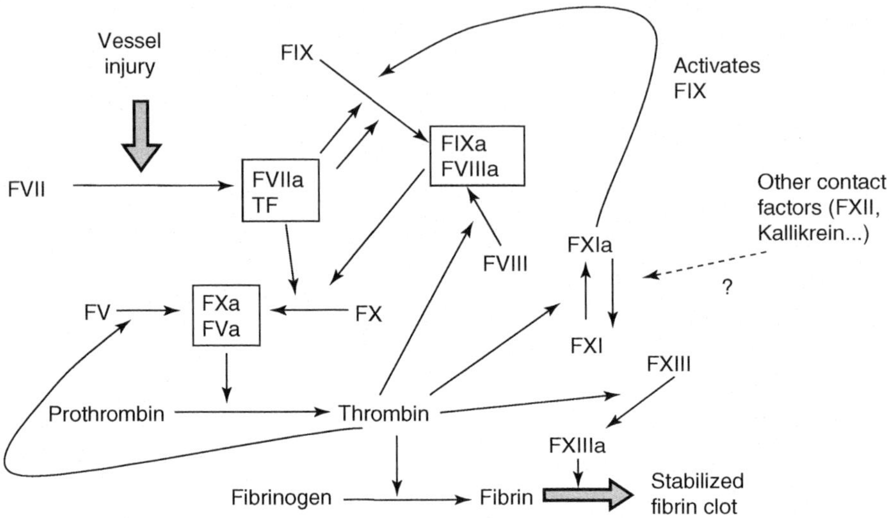

Fig. 4.1 Coagulation cascade (Endler et al. 2003)

In vitro, coagulation can be initiated via two pathways: the intrinsic system comprising factor XII, prekallikrein, high-molecular-weight kininogen, and factors XI, IX, and VIII at negatively charged surfaces and the extrinsic pathway via activation of factor VII in the presence of tissue factor. Both pathways converge at a common stage involving factors X, V, and II. As a result thrombin is formed, which then converts fibrinogen (factor I) to fibrin. The cross-linking of fibrin filaments by activated factor XIII increases clot stability and its resistance to fibrinolysis. Today, we know that the intrinsic and extrinsic coagulation systems are intertwined and the *in vivo* activation of the coagulation cascade is complex (Fig. 4.1).

4.6.1 Coagulation Factors, Gene Polymorphisms, and Thrombotic Disease

Thrombotic events in coronary arteries are central for the development of MI. Twin and sibling studies have shown that inherited risk factors contribute significantly to the development of coronary artery disease (CAD) and ischemic stroke (Brass et al. 1996, 1992). The formation of blood clots that occlude major arteries represents the leading complication of arteriosclerosis. In the 1990s it became evident that besides mutations in coding regions of genes, polymorphisms in regulatory elements of coagulation factor genes, particularly in the promoter regions, have an important impact on blood coagulation because of their effect on the concentration of the proteins (Franco and Reitsma 2001; Blake et al. 2001). For some coagulation factors, reproducible data exist regarding the contribution of various SNPs to the variability of plasma concentration and consequently an effect on the clinical phenotype.

4.6.2 Fibrinogen (Factor I), Factor XIII, and Fibrinolysis

Fibrinogen is a 340-kDa glycoprotein consisting of three nonidentical polypeptide chains (α, β, and γ) linked by disulfide bridges. The three genes encoding fibrinogen Bβ (*FGB*), Aα (*FGA*), and γ (*FGG*) are clustered in a region of ~50 kilobases (kb) on human chromosome 4. Each gene is separately transcribed and translated to produce the nascent Aα, Bβ, and γ-polypeptides, which assemble to form fibrinogen. Fibrinogen levels are subject to biological variation. They are upregulated during acute phase reactions via the activation of the IL-6 responsive elements in the promoter region of all three fibrinogen chains. High levels of fibrinogen have been associated with the development of arterial thrombosis. Prospective studies such as the Northwick Park Heart Study, the Gothenburg study, and the PROCAM study have related elevated fibrinogen levels to MI and stroke (Heinrich et al. 1994; Wilhelmsen et al. 1984). Fibrinogen levels cluster with other cardiovascular risk factors including hypertension, diabetes, smoking, and peripheral arterial disease (Lee et al. 1993; Endler and Mannhalter 2003).

Genetic factors contribute to the variability of fibrinogen plasma levels. The two most frequently studied polymorphisms are the −455G>A (rs1800790) polymorphism located in the promoter region of the fibrinogen beta chain (FGB) and the 312Thr>Ala (rs6050) polymorphism in the fibrinogen alpha chain (FGA). −455G>A is in complete linkage disequilibrium with the −148C>T polymorphism in the IL-6 responsive element and leads to increased plasma fibrinogen levels, while 312Thr>Ala is associated with modestly lower plasma fibrinogen levels. The FGA rs6050 single-nucleotide polymorphism (SNP) indicates a decreased risk, whereas the FGB −455G>A SNP seems to increase the risk of stroke. The risk of myocardial infarction does not seem to be altered by either SNP (Siegerink et al. 2009). Interestingly, an association between poststroke mortality and the 312Thr>Ala polymorphism in patients with atrial fibrillation had been observed by Carter et al. (1999). Thus, although the general risk for MI or stroke apparently is not significantly affected by fibrinogen polymorphisms, they may play a role in subgroups of patients.

Currently, the role of fibrinogen polymorphisms in thrombotic disease has still not been definitely proven. The available results suggest that plasma fibrinogen levels could be more important as risk factors for ischemic stroke than for myocardial infarction.

Fibrinogen is cleaved by thrombin leading to the generation of fibrin monomers which have to polymerize and be cross-linked by factor XIII. This cross-linking of the fibrin polymers is essential for clot stability and its resistance to fibrinolysis (Siebenlist et al. 2001). Fibrin structure and stability have been linked to thrombotic diseases. Biochemical studies of fibrinogens have examined the interactions that mediate the conversion of soluble fibrinogen to the insoluble fibrin network. These studies show that fibrin structure modulates the enzymatic lysis of the fibrin network. Thus, the molecular mechanisms that control structure also control stability. FXIII which catalyzes the formation of covalent ε(γ-glutamyl)-lysyl bonds between the γ and α chains of fibrin plays an important role in this process.

Factor XIII circulates in plasma as a heterotetramer consisting of two catalytic A-subunits and two carrier protein B-subunits (A2B2). Monocytes and megakaryocytes synthesize FXIII, and it has also been identified in platelets. In the F13 gene, a common G>T polymorphism at nt163 (rs5985, c.103G>T), leading to a valine (Val) to leucine (Leu) substitution at amino acid position 35 (nomenclature according to HGVS, three amino acids away from the thrombin activation site in exon 2 of the FXIII A-subunit gene), was described (Mikkola et al. 1994). Due to the close proximity of the thrombin activation site, this polymorphism is thought to impair FXIII activation and contribute to the pathogenesis of thrombotic disorders (Kohler and Grant 1999). The Val35Leu polymorphism leads to faster activation of FXIII. *In vitro*, the 35Leu allele confers an increased catalytic activity to FXIII but decreases clot stability and resistance to fibrinolysis by alterations of the clot structure (Ariens et al. 2000). A lower prevalence of the 35Leu allele has been found in subjects with MI or deep vein thrombosis compared to healthy controls (Corral et al. 2000; Bereczky and Muszbek 2011). In spite of some positive reports, the importance of the FXIII 35Leu genetic variant for premature coronary artery disease (CAD) and thrombotic events remains controversial. The evaluation of its effect on long-term clinical outcome defined as a composite of cardiovascular death, recurrent MI, and revascularization indicated no association of FXIII 35Leu with premature CAD or clinical outcome. However, carriage of factor XIII 35Leu leads to a stepwise decrease in the rate of fibrinolysis with a significant gene dose effect in patients (Silvain et al. 2011). The role of the FXIII Val35Leu polymorphism in ischemic stroke is still unclear. Our group found no differences in genotype distribution between patients with ischemic stroke and healthy controls. This suggests that an association of the FXIII Val35Leu polymorphism with a decreased risk of ischemic stroke or an increased risk of intracerebral hemorrhage is unlikely (Endler et al. 2003). The results were recently confirmed by Shemirani et al. (2010) who also found no association between the risk of acute ischemic stroke and FXIIIA 35Leu carriership in either gender. Thus, while there is increasing evidence that the 35Leu allele protects from MI, its role in the development of stroke needs to be clarified.

In mammalian blood the fibrinolytic system is essential for the dissolution of blood clots and the maintenance of a patent vascular system. Abnormalities in the fibrinolytic system have been implicated in the pathogenesis of atherosclerotic diseases such as myocardial infarction and stroke (Babu et al. 2012). Fibrinolysis is initiated by specific interactions between its main components, the inactive proenzyme plasminogen, which is converted to the active enzyme plasmin, which degrades fibrin. Two plasminogen activators exist in blood: tissue-type plasminogen activator (tPA) and urokinase. Activation of plasminogen by tPA is enhanced in the presence of fibrin or at the endothelial cell surface. Inhibition of fibrinolysis occurs either at the level of plasminogen activation by a plasminogen activator inhibitor or at the level of plasmin by α_2-antiplasmin.

Plasminogen activator inhibitor (PAI)-1 is the key regulator of the fibrinolytic system and has the greatest inhibitory effect. It plays a crucial role in various physiological processes including fibrinolysis, tissue repair, blood coagulation, thrombolysis, ovulation, embryogenesis, angiogenesis, and cell adhesion and migration (Fig. 4.2).

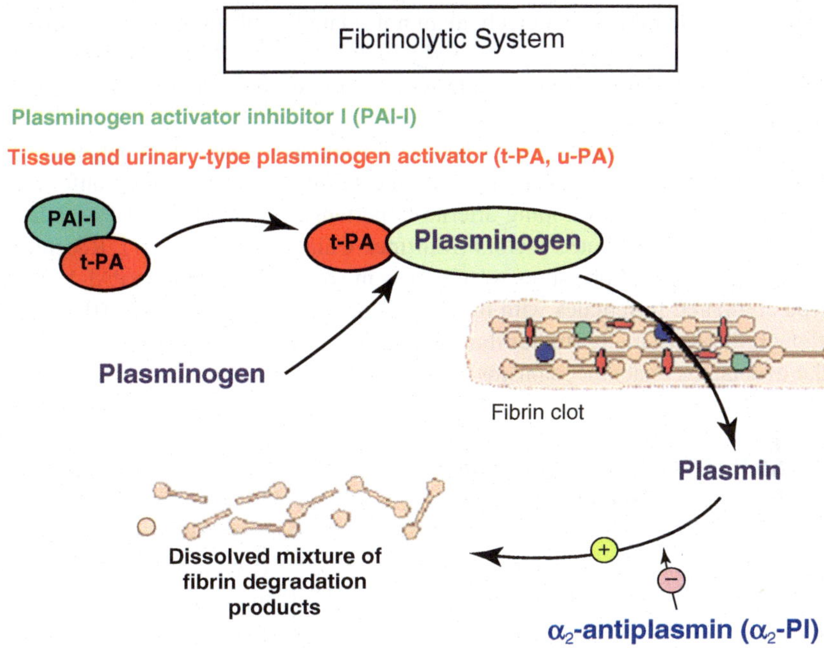

Fig. 4.2 Plasminogen activator inhibitor 1 (PAI-1)

A delicate interplay between tissue plasminogen activator (tPA) and its inhibitor PAI-1 regulates the propagation of thrombi. PAI-1 is synthesized locally in the vascular wall (in endothelial and smooth muscle cells). High levels of tPA and PAI-1 have been reported in patients with a history of stroke (Carter et al. 1998; Macko et al. 1999). Genetic variants of PAI-1 affect PAI-1 levels, and gene polymorphisms have been associated with myocardial infarction and stroke (Jood et al. 2005). The promoter region of the PAI-1 gene contains a common single guanine nucleotide insertion/deletion polymorphism (4G/5G) situated 675 bp from the transcription start site. The 5G allele contains a repressor site and binds both an enhancer and a repressor, while the 4G site binds only an enhancer. This leads to a lower transcription of PAI-1 in carriers of the 5G genotype (Erikson et al. 1995). A recently published meta-analysis comprising 41 studies including 12,461 cases and 14,993 controls identified the PAI-1 4G allele as weak risk factor for MI in Caucasian, Asian, or African populations. The MI risk for the 4G/4G genotype was increased compared to 5G/5G genotype and the 5G allele, with an odds ratio at 1.16 (Gong et al. 2012). Data on the association of PAI-1 and tPA levels with stroke as a function of PAI-1 4G/5G genotype is controversial. Elevation in PAI-1 and reduction in tPA levels were observed in stroke patients. Nevertheless, in some studies including ours, the 4G/4G genotype conferred a reduced risk of stroke, possibly due to a more stable clot and less likelihood of embolic events (Endler et al. 2000). In a study by Saidi et al. (2007), regression analysis demonstrated that 4G homozygosity was an independent predictor of a protective effect. The authors speculate that this may possibly be mediated through a mechanism not related to fibrinolysis. In contrast to these results, a nested case–control study in two independent Swedish cohorts reported an association of the 4G allele with an increased risk of future ischemic stroke (Wiklund et al. 2005).

The PAI-1 4G/5G polymorphism has also been associated with hypertension. Indeed, in 815 unrelated Spanish individuals, the 4G/4G PAI-1 genotype conferred an elevated relative risk of developing arterial hypertension, regardless of PAI-1 levels and other hypertension-related factors (Martínez-Calatrava et al. 2007). Recently, it was reported that central pressures were higher in women carrying the PAI-1 4G/4G genotype compared to females with the 5G/5G genotype (Björck et al. 2011). Thus, even though there are many studies that suggest a contribution of PAI-1 to the risk of vascular events, the exact mechanisms as well as the disease entity are unclear.

4.6.3 SNPs in Other Coagulation Factors: Factor XII, Factor VII, Factor V, Factor II, Factor VIII, and von Willebrand Factor

In vitro, clot formation is initiated by the contact of FXII with negatively charged surfaces. This leads to proteolytic cleavage of the FXII molecule and activation to FXIIa, which induces activation of factor XI. FXII also participates in the conversion of plasminogen to plasmin. Zito et al. showed that FXII and FXIIa levels are highly variable in the population (Zito et al. 2000). Even though it is known that FXII levels are influenced by hypertriglyceridemia, smoking, or estrogens, his data indicated that the FXII variability is also genetically determined. A common variant in the factor XII (F12) gene (46C>T, −4C>T, rs 1801020) resulting in decreased plasma levels was identified by Kanaji et al. (1998). The polymorphism destroys a Kozak consensus sequence, resulting in lower translation efficiency and a decrease in FXII plasma levels. There is general agreement that FXII and FXIIa levels are significantly lower in carriers of the T allele, e.g., 132 U/dl FXII in −4CC, 87 U/dl FXII in −4CT, and 53 U/dl FXII in −4TT carriers $p<0.001$. This polymorphism is frequent in Orientals and may explain the lower plasma FXII activity found in this population. The role of the polymorphism for the development and progression of coronary artery disease has been debated for more than 10 years. In Viennese patients the homozygous F12 −4TT genotype was less frequent in patients with an acute coronary syndrome (ACS) than in patients with stable CAD (Endler et al. 2001). However, data of a large retrospective study performed in Vienna indicated gradually increased hazard ratios for vascular mortality with decreasing FXII levels (from 1.0 in the >100 % FXII to 1.5 in the 80–90 % to ~4.7 in the 10–20 % FXII group). No significant increase in all-cause mortality was observed in FXII-deficient subjects (0–10 % FXII) (Endler et al. 2007). Bach et al. (2008) found no association of the FXII genotype with any clinical phenotype but could confirm the association of FXII levels with coronary risk. To shed light into the long discussion of an association between low factor XII levels or F12 gene variants and thrombotic outcomes, a recent review and meta-analysis searched MEDLINE, EMBASE, and HuGE Navigator for an association between F12 −4C>T and venous thromboembolism as well as myocardial infarction. Sixteen candidate gene studies (4,386 cases, 40,089 controls) were analyzed. None of the investigated effects reached statistical significance at $p<0.05$, apart from a very weak association with myocardial infarction for the TT+CT versus CC genotype (odds ratio 1.13). Overall, the evidence for an association between F12 −4C>T and venous thromboembolism and myocardial infarction was weak (Johnson et al. 2011a). The F12 −4C>T polymorphism apparently does not play a role in the development of ischemic stroke, except probably in patients with atherothrombotic stroke (Oguchi et al. 2000). However, according to a recent genome-wide linkage screen, the factor 12 gene seems to contribute to the susceptibility to venous thrombosis. The effect was only partly due to the F12 −4C>T polymorphism, and still unknown polymorphisms within the F12 gene appear to contribute (Endler and Mannhalter 2003). Thus, further studies regarding the role of SNPs in the F12 gene will be necessary before a final conclusion can be drawn.

The extrinsic coagulation pathway is initiated by activation of factor VII. Coagulation factor VII (FVII) is a vitamin K-dependent coagulation factor circulating in the blood as inactive zymogen. In the Northwick Park Heart Study, elevated FVII coagulant activity has been associated with fatal ischemic heart disease. These results could only partly be confirmed in the prospective PROCAM study (Meade et al. 1993; Heinrich et al. 1994). FVII was elevated in patients with coronary events; however, in a multivariate logistic regression model, it did not remain an independent risk factor. FVII activity and plasma triglyceride concentrations are closely linked which may in part explain the association between hypertriglyceridemia and arterial thrombotic disease (Grant and Humphries 1999). The gene for FVII contains five polymorphic sites, which account for ~30 % of the variation in FVII plasma levels (Bernardi et al. 1996; Campo et al. 2006). The contribution of the polymorphisms in the FVII gene to the development of arterial thrombosis is supported by several publications. Already in 1998 Iacoviello et al. reported a protective effect of the FVII 353Q allele (rs 6046) for the development of MI (Iacoviello et al. 1998). Girelli et al. (2000) described similar results for the 353R>Q polymorphism and the decanucleotide polymorphism at nt(−323). In 2006, Campo et al. published that FVII and TF antigen levels measured at admission, which partly correspond to polymorphisms in the respective gene, are independent predictors of mortality and reinfarction in patients with acute MI (Campo et al. 2006). In 2012, an investigation on the transcriptional regulation of F7 by epigenetic features indicated that the epigenetic regulation of the F7 promoter through methylation affects FVIIa plasma concentrations and is associated with coronary artery disease (Friso et al. 2012). Our group presented evidence that the FVII −402A allele in the promoter region which confers increased transcriptional activity resulting in higher FVIIa increases the risk of early ischemic cerebrovascular events, whereas the 353R>Q, the −401G>T, and the −323ins/del sequence variations, which are in close-linkage disequilibrium and have been associated with lower FVII activity, have no affect (Funk et al. 2006). Indications that individuals with the −323ins allele have an increased risk of primary intracerebral hemorrhage (PICH) could not be confirmed by Greisenegger et al. (2007).

The intrinsic and the extrinsic coagulation system converge at a common stage involving factors X, V, and II. Coagulation factor V (FV) acts as cofactor of FXa and plays an important role in the regulation of the coagulation process. It circulates in plasma as inactive pro-factor. FV is also found in the α-granules of platelets. To develop its procoagulatory activity, FV has to be activated either by thrombin or FXa by limited proteolysis of several peptide bonds (Monkovic and Tracy 1990). Activated FV (FVa) has to be inactivated by proteolytic cleavage at Arg506, Arg306, and Arg679 by activated protein C (APC) to maintain the hemostatic balance. In 1993 Dahlbäck et al. reported a family with poor anticoagulant response due to resistance to activated protein C (Dahlbäck et al. 1993). The resistance was shown to be caused by a point mutation in the F5 gene at nt1691, leading to an Arg>Gln amino acid exchange at amino acid position 506 (FV R506Q, FV Leiden, rs6025) (Bertina et al. 1994). The mutation is frequent in the Caucasian population (~5 %) and represents one of the most important risk factors for inherited venous thrombosis. Large meta-analyses investigated the role of the FV R506Q mutation in myocardial infarction and reported an odds ratio of ~1.3 for carriers of the FV 506Q allele (Juul et al. 2002). However, only in one study did the association reach statistical significance, probably due to the larger sample size. In a meta-analysis published by Wu and Tsongalis (2001) in unselected patients with CAD, no association with factor V Leiden was found. From these results, it can be concluded that screening for the FV R506Q polymorphism in unselected patients at risk for MI is not indicated. However, according to the literature, it still cannot be excluded that the FV 506Q mutation has a role in selected patient groups. Rosendaal et al. (1997) observed that the FV R506Q mutation represents a risk factor in young women with MI before the age of 45 years. In addition, a high prevalence of the FV 506Q allele in patients with MI without signs of coronary atheromatosis was reported (Van de Water et al. 2000). In this selected group of patients, vasospasms followed by coronary thrombosis seemed to trigger the development of MI. The analysis of the FV R506Q genotype could help with risk assessment in this patient subgroup. The contribution of this mutation to the risk for ischemic stroke in adults is still unclear. Wu and Tsongalis (2001) found an increased risk for cerebrovascular disease for carriers of the FV 506Q allele in their meta-analysis. In contrast, Juul et al. (2002) saw no association between FV R506Q and adult ischemic stroke in patients from the Copenhagen City Heart Study and from meta-analysis of available data. However, the statistical power of the meta-analysis was too low to rule out a 20 % risk increase. As reported by Lalouschek et al., FV Leiden confers a statistically significantly increased risk for stroke in female smokers (*OR* 8.8; $P=0.004$). No interaction between the mutation, smoking, and risk of stroke was observed in men (Lalouschek et al. 2005). While there is very good evidence for an increased risk of venous thrombosis in carriers of the FV Leiden mutation due to the impaired inactivation by activated protein C, a general role of this mutation in patients with cardiovascular or cerebrovascular thrombosis has not yet been proven. It appears that the mutation is only relevant in selected patient populations, e.g., those with vasospasms and patients with MI or stroke at young age (<55 years old) or without significant coronary stenosis.

Another important and well-studied procoagulant protein is prothrombin (FII). It is a vitamin K-dependent glycoprotein of about 70 kDa and represents the proenzyme of thrombin. Defects in the prothrombin gene can lead to inherited prothrombin deficiencies, which are associated with an increased bleeding tendency. In 1996, Poort et al. (1996) identified a single-nucleotide exchange (G>A transition) in the 3′-untranslated region of the prothrombin gene at position 20210, (rs1799963) which is associated with ~25 % increase in plasma thrombin activity (Franco and Reitsma 2001; Ceelie et al. 2004). The mutation, located 20 nt downstream of the poly A signal, increases the posttranslational 3′ end processing efficiency and leads to a higher transcription rate (Gehring et al. 2001). In many studies, an association between the 20210A allele and the risk for venous thrombosis was described. In contrast, the data on the effect of this variant on the development of arterial thrombosis and MI are contradictory. A meta-analysis performed by Franco et al. (1999) revealed a statistically significant association of the prothrombin 20210A variant with the development of MI (*OR* 2.5). A more recent meta-analysis testing the association of factor V Leiden, the prothrombin 20210G>A, and the MTHFR 677C>T (TT genotype) mutations with myocardial infarction, ischemic stroke, and peripheral vascular disease (Endler and Mannhalter 2003) showed a modest association of the factor V Leiden mutation (*OR* 1.21), the PT 20210G>A mutation (*OR* 1.32), and MTHFR TT mutation (*OR* 1.20) with arterial ischemic events. Subgroup analyses of younger patients (<55 years old) and of women revealed slightly stronger associations (Kim and Becker 2003). We investigated the prevalence of the prothrombin mutation in 468 patients with acute stroke or transient ischemic attack (TIA) before the age of 60 years. We found that the frequency of the F2 20210G>A mutation was significantly higher in male patients compared with controls (6 % versus 1 %; adjusted *OR* 6.1). Our data indicate an increased risk of stroke/TIA at young age in men who have the F2 20210G>A mutation (Lalouschek et al. 2005). The benefit of testing patients with arterial thrombotic diseases for factor V Leiden (FV) 1691G>A and/or F2 20210G>A polymorphisms is still disputed. However, recent data suggests that the F2 20210G>A polymorphism might be considered an important risk factor for acute myocardial infarction in aged patients (55–80 years old). The mutation seems to be involved as risk factor for MI in about 5 % of old subjects, justifying the opportunity of a genetic screening and an eventual preventive treatment, especially in old subjects in which other major risk factors, such as hypertension and atherosclerosis, are present (Forte et al. 2011).

For an efficient initiation and activation of the hemostatic system, von Willebrand factor and factor VIII are highly important. FVIII is a large glycoprotein which is synthesized in the liver. It is elevated in acute phase conditions like stress and inflammation. The F8 gene is located on the X chromosome, and mutations within the gene cause hemophilia A. In plasma, FVIII is bound to von Willebrand factor, a large plasma glycoprotein synthesized in endothelial cells and megakaryocytes. vWF mediates platelet adhesion to damaged vascular subendothelium and subsequent platelet aggregation. It is particularly important under conditions of high shear stress, where platelet adhesion and primary hemostasis are fully dependent on vWF (Savage et al. 1996; Alevriadou et al. 1993). vWF acts as stabilizer of FVIII in the circulation. Reduced levels of vWF are frequently accompanied by reduced levels of FVIII. Elevated plasma levels of FVIII and vWF have been associated with an increased risk for MI and stroke (Saito et al. 2000). In the prospective ARIC study, Folsom et al. (1999) reported that individuals within the fourth quartile of FVIII levels (median: 171 %) have an almost twofold increased risk to suffer from ischemic stroke. Until now, the causes for elevated FVIII remain unclear, but there is no doubt that high FVIII levels are not merely a consequence of inflammation but are partly genetically determined. Plasma VWF levels are, to a large extent, genetically determined. Numerous association studies have been performed to assess the effect of genetic variants in the VWF gene (VWF) on VWF antigen and activity levels and on the risk of arterial thrombosis (Kamphuisen et al. 1998). From currently available data, it is evident that sequence variations within the FVIII gene or the vWF gene explain a portion of the concentration variability. The SNP 92714C>G (rs1800291), a nonsynonymous SNP encoding the B-domain substitution D1241E (Asp1260Glu), is associated with FVIII levels (Viel et al. 2007). Furthermore, polymorphisms within the ABO blood group locus have been demonstrated to contribute to the wide population variability. Individuals who are carriers of blood group 0 have significantly lower vWF and FVIII levels than carriers of blood groups A, B, or AB (Souto et al. 2000; Ay et al. 2010). Recently an association of SNPs in STXBP5, STX2, TC2N, and CLEC4M genes with vWF levels and of SCARA5 and STAB2 genes with FVIII levels has been reported. Collectively, these genes explain ~10 % of the variability of vWF and FVIII levels (Antoni et al. 2011). Genetic variations in other regulators of VWF, including ADAMTS-13, thrombospondin-1, and SNARE protein genes, have also been investigated. A review of the current literature regarding associations between genetic variations in the VWF gene and the risk of arterial thrombosis has been performed. According to the results, vWF cannot yet be excluded as causal mediator of arterial thrombotic events, and further research into the relationship between VWF and arterial thrombosis is necessary and justified (van Schie et al. 2011).

4.6.4 Platelet Glycoproteins

The role of platelets in thrombotic complications of eroded or disrupted plaques is well defined (Davi and Patrono 2007). Occlusive thrombi are almost exclusively initiated by plaque rupture and adhesion of platelets to subendothelial surface. Collagen, one of the major proteins of subendothelial vasculature, gets exposed following endothelium denudement. Recent advances in the understanding of collagen-mediated platelet adhesion and aggregation have led to the identification of two prominent receptors: glycoprotein Ia/IIa (GPIa/IIa or integrin alpha(2)beta(1)) and glycoprotein VI (GPVI). Other important platelet receptors are glycoprotein (GP) Ib-alpha, GPIa/IIa ($\alpha2\beta1$), and GPIIb/IIIa ($\alpha IIb\beta3$). GPIa/IIa is the major collagen receptor and mediates adhesion of platelets to the exposed injured vessel wall. Tests with monoclonal and polyclonal antibodies, peptide inhibitors, knockout models, and collagen mimetics have revealed that the collagen receptor is an attractive target for new drugs to prevent intravascular thrombosis. Several studies have suggested that genetic variants affecting platelet receptors are involved in platelet adhesion and activation and may influence platelet reactivity and/or their interaction with endothelium, collagen, leukocytes, and platelets. The genetic variability of platelet receptors has been associated with the risk of arterial thrombosis (Nurden 1995). A polymorphism in the GPIa gene (ITGA2) at nt 807C>T (rs 1801106, Glu534Lys) has been related to expression density and the activity of GPIa. The 807T allele was found associated with a higher receptor surface expression density and is augmenting platelet deposition onto collagen under shear stress (Kunicki 2002). The study showed that a fourfold change in the concentration of GPIa/IIa resulted in significant differences in platelet adhesion to collagen. Some investigations suggested a correlation between the 807T allele and the risk of thrombotic events, but others did not (Lewandowski et al. 2005). Different frequencies of the GPIa 807C>T polymorphism among various ethnic groups may explain the contradictory results. In the Chinese Han ethnic population, the prevalence of carriers of the 807T allele was markedly higher in patients with unstable angina pectoris (UAP) (Zhao et al. 2011). In this population, the lag time to achieve a 30 % platelet aggregation was significantly longer in CC genotype than in TC genotype carriers, although there was no significant difference in maximal platelet aggregation among healthy subjects with either genotype. Possibly, the platelet membrane GPIa 807T allele confers a rapid initiation of collagen-induced platelet aggregation.

In diabetic patients a higher prevalence of the 807T allele appears to support the generation of more reactive platelets which presumably cause a persistently increased release of pro-atherogenic TGF-beta and platelet-derived growth factor that could accelerate the atherothrombotic disease (Pellitero et al. 2010). Previously, a study on young female patients revealed an association between the GPIa 807C>T polymorphism and ischemic stroke (Lopaciuk et al. 2001), indicating the possibility of gene-sex hormone interactions. A contribution of GPIa 807C>T to stroke risk in patient subgroups cannot be ruled out, even though a meta-analysis evaluating candidate genes for ischemic stroke did not consider the modest association between GPIa 807C>T and ischemic stroke as positive result because the existence of a publication bias was suspected (Xin et al. 2009).

In conclusion, although there is no doubt that genetic factors contribute significantly to the prothrombotic state, the data on mutations and polymorphisms in candidate genes are still inconclusive. We face the paradox that certain laboratory parameters correlate consistently with diseases (e.g., fibrinogen levels with MI), and polymorphisms in certain genes correlate with plasma levels of the respective proteins (e.g., fibrinogen −455G>A polymorphism with plasma fibrinogen levels), yet positive correlations between genotype and disease phenotype can frequently not be established. By and large, the intensive search for coagulation factor polymorphisms predisposing to arterial thrombosis has yielded disappointing results. This is not surprising as the pathogenesis of arterial thrombosis is a complex process in which environmental and genetic factors interact. The clinical manifestation of MI or stroke represents the common endpoint of a myriad of different processes (Reiner et al. 2008), and a single gene polymorphism contributes only a small percentage to the total risk and does not fully reflect the complex processes. Homogenous subgroups of patients may share a common "pathogenetic" genotype. For example, in the patient subgroup with MI caused by vasospastic angina with subsequent coronary thrombosis, the FV 506R>Q mutation is a highly significant risk factor (*OR* 2.9–4.7), while the contribution of the FV 506R>Q mutation to the overall risk for MI in unselected patients is modest (*OR* 1.30; 95 % *CI*: 1.08–1.39) (Endler and Mannhalter 2003). A suitable approach to identify genetic risk profiles for thrombosis may be the simultaneous testing for numerous polymorphisms or a genome-wide association analysis (Wang et al. 2009). Currently available statistical methods are still unable to identify combined risk genotypes unless very large patient collectives (>10,000 cases) are tested. An appropriate analysis algorithm of the data produced by new genotyping technologies needs to be developed, and this remains a demanding task for the years to come.

4.7 Conclusion and Perspectives

Although our knowledge is still far from being complete, it is important to ask how genetic findings can be translated to clinical practice. The data from candidate gene studies to genome-wide association studies and recent sequencing experiments have to be dissected to identify candidates for therapeutic approaches. There are good reasons to believe that in the future, new drugs will increasingly be developed based on genetic data. Even though a larger body of evidence is still lacking, well-designed studies are under way.

References

Abifadel M, Varret M, Rabès JP, Allard D, Ouguerram K, Devillers M, Cruaud C, Benjannet S, Wickham L, Erlich D, Derré A, Villéger L, Farnier M, Beucler I, Bruckert E, Chambaz J, Chanu B, Lecerf JM, Luc G, Moulin P, Weissenbach J, Prat A, Krempf M, Junien C, Seidah NG, Boileau C (2003) Mutations in PCSK9 cause autosomal dominant hypercholesterolemia. Nat Genet 34(2):154–156

Akhter Q, Masood A, Ashraf R, Majid S, Rasool S, Khan T, Rashid T, Sameer AS, Ganai BA (2012) Polymorphisms in the 3′UTR of the human leptin gene and their role in hypertension. Mol Med Rep 5(4):1058–1062

Albert MA, Pare G, Morris A, Rose L, Buring J, Ridker PM, Zee RY (2009) Candidate genetic variants in the fibrinogen, methylenetetrahydrofolate reductase, and intercellular adhesion molecule-1 genes and plasma levels of fibrinogen, homocysteine, and intercellular adhesion molecule-1 among various race/ethnic groups: data from the Women's Genome Health Study. Am Heart J 157(4):777–83.e1

Alevriadou BR, Moake JL, Turner NA, Ruggeri ZM, Folie BJ, Phillips MD, Schreiber AB, Hrinda ME, McIntire LV (1993) Real-time analysis of shear-dependent thrombus formation and its blockade by inhibitors of von Willebrand factor binding to platelets. Blood 81:1263–1276

Allison DB, Heo M, Faith MS, Pietrobelli A (1998) Meta-analysis of the association of the Trp64Arg polymorphism in the beta3 adrenergic receptor with body mass index. Int J Obes Relat Metab Disord 22(6):559–566

Altshuler D, Hirschhorn JN, Klannemark M, Lindgren CM, Vohl MC, Nemesh J, Lane CR, Schaffner SF, Bolk S, Brewer C, Tuomi T, Gaudet D, Hudson TJ, Daly M, Groop L, Lander ES (2000) The common PPARgamma Pro12Ala polymorphism is associated with decreased risk of type 2 diabetes. Nat Genet 26(1):76–80

Amin AS, Giudicessi JR, Tijsen AJ, Spanjaart AM, Reckman YJ, Klemens CA, Tanck MW, Kapplinger JD, Hofman N, Sinner MF, Müller M, Wijnen WJ, Tan HL, Bezzina CR, Creemers EE, Wilde AA, Ackerman MJ, Pinto YM (2012) Variants in the 3′ untranslated region of the KCNQ1-encoded Kv7.1 potassium channel modify disease severity in patients with type 1 long QT syndrome in an allele-specific manner. Eur Heart J 33(6):714–723

Andersen G, Wegner L, Yanagisawa K, Rose CS, Lin J, Glümer C, Drivsholm T, Borch-Johnsen K, Jørgensen T, Hansen T, Spiegelman BM, Pedersen O (2005) Evidence of an association between genetic variation of the coactivator PGC-1beta and obesity. J Med Genet 42(5):402–407

Andreasen CH, Stender-Petersen KL, Mogensen MS, Torekov SS, Wegner L, Andersen G, Nielsen AL, Albrechtsen A, Borch-Johnsen K, Rasmussen SS, Clausen JO, Sandbaek A, Lauritzen T, Hansen L, Jørgensen T, Pedersen O, Hansen T (2008) Low physical activity accentuates the effect of the FTO rs9939609 polymorphism on body fat accumulation. Diabetes 57:95–101

Antoni G, Oudot-Mellakh T, Dimitromanolakis A, Germain M, Cohen W, Wells P, Lathrop M, Gagnon F, Morange PE, Tregouet DA (2011) Combined analysis of three genome-wide association studies on vWF and FVIII plasma levels. BMC Med Genet 12:102

Arbustini E, Diegoli M, Morbini P, Dal Bello B, Banchieri N, Pilotto A, Magani F, Grasso M, Narula J, Gavazzi A, Viganò M, Tavazzi L (2000) Prevalence and characteristics of dystrophin defects in adult male patients with dilated cardiomyopathy. J Am Coll Cardiol 35(7): 1760–1768

Ariens RA, Philippou H, Nagaswami C, Weisel JW, Lane DA, Grant PJ (2000) The factor XIII V34L polymorphism accelerates thrombin activation of factor XIII and affects cross-linked fibrin structure. Blood 96:988–995

Arita Y, Kihara S, Ouchi N, Takahashi M, Maeda K, Miyagawa J, Hotta K, Shimomura I, Nakamura T, Miyaoka K, Kuriyama H, Nishida M, Yamashita S, Okubo K, Matsubara K, Muraguchi M, Ohmoto Y, Funahashi T, Matsuzawa Y (1999) Paradoxical decrease of an adipose-specific protein, adiponectin, in obesity. Biochem Biophys Res Commun 257(1):79–83

Arking DE, Pfeufer A, Post W, Kao WH, Newton-Cheh C, Ikeda M, West K, Kashuk C, Akyol M, Perz S, Jalilzadeh S, Illig T, Gieger C, Guo CY, Larson MG, Wichmann HE, Marbán E, O'Donnell CJ, Hirschhorn JN, Kääb S, Spooner PM, Meitinger T, Chakravarti A (2006) A common genetic variant in the NOS1 regulator NOS1AP modulates cardiac repolarization. Nat Genet 38(6):644–651

Ay C, Thom K, Abu-Hamdeh F, Horvath B, Quehenberger P, Male C, Mannhalter C, Pabinger I (2010) Determinants of factor VIII plasma levels in carriers of haemophilia A and in control women. Haemophilia 16(1):111–117

Babu MS, Prabha TS, Kaul S, Al-Hazzani A, Shafi G, Roy S, Balakrishna N, Jyothy A, Munshi A (2012) Association of genetic variants of fibrinolytic system with stroke and stroke subtypes. Gene 495(1):76–80

Bach J, Endler G, Winkelmann BR, Boehm BO, Maerz W, Mannhalter C, Hellstern P (2008) Coagulation factor XII (FXII) activity, activated FXII, distribution of FXII C46T gene polymorphism and coronary risk. J Thromb Haemost 6(2):291–296

Bender N, Allemann N, Marek D, Vollenweider P, Waeber G, Mooser V, Egger M, Bochud M (2011) Association between variants of the leptin receptor gene (LEPR) and overweight: a systematic review and an analysis of the CoLaus study. PLoS One 6(10):e26157

Benetti R, Copetti T, Dell'Orso S, Melloni E, Brancolini C, Monte M, Schneider C (2005) The calpain system is involved in the constitutive regulation of beta-catenin signaling functions. J Biol Chem 280(23):22070–22080

Benn M, Nordestgaard BG, Grande P, Schnohr P, Tybjaerg-Hansen A (2010) PCSK9 R46L, low-density lipoprotein cholesterol levels, and risk of ischemic heart disease: 3 independent studies and meta-analyses. J Am Coll Cardiol 55(25):2833–2842

Benzinou M, Creemers JW, Choquet H, Lobbens S, Dina C, Durand E, Guerardel A, Boutin P, Jouret B, Heude B, Balkau B, Tichet J, Marre M, Potoczna N, Horber F, Le Stunff C, Czernichow S, Sandbaek A, Lauritzen T, Borch-Johnsen K, Andersen G, Kiess W, Körner A, Kovacs P, Jacobson P, Carlsson LM, Walley AJ, Jørgensen T, Hansen T, Pedersen O, Meyre D, Froguel P (2008) Common nonsynonymous variants in PCSK1 confer risk of obesity. Nat Genet 40(8):943–945

Bereczky Z, Muszbek L (2011) Factor XIII and venous thromboembolism. Semin Thromb Hemost 37(3):305–314

Bernardi F, Marchetti G, Pinotti M, Arcieri P, Baroncini C, Papacchini M, Zepponi E, Ursicino N, Chiarotti F, Mariani G (1996) Factor VII gene polymorphisms contribute about one third of the factor VII level variation in plasma. Arterioscler Thromb Vasc Biol 16:72–76

Bertina RM, Koeleman BP, Koster T, Rosendaal FR, Dirven RJ, de Ronde H, van der Velden PA, Reitsma PH (1994) Mutation in blood coagulation factor V associated with resistance to activated protein C. Nature 369(6475):64–67

Bezzina CR, Verkerk AO, Busjahn A, Jeron A, Erdmann J, Koopmann TT, Bhuiyan ZA, Wilders R, Mannens MM, Tan HL, Luft FC, Schunkert H, Wilde AA (2003) A common polymorphism in KCNH2 (HERG) hastens cardiac repolarization. Cardiovasc Res 59(1):27–36

Björck HM, Eriksson P, Alehagen U, De Basso R, Ljungberg LU, Persson K, Dahlström U, Länne T (2011) Gender-specific association of the plasminogen activator inhibitor-1 4G/5G polymorphism with central arterial blood pressure. Am J Hypertens 24(7):802–808

Blake GJ, Schmitz C, Lindpaintner K, Ridker PM (2001) Mutation in the promoter region of the beta-fibrinogen gene and the risk of future myocardial infarction, stroke and venous thrombosis. Eur Heart J 22:2262–2266

Brass LM, Isaacsohn JL, Merikangas KR, Robinette CD (1992) A study of twins and stroke. Stroke 23:221–223

Brass LM, Hartigan PM, Page WF, Concato J (1996) Importance of cerebrovascular disease in studies of myocardial infarction. Stroke 27:1173–1176

Bray PF (2000) Platelet glycoprotein polymorphisms as risk factors for thrombosis. Curr Opin Hematol 7(5):284–289

Brown MS, Goldstein JL (1986) A receptor-mediated pathway for cholesterol homeostasis. Science 232:34–47

Büsst CJ, Bloomer LDS, Scurrah KJ, Ellis JA, Barnes TA, Charchar FJ, Braund P, Hopkins PN, Samani NJ, Hunt SC, Tomaszewski M, Harrap SB (2011) The epithelial sodium channel γ-subunit gene and blood pressure: family based association, renal gene expression, and physiological analyses. Hypertension 58:1073–1078

Calafell F, Almasy L, Sabater-Lleal M, Buil A, Mordillo C, Ramírez-Soriano A, Sikora M, Souto JC, Blangero J, Fontcuberta J, Soria JM (2010) Sequence variation and genetic evolution at the human F12 locus: mapping quantitative trait nucleotides that influence FXII plasma levels. Hum Mol Genet 19(3):517–525

Campo G, Valgimigli M, Ferraresi P, Malagutti P, Baroni M, Arcozzi C, Gemmati D, Percoco G, Parrinello G, Ferrari R, Bernardi F (2006) Tissue factor and coagulation factor VII levels during acute myocardial infarction: association with genotype and adverse events. Arterioscler Thromb Vasc Biol 26(12):2800–2806

Cappola TP, Li M, He J, Ky B, Gilmore J, Qu L, Keating B, Reilly M, Kim CE, Glessner J, Frackelton E, Hakonarson H, Syed F, Hindes A, Matkovich SJ, Cresci S, Dorn GW (2010) Common variants in HSPB7 and FRMD4B associated with advanced heart failure. Circ Cardiovasc Genet 3(2):147–154

Carter AM, Catto AJ, Grant PJ (1998) Determinants of tPA antigen and associations with coronary artery disease and acute cerebrovascular disease. Thromb Haemost 80(4):632–636

Carter AM, Catto AJ, Grant PJ (1999) Association of the alpha-fibrinogen Thr312Ala polymorphism with poststroke mortality in subjects with atrial fibrillation. Circulation 99: 2423–2426

Carty CL, Heagerty P, Heckbert SR, Jarvik GP, Lange LA, Cushman M, Tracy RP, Reiner AP (2010) Interaction between fibrinogen and IL-6 genetic variants and associations with cardiovascular disease risk in the Cardiovascular Health Study. Ann Hum Genet 74(1):1–10

Ceelie H, Spaargaren-van Riel CC, Bertina RM, Vos HL (2004) G20210A is a functional mutation in the prothrombin gene; effect on protein levels and 3'-end formation. J Thromb Haemost 2(1):119–127

Chandra V, Huang P, Hamuro Y, Raghuram S, Wang Y, Burris TP, Rastinejad F (2008) Structure of the intact PPAR-gamma-RXR- nuclear receptor complex on DNA. Nature 456:350–356

Chinchilla A, Daimi H, Lozano-Velasco E, Dominguez JN, Caballero R, Delpon E, Tamargo J, Cinca J, Hove-Madsen L, Aranega AE, Franco D (2011) Pitx2 insufficiency leads to atrial electrical and structural remodeling linked to arrhythmogenesis. Circ Cardiovasc Genet 4:269–279

Christoffels VM, Smits GJ, Kispert A, Moorman AF (2010) Development of the pacemaker tissues of the heart. Circ Res 106:240–254

Combs TP, Wagner JA, Berger J, Doebber T, Wang WJ, Zhang BB, Tanen M, Berg AH, O'Rahilly S, Savage DB, Chatterjee K, Weiss S, Larson PJ, Gottesdiener KM, Gertz BJ, Charron MJ, Scherer PE, Moller DE (2002) Induction of adipocyte complement-related protein of 30 kilodaltons by PPARgamma agonists: a potential mechanism of insulin sensitization. Endocrinology 143(3):998–1007

Corbo RM, Scacchi R (1999) Apolipoprotein E (APOE) allele distribution in the world. Is APOE}4 a'thrifty' allele? Ann Hum Genet 63(4):301–310

Corral J, Gonzalez-Conejero R, Iniesta JA, Rivera J, Martinez C, Vicente V (2000) The FXIII Val34Leu polymorphism in venous and arterial thromboembolism. Haematologica 85:293–297

Dahlbäck B, Carlsson M, Svensson PJ (1993) Familial thrombophilia due to a previously unrecognized mechanism characterized by poor anticoagulant response to activated protein C: prediction of a cofactor to activated protein C. Proc Natl Acad Sci USA 90:1004–1008

Davi G, Patrono C (2007) Platelet activation and atherothrombosis. N Engl J Med 357:2482–2494

Debette S, Seshadri S (2009) Genetics of atherothrombotic and lacunar stroke. Circ Cardiovasc Genet 2(2):191–198

Dehghan A, Yang Q, Peters A, Basu S, Bis JC, Rudnicka AR, Kavousi M, Chen MH, Baumert J, Lowe GD, McKnight B, Tang W, de Maat M, Larson MG, Eyhermendy S, McArdle WL, Lumley T, Pankow JS, Hofman A, Massaro JM, Rivadeneira F, Kolz M, Taylor KD, van Duijn CM, Kathiresan S, Illig T, Aulchenko YS, Volcik KA, Johnson AD, Uitterlinden AG, Tofler GH, Gieger C; Wellcome Trust Case Control Consortium, Psaty BM, Couper DJ, Boerwinkle E, Koenig W, O'Donnell CJ, Witteman JC, Strachan DP, Smith NL, Folsom AR (2009) Association of novel genetic loci with circulating fibrinogen levels: a genome-wide association study in 6 population-based cohorts. Circ Cardiovasc Genet 2(2):125–133

De la Peña-Díaz A, Izaguirre-Avila R, Anglés-Cano E (2000) Lipoprotein Lp(a) and atherothrombotic disease. Arch Med Res 31(4):353–359

Delluc A, Gourhant L, Lacut K, Mercier B, Audrezet MP, Nowak E, Oger E, Leroyer C, Mottier D, Le Gal G, Couturaud F (2010) Association of common genetic variations and idiopathic venous thromboembolism. Results from EDITh, a hospital-based case–control study. Thromb Haemost 103(6):1161–1169

DeKroon R, Robinette JB, Hjelmeland AB, Wiggins E, Blackwell M, Mihovilovic M, Fujii M, York J, Hart J, Kontos C, Rich J, Strittmatter WJ (2006) APOE4-VLDL inhibits the HDL-activated phosphatidylinositol 3-kinase/Akt pathway via the phosphoinositol phosphatase SHIP2. Circ Res 99:829–836

Diness JG, Sorensen US, Nissen JD, Al-Shahib B, Jespersen T, Grunnet M, Hansen RS (2010) Inhibition of small-conductance Ca2+–activated K+channels terminates and protects against atrial fibrillation. Circ Arrhythm Electrophysiol 3:380–390

Eisenach JH, Wittwer ED (2010) {beta}-Adrenoceptor gene variation and intermediate physiological traits: prediction of distant phenotype. Exp Physiol 95(7):757–764

Ellinor PT, Lunetta KL, Albert CM, Glazer NL, Ritchie MD, Smith AV, Arking DE, Müller-Nurasyid M, Krijthe BP, Lubitz SA, Bis JC, Chung MK, Dörr M, Ozaki K, Roberts JD, Smith JG, Pfeufer A, Sinner MF, Lohman K, Ding J, Smith NL, Smith JD, Rienstra M, Rice KM, Van

Wagoner DR, Magnani JW, Wakili R, Clauss S, Rotter JI, Steinbeck G, Launer LJ, Davies RW, Borkovich M, Harris TB, Lin H, Völker U, Völzke H, Milan DJ, Hofman A, Boerwinkle E, Chen LY, Soliman EZ, Voight BF, Li G, Chakravarti A, Kubo M, Tedrow UB, Rose LM, Ridker PM, Conen D, Tsunoda T, Furukawa T, Sotoodehnia N, Xu S, Kamatani N, Levy D, Nakamura Y, Parvez B, Mahida S, Furie KL, Rosand J, Muhammad R, Psaty BM, Meitinger T, Perz S, Wichmann HE, Witteman JC, Kao WH, Kathiresan S, Roden DM, Uitterlinden AG, Rivadeneira F, McKnight B, Sjögren M, Newman AB, Liu Y, Gollob MH, Melander O, Tanaka T, Stricker BH, Felix SB, Alonso A, Darbar D, Barnard J, Chasman DI, Heckbert SR, Benjamin EJ, Gudnason V, Kääb S (2012) Meta-analysis identifies six new susceptibility loci for atrial fibrillation. Nat Genet 44(6):670–675

Endler G, Mannhalter C (2003) Polymorphisms in coagulation factor genes and their impact on arterial and venous thrombosis. Clin Chim Acta 330:31–55

Endler G, Lalouschek W, Exner M, Mitterbauer G, Häring D, Mannhalter C (2000) The 4G/4G genotype at nucleotide position −675 in the promoter region of the plasminogen activator inhibitor 1 (PAI-1) gene is less frequent in young patients with minor stroke than in controls. Br J Haematol 110(2):469–471

Endler G, Mannhalter C, Sunder-Plassmann H, Lalouschek W, Kapiotis S, Exner M, Jordanova N, Meier S, Kunze F, Wagner O, Huber K (2001) Homozygosity for the C>T polymorphism at nucleotide 46 in the 5′untranslated region of the factor XII gene protects from development of acute coronary syndrome. Br J Haematol 115:1007–1009

Endler G, Funk M, Haering D, Lalouschek W, Lang W, Mirafzal M, Wagner O, Mannhalter C (2003) Is the factor XIII 34Val/Leu polymorphism a protective factor for cerebrovascular disease? Br J Haematol 120(2):310–314

Endler G, Marsik C, Jilma B, Schickbauer T, Quehenberger P, Mannhalter C (2007) Evidence of a U-shaped association between factor XII activity and overall survival. J Thromb Haemost 5(6):1143–1148

Erdmann J, Grosshennig A, Braund PS, König IR, Hengstenberg C, Hall AS, Linsel-Nitschke P, Kathiresan S, Wright B, Trégouët DA, Cambien F, Bruse P, Aherrahrou Z, Wagner AK, Stark K, Schwartz SM, Salomaa V, Elosua R, Melander O, Voight BF, O'Donnell CJ, Peltonen L, Siscovick DS, Altshuler D, Merlini PA, Peyvandi F, Bernardinelli L, Ardissino D, Schillert A, Blankenberg S, Zeller T, Wild P, Schwarz DF, Tiret L, Perret C, Schreiber S, El Mokhtari NE, Schäfer A, März W, Renner W, Bugert P, Klüter H, Schrezenmeir J, Rubin D, Ball SG, Balmforth AJ, Wichmann HE, Meitinger T, Fischer M, Meisinger C, Baumert J, Peters A, Ouwehand WH; Italian Atherosclerosis, Thrombosis, and Vascular Biology Working Group; Myocardial Infarction Genetics Consortium; Wellcome Trust Case Control Consortium; Cardiogenics Consortium, Deloukas P, Thompson JR, Ziegler A, Samani NJ, Schunkert H (2009) New susceptibility locus for coronary artery disease on chromosome 3q22.3. Nat Genet 41(3):280–282

Erikson P, Kallin B, Van'T Hooft FM, Bovenholm P, Hamsten A (1995) Allele-specific increase in basal transcription of plasminogen activator inhibitor 1 gene is associated with myocardial infarction. Proc Natl Acad Sci USA 92:1851–1855

Evans JC, Frayling TM, Cassell PG, Saker PJ, Hitman GA, Walker M, Levy JC, O'Rahilly S, Rao PV, Bennett AJ, Jones EC, Menzel S, Prestwich P, Simecek N, Wishart M, Dhillon R, Fletcher C, Millward A, Demaine A, Wilkin T, Horikawa Y, Cox NJ, Bell GI, Ellard S, McCarthy MI, Hattersley AT (2001) Studies of association between the gene for calpain-10 and type 2 diabetes mellitus in the United Kingdom. Am J Hum Genet 69:544–552

Feng T, Elston RC, Zhu X (2011) Detecting rare and common variants for complex traits: sibpair and odds ratio weighted sum statistics (SPWSS, ORWSS). Genet Epidemiol 35(5):398–409

Flanagan SE, Clauin S, Bellanne-Chantelot C, de Lonlay P, Harries LW, Gloyn AL, Ellard S (2009) Update of mutations in the genes encoding the pancreatic beta-cell KATP channel subunits Kir6.2 (KCNJ11) and sulfonylurea receptor 1 (ABCC8) in diabetes mellitus and hyperinsulinism. Hum Mutat 30:170–180

Folsom AR, Rosamond WD, Shahar E, Cooper LS, Aleksic N, Nieto FJ, Rasmussen ML, Wu KK (1999) Prospective study of markers of hemostatic function with risk of ischemic stroke. The Atherosclerosis Risk in Communities (ARIC) Study Investigators. Circulation 100:736–742

Forte GI, Vaccarino L, Palmeri M, Branzi A, Caldarera CM, Scola L, Caruso C, Licastro F, Lio D (2011) Analysis of polymorphisms Leiden Factor V G1691A and prothrombin G20210A as risk factors for acute myocardial infarction. Biogerontology 12(5):485–490

Franco RF, Trip MD, ten Cate H, van den Ende A, Prins MH, Kastelein JJ, Reitsma PH (1999) The 20210 G→A mutation in the 3′-untranslated region of the prothrombin gene and the risk for arterial thrombotic disease. Br J Haematol 104:50–54

Franco RF, Reitsma PH (2001) Gene polymorphisms of the haemostatic system and the risk of arterial thrombotic disease. Br J Haematol 115:491–506

Frank-Kamenetsky M, Grefhorst A, Anderson NN, Racie TS, Bramlage B, Akinc A, Butler D, Charisse K, Dorkin R, Fan Y, Gamba-Vitalo C, Hadwiger P, Jayaraman M, John M, Jayaprakash KN, Maier M, Nechev L, Rajeev KG, Read T, Röhl I, Soutschek J, Tan P, Wong J, Wang G, Zimmermann T, de Fougerolles A, Vornlocher HP, Langer R, Anderson DG, Manoharan M, Koteliansky V, Horton JD, Fitzgerald K (2008) Therapeutic RNAi targeting PCSK9 acutely lowers plasma cholesterol in rodents and LDL cholesterol in nonhuman primates. Proc Natl Acad Sci USA 105(33):11915–11920

Frayling TM, Timpson NJ, Weedon MN, Zeggini E, Freathy RM, Lindgren CM, Perry JR, Elliott KS, Lango H, Rayner NW, Shields B, Harries LW, Barrett JC, Ellard S, Groves CJ, Knight B, Patch AM, Ness AR, Ebrahim S, Lawlor DA, Ring SM, Ben-Shlomo Y, Jarvelin MR, Sovio U, Bennett AJ, Melzer D, Ferrucci L, Loos RJ, Barroso I, Wareham NJ, Karpe F, Owen KR, Cardon LR, Walker M, Hitman GA, Palmer CN, Doney AS, Morris AD, Smith GD, Hattersley AT, McCarthy MI (2007) A common variant in the FTO gene is associated with body mass index and predisposes to childhood and adult obesity. Science 316(5826):889–894

Friso S, Lotto V, Choi SW, Girelli D, Pinotti M, Guarini P, Udali S, Pattini P, Pizzolo F, Martinelli N, Corrocher R, Bernardi F, Olivieri O (2012) Promoter methylation in coagulation F7 gene influences plasma FVII concentrations and relates to coronary artery disease. J Med Genet 49(3):192–199

Fujisawa T, Ikegami H, Kawaguchi Y, Ogihara T (1998) Meta-analysis of the association of Trp64Arg polymorphism of beta 3-adrenergic receptor gene with body mass index. J Clin Endocrinol Metab 83(7):2441–2444

Funk M, Endler G, Lalouschek W, Hsieh K, Schillinger M, Lang W, Mannhalter C (2006) Factor VII gene haplotypes and risk of ischemic stroke. Clin Chem 52(6):1190–1192

Galasso G, Santulli G, Piscione F, De Rosa R, Trimarco V, Piccolo R, Cassese S, Iaccarino G, Trimarco B, Chiariello M (2010) The GPIIIA PlA2 polymorphism is associated with an increased risk of cardiovascular adverse events. BMC Cardiovasc Disord 10:41

Gehring NH, Frede U, Neu-Yilik G, Hundsdoerfer P, Vetter B, Hentze MW, Kulozik AE (2001) Increased efficiency of mRNA 3′ end formation: a new genetic mechanism contributing to hereditary thrombophilia. Nat Genet 28:389–392

Geller F, Reichwald K, Dempfle A, Illig T, Vollmert C, Herpertz S, Siffert W, Platzer M, Hess C, Gudermann T, Biebermann H, Wichmann HE, Schäfer H, Hinney A, Hebebrand J (2004) Melanocortin-4 receptor gene variant I103 is negatively associated with obesity. Am J Hum Genet 74(3):572–581

Girelli D, Russo C, Ferraresi P, Olivieri O, Pinotti M, Friso S, Manzato F, Mazzucco A, Bernardi F, Corrocher R (2000) Polymorphisms in the factor VII gene and the risk of myocardial infarction in patients with coronary artery disease. N Engl J Med 343:774–780

Gjesing AP, Andersen G, Borch-Johnsen K, Jørgensen T, Hansen T, Pedersen O (2008) Association of the beta3-adrenergic receptor Trp64Arg polymorphism with common metabolic traits: studies of 7605 middle-aged white people. Mol Genet Metab 94(1):90–97

Gloyn AL, Noordam K, Willemsen MA, Ellard S, Lam WW, Campbell IW, Midgley P, Shiota C, Buettger C, Magnuson MA, Matschinsky FM, Hattersley AT (2003) Insights into the biochemical and genetic basis of glucokinase activation from naturally occurring hypoglycemia mutations. Diabetes 52(9):2433–2440

Gong LL, Peng JH, Han FF, Zhu J, Fang LH, Wang YH, Du GH, Wang HY, Liu LH (2012) Association of tissue plasminogen activator and plasminogen activator inhibitor polymorphism with myocardial infarction: A meta-analysis. Thromb Res 130(3):e43–e51

Gouda HN, Sagoo GS, Harding A-J, Yates J, Sandhu MS, Higgins JPT (2010) The Association Between the Peroxisome Proliferator-Activated Receptor-g2 (PPARG2) Pro12Ala Gene Variant and Type 2 Diabetes Mellitus: A HuGE Review and Meta-Analysis. Am J Epidemiol 171:645–655

Grant PJ, Humphries S (1999) Genetic determinants of arterial thrombosis. Baillieres Best Pract Res Clin Haematol 12:505–532

Grant SF, Thorleifsson G, Reynisdottir I, Benediktsson R, Manolescu A, Sainz J, Helgason A, Stefansson H, Emilsson V, Helgadottir A, Styrkarsdottir U, Magnusson KP, Walters GB, Palsdottir E, Jonsdottir T, Gudmundsdottir T, Gylfason A, Saemundsdottir J, Wilensky RL, Reilly MP, Rader DJ, Bagger Y, Christiansen C, Gudnason V, Sigurdsson G, Thorsteinsdottir U, Gulcher JR, Kong A, Stefansson K (2006) Variant of transcription factor 7-like 2 (TCF7L2) gene confers risk of type 2 diabetes. Nat Genet 38(3):320–323

Greisenegger S, Weber M, Funk M, Endler G, Lang W, Ferrari J, Lalouschek W, Mannhalter C (2007) Polymorphisms in the coagulation factor VII gene and risk of primary intracerebral hemorrhage. Eur J Neurol 14(10):1098–1101

Gudbjartsson DF, Arnar DO, Helgadottir A, Gretarsdottir S, Holm H, Sigurdsson A, Jonasdottir A, Baker A, Thorleifsson G, Kristjansson K, Palsson A, Blondal T, Sulem P, Backman VM, Hardarson GA, Palsdottir E, Helgason A, Sigurjonsdottir R, Sverrisson JT, Kostulas K, Ng MC, Baum L, So WY, Wong KS, Chan JC, Furie KL, Greenberg SM, Sale M, Kelly P, MacRae CA, Smith EE, Rosand J, Hillert J, Ma RC, Ellinor PT, Thorgeirsson G, Gulcher JR, Kong A, Thorsteinsdottir U, Stefansson K (2007) Variants conferring risk of atrial fibrillation on chromosome 4q25. Nature 448:353–357

Guella I, Duga S, Ardissino D, Merlini PA, Peyvandi F, Mannucci PM, Asselta R (2011) Common variants in the haemostatic gene pathway contribute to risk of early-onset myocardial infarction in the Italian population. Thromb Haemost 106(4):655–664

Guevara J Jr, Spurlino J, Jan AY, Yang CY, Tulinsky A, Prasad BV, Gaubatz JW, Morrisett JD (1993) Proposed mechanisms for binding of apo[a] kringle type 9 to apo B-100 in human lipoprotein[a]. Biophys J 64(3):686–700

Hasan N, McColgan P, Bentley P, Edwards RJ, Sharma P (2012) Towards the identification of blood biomarkers for acute stroke in humans: a comprehensive systematic review. Br J Clin Pharmacol 74(2):230–240

Heid IM, Vollmert C, Hinney A, Döring A, Geller F, Löwel H, Wichmann HE, Illig T, Hebebrand J, Kronenberg F, Group KORA (2005) Association of the 103I MC4R allele with decreased body mass in 7937 participants of two population based surveys. J Med Genet 42(4):e21

Heike Münzberg H, Myers MG (2005) Molecular and anatomical determinants of central leptin resistance. Nat Neurosci 8(5):566–570

Heinrich J, Balleisen L, Schulte H, Assmann G, van de Loo J (1994) Fibrinogen and factor VII in the prediction of coronary risk. Results from the PROCAM study in healthy men. Arterioscler Thromb 14:54–59

Herman DS, Lam L, Taylor MR, Wang L, Teekakirikul P, Christodoulou D, Conner L, DePalma SR, McDonough B, Sparks E, Teodorescu DL, Cirino AL, Banner NR, Pennell DJ, Graw S, Merlo M, Di Lenarda A, Sinagra G, Bos JM, Ackerman MJ, Mitchell RN, Murry CE, Lakdawala NK, Ho CY, Barton PJ, Cook SA, Mestroni L, Seidman JG, Seidman CE (2012) Truncations of titin causing dilated cardiomyopathy. New Engl J Med 366:619–628

Hinney A, Hebebrand J (2008) Polygenic Obesity in Humans. Obes Facts 1:35–42

Hoekstra T, Geleijnse JM, Schouten EG, Kluft C (2004) Plasminogen activator inhibitor-type 1: its plasma determinants and relation with cardiovascular risk. Thromb Haemost 91(5): 861–872

Holm H, Gudbjartsson DF, Arnar DO, Thorleifsson G, Thorgeirsson G, Stefansdottir H, Gudjonsson SA, Jonasdottir A, Mathiesen EB, Njølstad I, Nyrnes A, Wilsgaard T, Hald EM, Hveem K, Stoltenberg C, Løchen ML, Kong A, Thorsteinsdottir U, Stefansson K (2010) Several common variants modulate heart rate, PR interval and QRS duration. Nat Genet 42(2):117–122

Horton JD, Cohen JC, Hobbs HH (2007) Molecular biology of PCSK9: its role in LDL metabolism. Trends Biochem Sci 32(2):71–77

Horton JD, Cohen JC, Hobbs HH (2009) PCSK9: a convertase that coordinates LDL catabolism. J Lipid Res 50(Suppl):S172–S177

Hsieh K, Lalouschek W, Schillinger M, Endler G, Reisinger M, Janisiw M, Lang W, Cheng S, Wagner O, Mannhalter C (2005) Impact of alphaENaC polymorphisms on the risk of ischemic cerebrovascular events: a multicenter case–control study. Clin Chem 51(6):952–956

Iacoviello L, Di Castelnuovo A, De Knijff P, D'Orazio A, Amore C, Arboretti R, Kluft C, Benedetta Donati M (1998) Polymorphisms in the coagulation factor VII gene and the risk of myocardial infarction. N Engl J Med 338:79–85

Iwabu M, Yamauchi T, Okada-Iwabu M, Sato K, Nakagawa T, Funata M, Yamaguchi M, Namiki S, Nakayama R, Tabata M, Ogata H, Kubota N, Takamoto I, Hayashi YK, Yamauchi N, Waki H, Fukayama M, Nishino I, Tokuyama K, Ueki K, Oike Y, Ishii S, Hirose K, Shimizu T, Touhara K, Kadowaki T (2010) Adiponectin and AdipoR1 regulate PGC-1alpha and mitochondria by Ca(2+) and AMPK/SIRT1. Nature 464(7293):1313–1319

Jeninga EH, Gurnell M, Kalkhoven E (2009) Functional implications of genetic variation in human PPARg. Trends Endocrinol Metab 20(8):380–387

Johnson CY, Tuite A, Morange PE, Tregouet DA, Gagnon F (2011a) The factor XII -4C>T variant and risk of common thrombotic disorders: A HuGE review and meta-analysis of evidence from observational studies. Am J Epidemiol 173(2):136–144

Johnson T, Gaunt TR, Newhouse SJ et al (2011b) Blood pressure loci identified with a gene-centric array. Am J Hum Genet 89(6):688–700

Jood K, Ladenvall P, Tjärnlund-Wolf A, Ladenvall C, Andersson M, Nilsson S, Blomstrand C, Jern C (2005) Fibrinolytic gene polymorphism and ischemic stroke. Stroke 36(10):2077–2081

Juul K, Tybjaerg-Hansen A, Steffensen R, Kofoed S, Jensen G, Nordestgaard BG (2002) Factor V Leiden: the Copenhagen City Heart Study and 2 meta-analyses. Blood 100:3–10

Kääb S, Crawford DC, Sinner MF, Behr ER, Kannankeril PJ, Wilde AA, Bezzina CR, Schulze-Bahr E, Guicheney P, Bishopric NH, Myerburg RJ, Schott JJ, Pfeufer A, Beckmann BM, Martens E, Zhang T, Stallmeyer B, Zumhagen S, Denjoy I, Bardai A, Van Gelder IC, Jamshidi Y, Dalageorgou C, Marshall V, Jeffery S, Shakir S, Camm AJ, Steinbeck G, Perz S, Lichtner P, Meitinger T, Peters A, Wichmann HE, Ingram C, Bradford Y, Carter S, Norris K, Ritchie MD, George AL Jr, Roden DM (2012) A large candidate gene survey identifies the KCNE1 D85N polymorphism as a possible modulator of drug-induced torsades de pointes. Circ Cardiovasc Genet 5(1):91–99

Kamphuisen PW, Houwing-Duistermaat JJ, van Houwelingen HC, Eikenboom JC, Bertina RM, Rosendaal FR (1998) Familial clustering of factor VIII and von Willebrand factor levels. Thromb Haemost 79:323–327

Kamstrup PR (2010) Lipoprotein(a) and ischemic heart disease – a causal association? A review. Atherosclerosis 211(1):15–23

Kanaji T, Okamura T, Osaki K, Kuroiwa M, Shimoda K, Hamasaki N, Niho Y (1998) A common genetic polymorphism (46C to T substitution) in the 5′-untranslated region of the coagulation factor XII gene is associated with low translation efficiency and decrease in plasma factor XII level. Blood 91:2010–2014

Karelis AD, Brochu M, Rabasa-Lhoret R (2004) Can we identify metabolically healthy but obese individuals (MHO)? Diabetes Metab 30(6):569–572

Kathiresan S, Willer CJ, Peloso GM, Demissie S, Musunuru K, Schadt EE, Kaplan L, Bennett D, Li Y, Tanaka T, Voight BF, Bonnycastle LL, Jackson AU, Crawford G, Surti A, Guiducci C, Burtt NP, Parish S, Clarke R, Zelenika D, Kubalanza KA, Morken MA, Scott LJ, Stringham HM, Galan P, Swift AJ, Kuusisto J, Bergman RN, Sundvall J, Laakso M, Ferrucci L, Scheet P, Sanna S, Uda M, Yang Q, Lunetta KL, Dupuis J, de Bakker PI, O'Donnell CJ, Chambers JC, Kooner JS, Hercberg S, Meneton P, Lakatta EG, Scuteri A, Schlessinger D, Tuomilehto J, Collins FS, Groop L, Altshuler D, Collins R, Lathrop GM, Melander O, Salomaa V, Peltonen L, Orho-Melander M, Ordovas JM, Boehnke M, Abecasis GR, Mohlke KL, Cupples LA (2009) Common variants at 30 loci contribute to polygenic dyslipidemia. Nat Genet 41(1):56–65

Ken-Dror G, Drenos F, Humphries SE, Talmud PJ, Hingorani AD, Kivimäki M, Kumari M, Bauer KA, Morrissey JH, Ireland HA (2010) Haplotype and genotype effects of the F7 gene on

circulating factor VII, coagulation activation markers and incident coronary heart disease in UK men. J Thromb Haemost 8(11):2394–2403
Kim J, Basak JM, Holtzman DM (2009) The role of apolipoprotein E in Alzheimer's disease. Neuron 63(3):287–303
Kim RJ, Becker RC (2003) Association between factor V Leiden, prothrombin G20210A, and methylenetetrahydrofolate reductase C677T mutations and events of the arterial circulatory system: a meta-analysis of published studies. Am Heart J 146(6):948–957
Kirchhof P, Kahr PC, Kaese S, Piccini I, Vokshi I, Scheld HH, Rotering H, Fortmueller L, Laakmann S, Verheule S, Schotten U, Fabritz L, Brown NA (2011) PITX2c is expressed in the adult left atrium, and reducing Pitx2c expression promotes atrial fibrillation inducibility and complex changes in gene expression. Circ Cardiovasc Genet 4:123–133
Kjolby M, Andersen OM, Breiderhoff T, Fjorback AW, Pedersen KM, Madsen P, Jansen P, Heeren J, Willnow TE, Nykjaer A (2010) Sort1, encoded by the cardiovascular risk locus 1p13.3, is a regulator of hepatic lipoprotein export. Cell Metab 12(3):213–223
Knowles JW, Wang H, Itakura H, Southwick A, Myers RM, Iribarren C, Fortmann SP, Go AS, Quertermous T, Hlatky MA (2007) Association of polymorphisms in platelet and hemostasis system genes with acute myocardial infarction. Am Heart J 154(6):1052–1058
Kohler HP, Grant PJ (1999) The role of factor XIII Val34Leu in cardiovascular disease. QJM 92:67–72
Kolder IC, Tanck MW, Bezzina CR (2012) Common genetic variation modulating cardiac ECG parameters and susceptibility to sudden cardiac death. J Mol Cell Cardiol 52:620–629
Kraft HG, Lingenhel A, Köchl S, Hoppichler F, Kronenberg F, Abe A, Mühlberger V, Schönitzer D, Utermann G (1996a) Apolipoprotein(a) kringle IV repeat number predicts risk for coronary heart disease. Arterioscler Thromb Vasc Biol 16(6):713–719
Kraft HG, Lingenhel A, Pang RW, Delport R, Trommsdorff M, Vermaak H, Janus ED, Utermann G (1996b) Frequency distributions of apolipoprotein(a) kringle IV repeat alleles and their effects on lipoprotein(a) levels in Caucasian, Asian, and African populations: the distribution of null alleles is non-random. Eur J Hum Genet 4(2):74–87
Kucharska-Newton AM, Monda KL, Campbell S, Bradshaw PT, Wagenknecht LE, Boerwinkle E, Wasserman BA, Heiss G (2011) Association of the platelet GPIIb/IIIa polymorphism with atherosclerotic plaque morphology: the Atherosclerosis Risk in Communities (ARIC) Study. Atherosclerosis 216(1):151–156
Kudaravalli R, T T, Pinotti M, Ratti A, Santacroce R, Margaglione M, Dallapiccola B, Bernardi F, Fortina P, Devoto M, Pollak ES (2002) Polymorphic changes in the 5¢ flanking region of factor VII have a combined effect on promoter strength. Thromb Haemost 88:763–767
Kunes J, Zicha J (2009) The interaction of genetic and environmental factors in the etiology of hypertension. Physiol Res 58:S33–S41
Kunicki TJ (2002) The influence of platelet collagen receptor polymorphisms in hemostasis and thrombotic disease. Arterioscler Thromb Vasc Biol 22:14–20
Kunicki TJ, Williams SA, Salomon DR, Harrison P, Crisler P, Nakagawa P, Mondala TS, Head SR, Nugent DJ (2009) Genetics of platelet reactivity in normal, healthy individuals. J Thromb Haemost 7(12):2116–2122
Kunicki TJ, Williams SA, Diaz D, Farndale RW, Nugent DJ (2012) Platelet adhesion to decorin but not collagen I correlates with the integrin $\alpha 2$ dimorphism E534K, the basis of the human platelet alloantigen (HPA)-5 system. Haematologica 97(5):692–695
Kurokawa N, Nakai K, Kameo S, Liu ZM, Satoh H (2001) Association of BMI with the beta3-adrenergic receptor gene polymorphism in Japanese: meta-analysis. Obes Res 9(12):741–745
Kurokawa N, Young EH, Oka Y, Satoh H, Wareham NJ, Sandhu MS, Loos RJ (2008) The ADRB3 Trp64Arg variant and BMI: a meta-analysis of 44 833 individuals. Int J Obes (Lond) 32(8):1240–1249
Lalouschek W, Schillinger M, Hsieh K, Endler G, Tentschert S, Lang W, Cheng S, Mannhalter C (2005) Matched case–control study on factor V Leiden and the prothrombin G20210A mutation in patients with ischemic stroke/transient ischemic attack up to the age of 60 years. Stroke 36(7):1405–1409

Lanktree MB, Dichgans M, Hegele RA (2010) Advances in genomic analysis of stroke: what have we learned and where are we headed? Stroke 41(4):825–832

Lee AJ, Lowe GD, Woodward M, Tunstall-Pedoe H (1993) Fibrinogen in relation to personal history of prevalent hypertension, diabetes, stroke, intermittent claudication, coronary heart disease, and family history: the Scottish Heart Health Study. Br Heart J 69:338–342

Lehmann ED, Hopkins KD, Parker JR, Gosling RS (1995) Hyperlipidaemia, hypertension, and coronary heart disease. Lancet 345(8953):863

Lewandowski K, Swierczyńska A, Kwaśnikowski P, Elikowski W, Rzeźniczak M (2005) The prevalence of C807T mutation of glycoprotein Ia gene among young male survivors of myocardial infarction: a relation with coronary angiography results. Kardiol Pol 63:107–113

Li D, Tapscoft T, Gonzalez O, Burch PE, Quiñones MA, Zoghbi WA, Hill R, Bachinski LL, Mann DL, Roberts R (1999) Desmin mutation responsible for idiopathic dilated cardiomyopathy. Circulation 100(5):461–464

Li H, Chen Y, Zhou B, Peng Y, Sheng Y, Rao L (2011) Polymorphisms of presenilin-1 gene associate with dilated cardiomyopathy susceptibility. Mol Cell Biochem 358(1–2):31–36

Lingrel JB (2010) The physiological significance of the cardiotonic steroid/ouabain-binding site of the Na, K-ATPase. Ann Rev Physiol 72:395–412

Linsel-Nitschke P, Heeren J, Aherrahrou Z, Bruse P, Gieger C, Illig T, Prokisch H, Heim K, Doering A, Peters A, Meitinger T, Wichmann HE, Hinney A, Reinehr T, Roth C, Ortlepp JR, Soufi M, Sattler AM, Schaefer J, Stark K, Hengstenberg C, Schaefer A, Schreiber S, Kronenberg F, Samani NJ, Schunkert H, Erdmann J (2010) Genetic variation at chromosome 1p13.3 affects sortilin mRNA expression, cellular LDL-uptake and serum LDL levels which translates to the risk of coronary artery disease. Atherosclerosis 208(1):183–189

Liu X, Wang F, Knight AC, Zhao J, Xiao J (2012) Common variants for atrial fibrillation: results from genome-wide association studies. Hum Genet 131(1):33–39

Lopaciuk S, Bykowska K, Kwiecinski H, Mickielewicz A, Czlonkowska A, Mendel T, Kuczynska-Zardzewialy A, Szelagowska D, Windyga J, Schröder W, Herrmann FH, Jedrzejowska H (2001) Factor V Leiden, prothrombin gene G20210A variant, and methylenetetrahydrofolate reductase C677T genotype in young adults with ischemic stroke. Clin Appl Thromb Hemost 7:346–350

Loscalzo J, Weinfeld M, Fless GM, Scanu AM (1990) Lipoprotein(a), fibrin binding, and plasminogen activation. Arteriosclerosis 10(2):240–245

Lotta LA, Wang M, Yu J, Martinelli I, Yu F, Passamonti SM, Consonni D, Pappalardo E, Menegatti M, Scherer SE, Lewis LL, Akbar H, Wu Y, Bainbridge MN, Muzny DM, Mannucci PM, Gibbs RA, Peyvandi F (2012) Identification of genetic risk variants for deep vein thrombosis by multiplexed next-generation sequencing of 186 hemostatic/pro-inflammatory genes. BMC Med Genomics 5:7

Lovely RS, Yang Q, Massaro JM, Wang J, D'Agostino RB Sr, O'Donnell CJ, Shannon J, Farrell DH (2011) Assessment of genetic determinants of the association of γ' fibrinogen in relation to cardiovascular disease. Arterioscler Thromb Vasc Biol 31(10):2345–2352

Luke MM, Kane JP, Liu DM, Rowland CM, Shiffman D, Cassano J, Catanese JJ, Pullinger CR, Leong DU, Arellano AR, Tong CH, Movsesyan I, Naya-Vigne J, Noordhof C, Feric NT, Malloy MJ, Topol EJ, Koschinsky ML, Devlin JJ, Ellis SG (2007) A polymorphism in the protease-like domain of apolipoprotein(a) is associated with severe coronary artery disease. Arterioscler Thromb Vasc Biol 27(9):2030–2036

Macko RF, Kittner SJ, Epstein A, Cox DK, Wozniak MA, Wityk RJ, Stern BJ, Sloan MA, Sherwin R, Price TR, McCarter RJ, Johnson CJ, Earley CJ, Buchholz DW, Stolley PD (1999) Elevated tissue plasminogen activator antigen and stroke risk. The stroke prevention in young women study. Stroke 30:7–11

Maes HH, Neale MC, Eaves LJ (1997) Genetic and environmental factors in relative body weight and human obesity. Behav Genet 27:325–351

Maitland-van der Zee AH, Peters BJ, Lynch AI, Boerwinkle E, Arnett DK, Cheng S, Davis BR, Leiendecker-Foster C, Ford CE, Eckfeldt JH (2009) The effect of nine common polymorphisms in coagulation factor genes (F2, F5, F7, F12 and F13) on the effectiveness of statins: the GenHAT study. Pharmacogenet Genomics 19(5):338–344

Männikkö R, Overend G, Perrey C, Gavaghan CL, Valentin JP, Morten J, Armstrong M, Pollard CE (2010) Pharmacological and electrophysiological characterization of nine, single nucleotide polymorphisms of the hERG-encoded potassium channel. Br J Pharmacol 159(1):102–114

Martínez-Calatrava MJ, González-Sánchez JL, Zabena C, Martínez-Larrad MT, Luque-Otero M, Serrano-Ríos M (2007) Is the plasminogen activator inhibitor-1 gene a candidate gene predisposing to hypertension? Results from a population-based study in Spain. J Hypertens 25(4):773–777

Masuzaki H, Tanaka T, Ebihara K, Hosoda K, Nakao K (2009) Hypothalamic melanocortin signaling and leptin resistance–perspective of therapeutic application for obesity-diabetes syndrome. Peptides 30(7):1383–1386

Matkovich SJ, Van Booven DJ, Hindes A, Kang MY, Druley TE, Vallania FL, Mitra RD, Reilly MP, Cappola TP, Dorn GW 2nd (2010) Cardiac signaling genes exhibit unexpected sequence diversity in sporadic cardiomyopathy, revealing HSPB7 polymorphisms associated with disease. J Clin Invest 120:280–289

Meade TW, Ruddock V, Stirling Y, Chakrabarti R, Miller GJ (1993) Fibrinolytic activity, clotting factors, and long-term incidence of ischaemic heart disease in the Northwick Park Heart Study. Lancet 342:1076–1079

Mikkola H, Syrjälä M, Rasi V, Vahtera E, Hämäläinen E, Peltonen L, Palotie A (1994) Deficiency in the A-subunit of coagulation factor XIII: two novel point mutations demonstrate different effects on transcript levels. Blood 84:517–525

Mohana VU, Swapna N, Surender RS, Vishnupriya S, Padma T (2012) Gender-related association of AGT gene variants (M235T and T174M) with essential hypertension–a case–control study. Clin Exp Hypertens 34(1):38–44

Monkovic DD, Tracy PB (1990) Activation of human factor V by factor Xa and thrombin. Biochemistry 29:1118–1128

Morandi A, Maffeis C, Lobbens S, Bouatia-Naji N, Heude B, Pinelli L, Meyre D, Froguel P (2010) Early detrimental metabolic outcomes of rs17300539-A allele of ADIPOQ gene despite higher adiponectinemia. Obesity (Silver Spring) 18(7):1469–1473

Morris BJ (2011) A "SNP in time" for SCNN1G to join the "highly likely" list of genes for essential hypertension. Hypertension 58(6):996–997

Mottillo S, Filion KB, Genest J, Joseph L, Pilote L, Poirier P, Rinfret S, Schiffrin EL, Eisenberg MJ (2010) The metabolic syndrome and cardiovascular risk a systematic review and meta-analysis. J Am Coll Cardiol 56:1113–1132

Mulholland DJ, Dedhar S, Coetzee GA, Nelson CC (2005) Interaction of nuclear receptors with the Wnt/beta-catenin/Tcf signaling axis: Wnt you like to know? Endocr Rev 26(7): 898–915

Münzberg H, Björnholm M, Bates SH, Myers MG (2005) Leptin receptor action and mechanisms of leptin resistance. Cell Mol Life Sci 62:642–652

Musunuru K, Strong A, Frank-Kamenetsky M, Lee NE, Ahfeldt T, Sachs KV, Li X, Li H, Kuperwasser N, Ruda VM, Pirruccello JP, Muchmore B, Prokunina-Olsson L, Hall JL, Schadt EE, Morales CR, Lund-Katz S, Phillips MC, Wong J, Cantley W, Racie T, Ejebe KG, Orho-Melander M, Melander O, Koteliansky V, Fitzgerald K, Krauss RM, Cowan CA, Kathiresan S, Rader DJ (2010) From noncoding variant to phenotype via SORT1 at the 1p13 cholesterol locus. Nature 466(7307):714–719

Ned RM, Yesupriya A, Imperatore G, Smelser DT, Moonesinghe R, Chang MH, Dowling NF (2012) The ACE I/D polymorphism in US adults: limited evidence of association with hypertension-related traits and sex-specific effects by race/ethnicity. Am J Hypertens 25(2): 209–215

Nielsen EM, Hansen L, Echwald SM, Drivsholm T, Borch-Johnsen K, Ekstrøm CT, Hansen T, Pedersen O (2003) Evidence for an association between the Leu162Val polymorphism of the PPARalpha gene and decreased fasting serum triglyceride levels in glucose tolerant subjects. Pharmacogenetics 13(7):417–423

Nielsen MS, Madsen P, Christensen EI, Nykjaer A, Gliemann J, Kasper D, Pohlmann R, Petersen CM (2001) The sortilin cytoplasmic tail conveys Golgi-endosome transport and binds the VHS domain of the GGA2 sorting protein. EMBO J 20(9):2180–2190

Nishio Y, Makiyama T, Itoh H, Sakaguchi T, Ohno S, Gong YZ, Yamamoto S, Ozawa T, Ding WG, Toyoda F, Kawamura M, Akao M, Matsuura H, Kimura T, Kita T, Horie M (2009) D85N, a KCNE1 polymorphism, is a disease-causing gene variant in long QT syndrome. J Am Coll Cardiol 54:812–819

Norton N, Li D, Rieder MJ, Siegfried JD, Rampersaud E, Züchner S, Mangos S, Gonzalez-Quintana J, Wang L, McGee S, Reiser J, Martin E, Nickerson DA, Hershberger RE (2011) Genome-wide studies of copy number variation and exome sequencing identify rare variants in BAG3 as a cause of dilated cardiomyopathy. Am J Hum Genet 88(3):273–282

Nurden AT (1995) Polymorphisms of human platelet membrane glycoproteins: structure and clinical significance. Thromb Haemost 74:345–351

Oguchi S, Ito D, Murata M, Yoshida T, Tanahashi N, Fukuuchi Y, Ikeda Y, Watanabe K (2000) Genotype distribution of the 46C/T polymorphism of coagulation factor XII in the Japanese population: absence of its association with ischemic cerebrovascular disease. Thromb Haemost 83:178–179

Olson TM, Terzic A (2010) Human KATP channelopathies: diseases of metabolic homeostasis. Pflugers Arch 460:295–306

Paavonen KJ, Chapman H, Laitinen PJ, Fodstad H, Piippo K, Swan H, Toivonen L, Viitasalo M, Kontula K, Pasternack M (2003) Functional characterization of the common amino acid 897 polymorphism of the cardiac potassium channel KCNH2 (HERG). Cardiovasc Res 59(3):603–611

Papadopoulou S, Edlund H (2005) Attenuated Wnt signaling perturbs pancreatic growth but not pancreatic function. Diabetes 54(10):2844–2851

Parson W, Kraft HG, Niederstätter H, Lingenhel AW, Köchl S, Fresser F, Utermann G (2004) A common nonsense mutation in the repetitive Kringle IV-2 domain of human apolipoprotein(a) results in a truncated protein and low plasma Lp(a). Hum Mutat 24(6):474–480

Pellitero S, Reverter JL, Tàssies D, Pizarro E, Monteagudo J, Salinas I, Aguilera E, Sanmartí A, Reverter JC (2010) Polymorphisms in platelet glycoproteins Ia and IIIa are associated with arterial thrombosis and carotid atherosclerosis in type 2 diabetes. Thromb Haemost 103(3):630–637

Penas-Steinhardt A, Tellechea ML, Gomez-Rosso L, Brites F, Frechtel GD, Poskus E (2011) Association of common variants in JAK2 gene with reduced risk of metabolic syndrome and related disorders. BMC Med Genet 12:166

Periasamy M, Bhupathy P, Babu GJ (2008) Regulation of sarcoplasmic reticulum Ca2+ ATPase pump expression and its relevance to cardiac muscle physiology and pathology. Cardiovasc Res 77:265–273

Perry JR, Frayling TM (2008) New gene variants alter type 2 diabetes risk predominantly through reduced beta-cell function. Curr Opin Clin Nutr Metab Care 11(4):371–377

Poelmann RE, Jongbloed MR, Gittenberger-de Groot AC (2008) Pitx2: a challenging teenager. Circ Res 102:749–751

Poort SR, Rosendaal FR, Reitsma PH, Bertina RM (1996) A common genetic variation in the 3′-untranslated region of the prothrombin gene is associated with elevated plasma prothrombin levels and an increase in venous thrombosis. Blood 88:3698–3703

Prunier C, Hocevar BA, Howe PH (2004) Wnt signaling: physiology and pathology. Growth Factors 22(3):141–150

Raffai RL, Dong LM, Farese RV Jr, Weisgraber KH (2001) Introduction of human apolipoprotein E4 “domain interaction” into mouse apolipoprotein E. Proc Natl Acad Sci USA A98(20): 11587–11591

Rask-Andersen M, Almén MS, Olausen HR, Olszewski PK, Eriksson J, Chavan RA, Levine AS, Fredriksson R, Schiöth HB (2011) Functional coupling analysis suggests link between the obesity gene FTO and the BDNF-NTRK2 signaling pathway. BMC Neurosci 12:117

Reaven GM (1988) (1988) Banting lecture. Role of insulin resistance in human disease. Diabetes 37:1595–1607

Reiner AP, Carlson CS, Thyagarajan B, Rieder MJ, Polak JF, Siscovick DS, Nickerson DA, Jacobs DR Jr, Gross MD (2008) Soluble P-selectin, SELP polymorphisms, and atherosclerotic risk in

European-American and African-African young adults: the Coronary Artery Risk Development in Young Adults (CARDIA) Study. Arterioscler Thromb Vasc Biol 28(8):1549–1555

Reiner AP, Lange LA, Smith NL, Zakai NA, Cushman M, Folsom AR (2009) Common hemostasis and inflammation gene variants and venous thrombosis in older adults from the Cardiovascular Health Study. J Thromb Haemost 7(9):1499–1505

Rienstra M, Lubitz SA, Mahida S, Magnani JW, Fontes JD, Sinner MF, Van Gelder IC, Ellinor PT, Benjamin EJ (2012) Symptoms and functional status of patients with atrial fibrillation: state of the art and future research opportunities. Circulation 125(23):2933–2943

Rosendaal FR, Siscovick DS, Schwartz SM, Beverly RK, Psaty BM, Longstreth WT Jr, Raghunathan TE, Koepsell TD, Reitsma PH (1997) Factor V Leiden (resistance to activated protein C) increases the risk of myocardial infarction in young women. Blood 89:2817–2821

Rulifson IC, Karnik SK, Heiser PW, ten Berge D, Chen H, Gu X, Taketo MM, Nusse R, Hebrok M, Kim SK (2007) Wnt signaling regulates pancreatic beta cell proliferation. Proc Natl Acad Sci USA 104(15):6247–6252

Saidi S, Slamia LB, Mahjoub T, Ammou SB, Almawi WY (2007) Association of PAI-1 4G/5G and −844G/A gene polymorphism and changes in PAI-1/tPA levels in stroke: a case–control study. J Stroke Cerebrovasc Dis 16(4):153–159

Saito I, Folsom AR, Brancati FL, Duncan BB, Chambless LE, McGovern PG (2000) Nontraditional risk factors for coronary heart disease incidence among persons with diabetes: the Atherosclerosis Risk in Communities (ARIC) Study. Ann Intern Med 133:81–91

Sanna S, Li B, Mulas A, Sidore C, Kang HM, Jackson AU, Piras MG, Usala G, Maninchedda G, Sassu A, Serra F, Palmas MA, Wood WH 3rd, Njølstad I, Laakso M, Hveem K, Tuomilehto J, Lakka TA, Rauramaa R, Boehnke M, Cucca F, Uda M, Schlessinger D, Nagaraja R, Abecasis GR (2011) Fine mapping of five loci associated with low-density lipoprotein cholesterol detects variants that double the explained heritability. PLoS Genet 7(7):e1002198

Satra M, Samara M, Wozniak G, Tzavara C, Kontos A, Valotassiou V, Vamvakopoulos NK, Tsougos I, Aleporou-Marinou V, Patrinos GP, Kollia P, Georgoulias P (2011) Sequence variations in the FII, FV, F13A1, FGB and PAI-1 genes are associated with differences in myocardial perfusion. Pharmacogenomics 12(2):195–203

Savage B, Saldivar E, Ruggeri Z (1996) Initiation of platelet adhesion by arrest onto fibrinogen or translocation on von Willebrand factor. Cell 84:289–297

Schiffrin EL (2005) Peroxisome proliferator-activated receptors and cardiovascular remodeling. Am J Physiol Heart Circ Physiol 288(3):H1037–H1043

Schild L (2010) The epithelial sodium channel and the control of sodium balance. Biochim Biophys Acta 1802(12):1159–1165

Schott JJ, Alshinawi C, Kyndt F, Probst V, Hoorntje TM, Hulsbeek M, Wilde AA, Escande D, Mannens MM, Le Marec H (1999) Cardiac conduction defects associate with mutations in SCN5A. Nat Genet 23(1):20–21

Schunkert H, Erdmann J, Samani NJ (2010) Genetics of myocardial infarction: a progress report. Eur Heart J 31(8):918–925

Scuteri A, Sanna S, Chen WM, Uda M, Albai G, Strait J, Najjar S, Nagaraja R, Orrú M, Usala G, Dei M, Lai S, Maschio A, Busonero F, Mulas A, Ehret GB, Fink AA, Weder AB, Cooper RS, Galan P, Chakravarti A, Schlessinger D, Cao A, Lakatta E, Abecasis GR (2007) Genome-wide association scan shows genetic variants in the FTO gene are associated with obesity-related traits. PLoS Genet 3:e115

Shemirani AH, Pongrácz E, Antalfi B, Adány R, Muszbek L (2010) Factor XIII A subunit Val34Leu polymorphism in patients suffering atherothrombotic ischemic stroke. Thromb Res 126(2):159–162

Shinlapawittayatorn K, Du XX, Liu H, Ficker E, Kaufman ES, Deschênes I (2011) A common SCN5A polymorphism modulates the biophysical defects of SCN5A mutations. Heart Rhythm 8(3):455–462

Shugart YY, Chen L, Day IN, Lewis SJ, Timpson NJ, Yuan W, Abdollahi MR, Ring SM, Ebrahim S, Golding J, Lawlor DA, Davey-Smith G (2009) Two British women studies replicated the association between the Val66Met polymorphism in the brain-derived neurotrophic factor (BDNF) and BMI. Eur J Hum Genet 17(8):1050–1055

Siebenlist KR, Meh DA, Mosesson MW (2001) Protransglutaminase (factor XIII) mediated cross-linking of fibrinogen and fibrin. Thromb Haemost 86(5):1221–1228
Siegerink B, Rosendaal FR, Algra A (2009) Genetic variation in fibrinogen; its relationship to fibrinogen levels and the risk of myocardial infarction and ischemic stroke. J Thromb Haemost 7(3):385–390
Silvain J, Pena A, Vignalou JB, Hulot JS, Galier S, Cayla G, Bellemain-Appaix A, Barthélémy O, Beygui F, Bal-dit-Sollier C, Drouet L, Weisel JW, Montalescot G, Collet JP (2011) FXIII-A Leu34 genetic variant in premature coronary artery disease: a genotype–phenotype case control study. Thromb Haemost 106(3):511–520
Sinner MF, Pfeufer A, Akyol M, Beckmann BM, Hinterseer M, Wacker A, Perz S, Sauter W, Illig T, Näbauer M, Schmitt C, Wichmann HE, Schömig A, Steinbeck G, Meitinger T, Kääb S (2008) The non-synonymous coding IKr-channel variant KCNH2-K897T is associated with atrial fibrillation: results from a systematic candidate gene-based analysis of KCNH2 (HERG). Eur Heart J 29(7):907–914
Souto JC, Almasy L, Muñiz-Diaz E, Soria JM, Borrell M, Bayén L, Mateo J, Madoz P, Stone W, Blangero J, Fontcuberta J (2000) Functional effects of the ABO locus polymorphism on plasma levels of von Willebrand factor, factor VIII, and activated partial thromboplastin time. Arterioscler Thromb Vasc Biol 20:2024–2028
Slowik A, Dziedzic T, Pera J, Figlewicz DA, Szczudlik A (2005) Coagulation factor XIII Vai34Leu polymorphism in patients with small vessel disease or primary intracerebral hemorrhage. Cerebrovasc Dis 19:165–170
Stark K, Esslinger UB, Reinhard W, Petrov G, Winkler T, Komajda M, Isnard R, Charron P, Villard E, Cambien F, Tiret L, Aumont MC, Dubourg O, Trochu JN, Fauchier L, Degroote P, Richter A, Maisch B, Wichter T, Zollbrecht C, Grassl M, Schunkert H, Linsel-Nitschke P, Erdmann J, Baumert J, Illig T, Klopp N, Wichmann HE, Meisinger C, Koenig W, Lichtner P, Meitinger T, Schillert A, König IR, Hetzer R, Heid IM, Regitz-Zagrosek V (2010) Hengstenberg C (2010) Genetic association study identifies HSPB7 as a risk gene for idiopathic dilated cardiomyopathy. PLoS Genet 6(10):e1001167
Stein EA, Gipe D, Bergeron J, Gaudet D, Weiss R, Dufour R, Wu R, Pordy R (2012) Effect of a monoclonal antibody to PCSK9, REGN727/SAR236553, to reduce low-density lipoprotein cholesterol in patients with heterozygous familial hypercholesterolaemia on stable statin dose with or without ezetimibe therapy: a phase 2 randomised controlled trial. Lancet 380(9836):29–36
Stumvoll M, Tschritter O, Fritsche A, Staiger H, Renn W, Weisser M, Machicao F, Häring H (2002) Association of the T-G polymorphism in adiponectin (exon 2) with obesity and insulin sensitivity: interaction with family history of type 2 diabetes. Diabetes 51(1):37–41
Stutzmann F, Vatin V, Cauchi S, Morandi A, Jouret B, Landt O, Tounian P, Levy-Marchal C, Buzzetti R, Pinelli L, Balkau B, Horber F, Bougnères P, Froguel P, Meyre D (2007) Non-synonymous polymorphisms in melanocortin-4 receptor protect against obesity: the two facets of a Janus obesity gene. Hum Mol Genet 16(15):1837–1844
Suzuki K, Kuriyama M, Saito T, Ichinose A (1997) Plasma lipoprotein(a) levels and expression of the apolipoprotein(a) gene are dependent on the nucleotide polymorphisms in its 5'-flanking region. J Clin Invest 99(6):1361–1366
Takeda N (2003) Cardiomyopathy: molecular and immunological aspects (review). Int J Mol Med 11(1):13–16
Takeuchi F, Yamamoto K, Katsuya T, Sugiyama T, Nabika T, Ohnaka K, Yamaguchi S, Takayanagi R, Ogihara T, Kato N (2012) Reevaluation of the association of seven candidate genes with blood pressure and hypertension: a replication study and meta-analysis with a larger sample size. Hypertens Res 35(8):825–831
Taylor MR, Slavov D, Ku L, Di Lenarda A, Sinagra G, Carniel E, Haubold K, Boucek MM, Ferguson D, Graw SL, Zhu X, Cavanaugh J, Sucharov CC, Long CS, Bristow MR, Lavori P, Mestroni L, ; Familial Cardiomyopathy Registry; BEST (Beta-Blocker Evaluation of Survival Trial) DNA Bank (2007) Prevalence of desmin mutations in dilated cardiomyopathy. Circulation 115(10):1244–1251
Temelkova-Kurktschiev T, Stefanov T (2012) Lifestyle and genetics in obesity and type 2 diabetes. Exp Clin Endocrinol Diabetes 120(1):1–6

Tessari A, Pietrobon M, Notte A, Cifelli G, Gage PJ, Schneider MD, Lembo G, Campione M (2008) Myocardial Pitx2 differentially regulates the left atrial identity and ventricular asymmetric remodeling programs. Circ Res 102:813–822

Tomaiuolo R, Bellia C, Caruso A, Di Fiore R, Quaranta S, Noto D, Cefalù AB, Di Micco P, Zarrilli F, Castaldo G, Averna MR, Ciaccio M (2012) Prothrombotic gene variants as risk factors of acute myocardial infarction in young women. J Transl Med 10:235

Tomaszewski M, Charchar FJ, Barnes T, Gawron-Kiszka M, Sedkowska A, Podolecka E, Kowalczyk J, Rathbone W, Kalarus Z, Grzeszczak W, Goodall AH, Samani NJ, Zukowska-Szczechowska E (2009) A common variant in low density lipoprotein receptor-related protein 6 gene (LRP6) is associated with LDL-cholesterol. Arterioscler Thromb Vasc Biol 29(9):1316–1321

Torekov SS, Larsen LH, Andersen G, Albrechtsen A, Glümer C, Borch-Johnsen K, Jørgensen T, Hansen T, Pedersen O (2006) Variants in the 5′ region of the neuropeptide Y receptor Y2 gene (NPY2R) are associated with obesity in 5,971 white subjects. Diabetologia 49(11):2653–2658

Trégouët DA, König IR, Erdmann J, Munteanu A, Braund PS, Hall AS, Grosshennig A, Linsel-Nitschke P, Perret C, DeSuremain M, Meitinger T, Wright BJ, Preuss M, Balmforth AJ, Ball SG, Meisinger C, Germain C, Evans A, Arveiler D, Luc G, Ruidavets JB, Morrison C, van der Harst P, Schreiber S, Neureuther K, Schäfer A, Bugert P, El Mokhtari NE, Schrezenmeir J, Stark K, Rubin D, Wichmann HE, Hengstenberg C, Ouwehand W; Wellcome Trust Case Control Consortium; Cardiogenics Consortium, Ziegler A, Tiret L, Thompson JR, Cambien F, Schunkert H, Samani NJ (2009) Genome-wide haplotype association study identifies the SLC22A3-LPAL2-LPA gene cluster as a risk locus for coronary artery disease. Nat Genet 41(3):283–285

Tsantes AE, Nikolopoulos GK, Bagos PG, Bonovas S, Kopterides P, Vaiopoulos G (2008) The effect of the plasminogen activator inhibitor-1 4G/5G polymorphism on the thrombotic risk. Thromb Res 122(6):736–742

Van de Water NS, French JK, Lund M, Hyde TA, White HD, Browett PJ (2000) Prevalence of factor V Leiden and prothrombin variant G20210A in patients age <50 years with no significant stenoses at angiography three to four weeks after myocardial infarction. J Am Coll Cardiol 36:717–722

Van der Zwaag PA, van Rijsingen IA, Asimaki A, Jongbloed JD, van Veldhuisen DJ, Wiesfeld AC, Cox MG, van Lochem LT, de Boer RA, Hofstra RM, Christiaans I, van Spaendonck-Zwarts KY, Lekannedit Deprez RH, Judge DP, Calkins H, Suurmeijer AJ, Hauer RN, Saffitz JE, Wilde AA, van den Berg MP, van Tintelen JP (2012) Phospholamban R14del mutation in patients diagnosed with dilated cardiomyopathy or arrhythmogenic right ventricular cardiomyopathy: evidence supporting the concept of arrhythmogenic cardiomyopathy. Eur J Heart Fail 14(11):1199–1207

Van Schie MC, van Loon JE, de Maat MP, Leebeek FW (2011) Genetic determinants of von Willebrand factor levels and activity in relation to the risk of cardiovascular disease: a review. J Thromb Haemost 9(5):899–908

Vehaskari VM (2009) Heritable forms of hypertension. Pediatr Nephrol 24:1929–1937

Viel KR, Machiah DK, Warren DM, Khachidze M, Buil A, Fernstrom K, Souto JC, Peralta JM, Smith T, Blangero J, Porter S, Warren ST, Fontcuberta J, Soria JM, Flanders WD, Almasy L, Howard TE (2007) A sequence variation scan of the coagulation factor VIII (FVIII) structural gene and associations with plasma FVIII activity levels. Blood 109(9):3713–3724

Villard E, Perret C, Gary F, Proust C, Dilanian G, Hengstenberg C, Ruppert V, Arbustini E, Wichter T, Germain M, Dubourg O, Tavazzi L, Aumont MC, DeGroote P, Fauchier L, Trochu JN, Gibelin P, Aupetit JF, Stark K, Erdmann J, Hetzer R, Roberts AM, Barton PJ, Regitz-Zagrosek V; Cardiogenics Consortium, Aslam U, Duboscq-Bidot L, Meyborg M, Maisch B, Madeira H, Waldenström A, Galve E, Cleland JG, Dorent R, Roizes G, Zeller T, Blankenberg S, Goodall AH, Cook S, Tregouet DA, Tiret L, Isnard R, Komajda M, Charron P, Cambien F (2011) A genome-wide association study identifies two loci associated with heart failure due to dilated cardiomyopathy. Eur Heart J 32(9):1065–1076

Wang X, Cheng S, Brophy VH, Erlich HA, Mannhalter C, Berger K, Lalouschek W, Browner WS, Shi Y, Ringelstein EB, Kessler C, Luedemann J, Lindpaintner K, Liu L, Ridker PM, Zee RY, Cook NR, ; RMS Stroke SNP Consortium (2009) A meta-analysis of candidate gene polymorphisms and ischemic stroke in 6 study populations: association of lymphotoxin-alpha in nonhypertensive patients. Stroke 40:683–695

Wang J, Klysik E, Sood S, Johnson RL, Wehrens XH, Martin JF (2010a) Pitx2 prevents susceptibility to atrial arrhythmias by inhibiting left-sided pacemaker specification. Proc Natl Acad Sci USA 107:9753–9758

Wang Y, Fu W, Xie F, Wang Y, Chu X, Wang H, Shen M, Wang Y, Wang Y, Sun W, Lei R, Yang L, Wu H, Foo J, Liu J, Jin L, Huang W (2010b) Common polymorphisms in ITGA2, PON1 and THBS2 are associated with coronary atherosclerosis in a candidate gene association study of the Chinese Han population. J Hum Genet 55(8):490–494

Wiklund PG, Nilsson L, Ardnor SN, Eriksson P, Johansson L, Stegmayr B, Hamsten A, Holmberg D, Asplund K (2005) Plasminogen activator inhibitor-1 4G/5G polymorphism and risk of stroke: replicated findings in two nested case–control studies based on independent cohorts. Stroke 36(8):1661–1665

Wilhelmsen L, Svardsudd K, Korsan-Bengtsen K, Larsson B, Welin L, Tibblin G (1984) Fibrinogen as a risk factor for stroke and myocardial infarction. N Engl J Med 311:501–505

Willer CJ, Sanna S, Jackson AU, Scuteri A, Bonnycastle LL, Clarke R, Heath SC, Timpson NJ, Najjar SS, Stringham HM, Strait J, Duren WL, Maschio A, Busonero F, Mulas A, Albai G, Swift AJ, Morken MA, Narisu N, Bennett D, Parish S, Shen H, Galan P, Meneton P, Hercberg S, Zelenika D, Chen WM, Li Y, Scott LJ, Scheet PA, Sundvall J, Watanabe RM, Nagaraja R, Ebrahim S, Lawlor DA, Ben-Shlomo Y, Davey-Smith G, Shuldiner AR, Collins R, Bergman RN, Uda M, Tuomilehto J, Cao A, Collins FS, Lakatta E, Lathrop GM, Boehnke M, Schlessinger D, Mohlke KL, Abecasis GR (2008) Newly identified loci that influence lipid concentrations and risk of coronary artery disease. Nat Genet 40(2):161–169

Wilson CP, McNulty H, Scott JM, Strain JJ, Ward M (2011) Postgraduate Symposium: The MTHFR C677T polymorphism, B-vitamins and blood pressure. Proc Nutr Soc 69(1):156–165

Wu AH, Tsongalis GJ (2001) Correlation of polymorphisms to coagulation and biochemical risk factors for cardiovascular diseases. Am J Cardiol 87:1361–1366

Xu Y, Tuteja D, Zhang Z, Xu D, Zhang Y, Rodriguez J, Nie L, Tuxson HR, Young JN, Glatter KA, Vazquez AE, Yamoah EN, Chiamvimonvat N (2003) Molecular identification and functional roles of a Ca(2+)-activated K+channel in human and mouse hearts. J Biol Chem 278:49085–49094

Xin XY, Song YY, Ma JF, Fan CN, Ding JQ, Yang GY, Chen SD (2009) Gene polymorphisms and risk of adult early-onset ischemic stroke: A meta-analysis. Thromb Res 124(5): 619–624

Yang Q, Kathiresan S, Lin JP, Tofler GH, O'Donnell CJ (2007a) Genome-wide association and linkage analyses of hemostatic factors and hematological phenotypes in the Framingham Heart Study. BMC Med Genet 8(Suppl 1):S12

Yang WS, Yang YC, Chen CL, Wu IL, Lu JY, Lu FH, Tai TY, Chang CJ (2007b) Adiponectin SNP276 is associated with obesity, the metabolic syndrome, and diabetes in the elderly. Am J Clin Nutr 86(2):509–513

Ye Z, Liu EH, Higgins JP, Keavney BD, Lowe GD, Collins R, Danesh J (2006) Seven haemostatic gene polymorphisms in coronary disease: meta-analysis of 66,155 cases and 91,307 controls. Lancet 367(9511):651–658

Young EH, Wareham NJ, Farooqi S, Hinney A, Hebebrand J, Scherag A, O'rahilly S, Barroso I, Sandhu MS (2007) The V103I polymorphism of the MC4R gene and obesity: population based studies and meta-analysis of 29 563 individuals. Int J Obes (Lond) 31(9):1437–1441

Zee RY, Glynn RJ, Cheng S, Steiner L, Rose L, Ridker PM (2009) An evaluation of candidate genes of inflammation and thrombosis in relation to the risk of venous thromboembolism: The Women's Genome Health Study. Circ Cardiovasc Genet 2(1):57–62

Zhang H, Wu LM, Wu J (2011) Cross-talk between apolipoprotein E and cytokines. Mediators Inflamm 2011:949072

Zhang X, Lynch AI, Davis BR, Ford CE, Boerwinkle E, Eckfeldt JH, Leiendecker-Foster C, Arnett DK (2012) Pharmacogenetic association of NOS3 variants with cardiovascular disease in patients with hypertension: the GenHAT study. PLoS One 7(3):e34217

Zhao YH, Xu Y, Gu YY, Li Y, Zhang JY, Su X (2011) Functional effect of platelet membrane glycoprotein Ia gene polymorphism in the pathogenesis of unstable angina pectoris. J Int Med Res 39(2):541–548

Ziemke F, Mantzoros CS (2010) Adiponectin in insulin resistance: lessons from translational research. Am J Clin Nutr 91(suppl):258S–261S

Zito F, Drummond F, Bujac SR, Esnouf MP, Morrissey JH, Humphries SE, Miller GJ (2000) Epidemiological and genetic associations of activated factor XII concentration with factor VII activity, fibrinopeptide A concentration, and risk of coronary heart disease in men. Circulation 102:2058–2062

Zipes DP, Wellens HJ (1998) Sudden cardiac death. Circulation 98(21):2334–2351

Zobel DP, Andreasen CH, Burgdorf KS, Andersson EA, Sandbaek A, Lauritzen T, Borch-Johnsen K, Jørgensen T, Maeda S, Nakamura Y, Eiberg H, Pedersen O, Hansen T (2009) Variation in the gene encoding Krüppel-like factor 7 influences body fat: studies of 14 818 Danes. Eur J Endocrinol 160(4):603–609

Zorio E, Gilabert-Estellés J, España F, Ramón LA, Cosín R, Estellés A (2008) Fibrinolysis: the key to new pathogenetic mechanisms. Curr Med Chem 15(9):923–929

Diet and Nutrition to Prevent and Treat Cardiovascular Diseases

5

Hiroshi Masuda

Abstract

In this chapter, the effects of nutrients on the human cardiovascular system and the molecular mechanisms of their protective effects are summarized. The etiology and pathomechanisms of cardiovascular diseases (CVD) are complex with numerous of gene-environment interaction. Importantly, coronary heart disease (CHD) is the most prevalent cause; about 60 % is attributed to dietary patterns. Numerous epidemiologic studies demonstrated the necessity to reduce total calories, the consumption of high glycemic index (GI) foods, sodium, saturated fats, and trans-fats and to increase the consumption of dietary fiber, complex carbohydrates, fruits, and vegetables. Major cardioprotective mechanisms of nutrients are classified as follows: (1) activating sirtuins (caloric restriction), (2) ameliorating serum lipid profiles (MUFA, n-3 PUFA, carotenoids, dietary fiber), (3) decreasing insulin resistances (MUFA, low GI diet), (4) anti-inflammatory effects (n-3 PUFA, n-6 PUFA, niacin, vitamin D, low GI diet), (5) antioxidants (niacin, vitamin C, vitamin E, carotenoids, low GI diet), (6) antithrombogenic actions (MUFA, n-3 PUFA, niacin, vitamin D, vitamin E), (7) improving vascular functions (n-3 PUFA, n-6 PUFA, vitamin C, vitamin D, vitamin E), (8) antiarrhythmic actions (n-3 PUFA), and (9) lowering homocysteine levels (B vitamins). Although changes in the quality of food is as important as restricting calorie intake, more information is needed to clarify the relation between the intake of defined nutrients and the risk of heart diseases.

Keywords

Cardiovascular disease • Caloric restriction • Dietary sodium • Saturated fatty acid • Monounsaturated fatty acid • n-3 polyunsaturated fatty acid • n-6 polyunsaturated fatty acid • Trans-fatty acid • B vitamins • Vitamin C • Vitamin D • Vitamin E • Carotenoid • Carbohydrate • Dietary fiber

H. Masuda
Department of Internal Medicine,
Nishinomiya Kaisei Hospital, 1-4 Ohama-cho, Nishinomiya, Hyogo 662-0957, Japan
e-mail: kaisei-incho@amail.plala.or.jp

I. Wakabayashi, K. Groschner (eds.), *Interdisciplinary Concepts in Cardiovascular Health*,
DOI 10.1007/978-3-7091-1334-9_5,

5.1 Introduction

Cardiovascular disease (CVD) is important as a lifestyle-related disease, although the inherited character to CVD has been documented. Among many lifestyle factors in CVD, diet factor has been a major focus of health research for almost half a century. Although epidemiological studies and randomized controlled trials (RCTs) demonstrate protective or hazardous effects of various nutrients for cardiovascular system, little information is actually available on how these effects are mediated at the molecular level. This chapter focuses on health research between diet and CVD and attempts to highlight the molecular targets of natural compounds of food in the cardiovascular system.

5.2 Caloric Restriction

Since the initial rodent study of 1935 (MaCay et al. 1935), hundreds of reports demonstrated that caloric restriction (CR) with adequate nutrition slows aging and extends lifespan in various species. CR has been also shown to delay the onset of age-related cardiac changes and ameliorate all known CVD risk factors in experimental animals (Kemi et al. 2000). In humans, several studies found that CR has a protective effect against atherosclerosis and cardiac dysfunction (Fontana et al. 2004; Meyer et al. 2006).

The sirtuin family (SIRT1-7) consists of proteins found within numerous compartments which are NAD^+ (nicotinamide adenine dinucleotide, oxidized form)-dependent class III histone deacetylase and adenosine diphosphate (ADP)-ribosyltransferases in mammals. The levels of SIRT1, the most extensively studied of the seven mammalian sirtuins, have been reported to increase in rodent and human tissues in response to CR (Cohen et al. 2004; Nisoli et al. 2005; Civitarese et al. 2007). The enzymatic activity of most sirtuins has been shown to be dependent on shifts in the redox potential that modulate the cellular NAD^+/NADH (NAD, reduced form) ratio. Increases in the NAD^+/NADH ratio have been shown to stimulate sirtuin activity (Lin et al. 2004). As CR decreases intracellular NADH levels, CR may activate sirtuins by increasing the NAD^+/NADH ratio.

Hence, sirtuins have been suggested as mediators of beneficial aspects of CR including the decreased incidence of age-related disorders such as CVD in genetic and transcriptional regulation (Bordone and Guarente 2005). Furthermore, the potential benefits of sirtuin activity on cardiovascular health lie not only in their direct actions on cells of the cardiovascular system but in the favorable metabolic profile created through their actions on non-cardiovascular tissues.

The function of many important proteins involved in metabolism is regulated by sirtuins, including peroxisome proliferator-activated receptor γ (PPARγ), PPARγ coactivator 1α (PGC-1α), nuclear receptor corepressor (NCoR), silencing mediator of retinoid and thyroid hormone receptor (SMRT), nuclear factor κB (NFκB), liver X receptor α (LXRα), forkhead box O1 (FoxO1), protein tyrosine phosphatase 1B (PTB1B), uncoupling protein 2 (UCP2), glutamate dehydrogenase (GDH), p53, angiotensin II type 1 receptor (AT_1R), and endothelial nitric oxide synthase (eNOS).

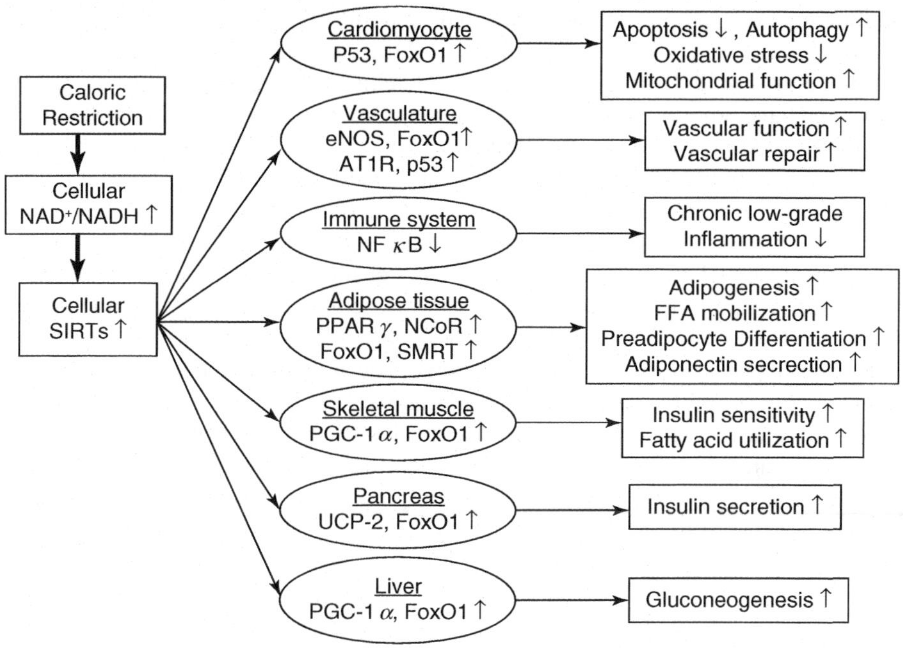

Fig. 5.1 Proposed roles and targets of SIRTs in cardioprotection

Many of these molecules have been found to play a direct or indirect role in cardioprotection. For example, at the cellular level, cardioprotection by CR is mediated by various mechanisms, among which attenuation of oxidative stress, mitochondrial dysfunction and inflammation, and a favorable modulation of apoptosis and autophagy are prominent contributors (Marzetti et al. 2009; Borradaile and Pickering 2009).

The proposed roles and targets of SIRTs in cardioprotective effects of CR are summarized in Fig. 5.1.

5.3 Dietary Sodium

Sodium, primarily consumed as salt (sodium chloride), is an essential nutrient because sodium and chloride are required to maintain extracellular fluid volume and plasma osmolality. The role of dietary sodium in health and disease has been a topic of great interest and debate for many years.

Experimental, epidemiological, and intervention studies have shown a positive relation of habitual dietary sodium intake and blood pressure (Elliott et al. 1996; Sacks et al. 2001). Equally well established is the association of increasing blood pressure and CVD morbidity and mortality (Gordon and Kannel 1982). However, findings from previous studies, independently from blood pressure, evaluating the direct relationship between sodium intake and cardiovascular events, have been conflicting. For example, although some have reported a positive association between sodium intake and cardiovascular outcomes (Strazzullo et al. 2009; Umesawa et al. 2008; Cook et al. 2007a), others have not (Geleijnse et al. 2007; Yang et al. 2011), and some have reported an inverse association (Cohen et al. 2006; Stolarz-Skrzypek et al. 2011). A meta-analysis (Cochrane review) of seven RCTs evaluating reduced sodium intake with follow-up of at least 6 months did not detect a significant reduction in CVD mortality or morbidity (Taylor et al. 2011). Furthermore, salt restriction increased the risk of all-cause mortality in those with heart failure. Recent observational analyses of participants in two hypertension treatment trials revealed a "J-shaped" relation of sodium intake to cardiovascular outcomes (O'Donnell et al. 2011). CVD was lowest between 4.0 and 5.9 g/day, with continuous increases in risk with sodium levels above and below that range. The authors concluded that CVD risk increases meaningfully when daily sodium intake falls below 3 g or rises to above 6 g/day (Alderman and Cohen 2012). The mechanisms responsible for an increased CVD associated with low sodium intake cannot be established by previous studies. A number of the possible mechanisms are proposed: increases in sympathetic nerve activity, plasma renin activity and aldosterone, serum cholesterol and triglyceride levels, insulin resistance, and adrenaline secretion have all been confirmed by multiple RCTs (Graudal et al. 2012; Petrie et al. 1998; Grassi et al. 2002). These biological explanations for harm at low sodium intake need further explication.

While blood pressure in the population is only modestly responsive to alterations in sodium intake, some individuals manifest large blood pressure changes in response to acute or chronic salt depletion or repletion and are termed "salt sensitive". Weinberger et al. (1986) demonstrated that salt sensitivity existed not only in hypertensive persons but also in normotensive persons. In addition, Weinberger et al. (2001) reported that salt sensitivity in both normotensive and hypertensive persons has been associated with increased CVD morbidity and total mortality, which were independent of alterations of blood pressure. A number of recent studies, using modern molecular genetic techniques, indicate that blood pressure salt sensitivity is under the control of many genes each of which has only relatively mild effects. Candidate genes that regulate the salt sensitivity have implicated aberrations in ion channels, ion channel regulation, aldosterone signaling, vasoconstriction, and inflammation (Sanada et al. 2011; Kanbay et al. 2011; see Table 5.1). Even people with normal blood pressure as well as hypertension that are salt sensitive need to take steps to reduce their salt intake in order to prevent salt-induced blood vessel stiffness, hypertension, and CVD.

5.4 Saturated Fatty Acids (SFA)

Saturated fatty acids (SFA) are long-chain carboxylic acids that usually have between 12 and 24 carbon atoms and have no double bonds. The most prevailing SFA in the diet are lauric acid (12:0), myristic acid (14:0), palmitic acid (16:0), and stearic acid (18:0). SFA are present in a type of single-bond animal or vegetable fat, as that found in butter, meat, egg yolks, and coconut or palm oil.

Table 5.1 Candidate genes involved in salt sensitivity

1. Genes of the renin-angiotensin system
Angiotensin-converting enzyme
Angiotensinogen
Angiotensin type 1 receptor
2. Genes related to aldosterone and other mineralocorticoids
Aldosterone synthase
Serum-/glucocorticoid-regulated kinase 1
11-β-hydroxysteroid dehydrogenase
3. Genes of the sympathetic nervous system
Amiloride-sensitive epithelial sodium channel
β-3 Subunit of G-protein (GNB3)
Renal chloride channels
4. Genes expressing proteins that normally do not decrease renal sodium transport
α-Adducin
5. Genes expressing proteins that normally decrease renal sodium transport
Adrenomedullin
Dopamine, dopamine receptors, and GRK4
Endothelial nitric oxide synthase
Eicosanoids
Kallikrein-kinin system
Natriuretic peptide precursor A/B
6. Genes of the vasculature
Endothelin
Endothelial nitric oxide synthase (eNOS)
Transforming growth factor-β (TGF-β)

Ecological data from the Seven Countries Study showed a strong positive association ($r=0.73$) between SFA intake and coronary heart disease (CHD) incidence (Keys and Aravanis 1980). Dietary saturated fat intake has been shown to increase low-density lipoprotein (LDL) cholesterol and therefore to be associated with increased risk of CVD. SFA inhibit the activity of the LDL receptor in liver. Thus, LDL cholesterol and total cholesterol in the blood are increased by SFA. In addition, high intake of dietary SFA is positively associated with the development of insulin resistance. Dietary SFA upregulated the mRNA levels of resistin, downregulated adiponectin and glucose transporter 4 (GLUT4), and upregulated lipoprotein lipase (LPL) (Saravanan et al. 2005). These findings suggest that dietary SFA alter the expression of different genes associated with insulin sensitivity in adipose tissue.

5.5 Monounsaturated Fatty Acids (MUFA)

MUFA are distinguished from the other fatty acid classes on the basis of having only one double bond. MUFA are not nutritionally essential as they can be synthesized from other (saturated) fatty acids and from carbohydrates. Common sources of MUFA are olive oil (contains about 79 % MUFA), canola oil (contains about 54 % MUFA), avocadoes, and nuts.

The typical diet of populations living in Mediterranean countries is high in olive oil, which provides 14–40 % of calories, and consequently is high in MUFA (16–29 % of calories) (Kris-Etherton 1999). Evidence suggesting the beneficial health effects of the "Mediterranean diet" has emerged from the classic studies of Keys and Aravanis (1980), which indicate that the consumption of diets enriched in MUFA reduced incidence of CVD (Keys and Aravanis 1980; Martinez-Gonzalez et al. 2002; Knoops et al. 2004). These cardioprotective effects can be explained by several mechanisms. Foods containing high MUFA reduce total and LDL cholesterol while did not alter high-density lipoprotein (HDL) cholesterol and triglycerides (TG) (Kris-Etherton 1999). High-MUFA diet significantly improved fasting glucose and insulin sensitivity and increased plasma glucagon-like peptide-1 (GLP-1) levels but has no effect on insulin secretion (Vessby et al. 2001). Several data suggest that MUFA may decrease platelet aggregation (Sirtori et al. 1986), increase bleeding time (McDonald et al. 1989), and increase fibrinolysis (Lopez-Segura et al. 1996), thereby protecting against thrombogenesis.

Two recent studies have reported interactions between genetic variants and MUFA for obesity (Warodomwichit et al. 2009; Lai et al. 2009). Variants of the adipocyte-derived protein adiponectin have been previously linked to obesity and insulin resistance, but association studies among different ethnic groups have not been consistent. In a population of 1,083 White Americans, the A allele of the adiponectin 11,391 G>A variant was protective against obesity independently of dietary intake and also under conditions of high MUFA intake, but this relationship was reversed in subjects with low intake of MUFA (Warodomwichit et al. 2009). MUFA can affect gene expression patterns of various proteins and accordingly may regulate pathways related to CVD pathophysiology.

5.6 N-3 Polyunsaturated Fatty Acids (n-3 PUFA)

Animals and humans cannot synthesize PUFA of the n-3 series, which contain double bonds at C-3, from the methyl end of the molecule. N-3 PUFA include the plant-derived α-linolenic acid (ALA, 18:3n-3) and the fish-oil-derived eicosapentaenoic acid (EPA, 20:5n-3) and docosahexaenoic acid (DHA, 22:6n-3). PUFA represent the fundamental components of phospholipids in cellular membranes where they maintain homeostasis for correct membrane protein function and influence membrane fluidity, thus regulating cell signaling processes, cellular functions, and gene expression (Das 2006).

In the 1970s, Dyerberg et al. (1978) reported that the rate of ischemic heart disease was markedly lower in Greenland Inuits, a population consuming a high-fat diet but rich in n-3 PUFA, than in age-matched population of Denmark. Inuits have high levels of EPA and low levels of AA, and they also have a tendency to bleed. It is possible that dietary enrichment with EPA will protect against thrombosis. The following prospective observational studies and adequately powered RCTs from many countries have shown that consumption of fish oil or n-3 PUFA decreases the risk of major cardiovascular events, such as myocardial infarction, sudden cardiac death, coronary heart disease (CHD), atrial fibrillation, and death in patients with heart failure (Lee et al. 2008b; Kromhout et al. 1985; Albert et al. 1998; Jialal et al. 1999; Yokoyama et al. 2007; Tavazzi et al. 2008).

N-3 PUFA regulate a wide variety of biological functions, depending on the location of the last double bond, which range from blood pressure and blood clotting to the correct development and functioning of the brain and nervous system. These include prevention of arrhythmias as well as lowering resting heart rate and blood pressure, decreasing platelet aggregation, and lowering triglyceride levels and might also include ameliorating myocardial efficiency and improving vascular function (Reiffel and McDonald 2006; O'Keefe et al. 2006; Harris et al. 2008).

N-3 PUFA affect a myriad of molecular pathways, including alteration of physical and chemical properties of cellular membranes, direct interaction with and modulation of membrane channels and proteins, regulation of gene expression via nuclear receptors and transcription factors, changes in eicosanoid profiles, and conversion of n-3 PUFA to bioactive metabolites. In addition, lipid mediators generated from long-chain PUFA (AA) in the n-6 series and EPA and DHA in the n-3 series have important roles in immune regulation and inflammation (Calder 2009a). Classically, eicosanoids derived from n-6 PUFA (AA) are pro-inflammatory, while eicosanoids derived from n-3 PUFA (EPA and DHA) are anti-inflammatory. Thus, high intake of n-6 PUFA, along with low intakes of n-3 PUFA, shifts the physiological state to one that is pro-inflammatory and prothrombotic.

Latest research has shown that by increasing the ratio of n-3 to n-6 fatty acids in the diet, and consequently favoring the production of EPA in the body, or by increasing the dietary intake of EPA and DHA through consumption of fatty fish or fish-oil supplements, reductions may be achieved in the incidence of many chronic diseases that involve inflammatory processes including CVD (Wall et al. 2010).

Recent years have seen the discovery of a new class of inflammation-dampening and resolution-promoting n-3 PUFA-derived lipid mediators called resolvins and protectins that are generated within resolving exudates during resolution of acute inflammation. These compounds are hydroxylated derivatives of the parent n-3 PUFA, EPA for the E-resolvins, and DHA for the D-resolvins and protectin D1 (Serhan 2007). Resolvin E1 and protectin D1 promote phagocyte removal during acute inflammation, via regulating leukocyte infiltration, increasing macrophage ingestion of apoptotic polymorphonuclear leukocyte *in vivo* and *in vitro*, and enhancing the appearance of phagocytes carrying engulfed zymosan in lymph nodes and spleen.

The role of n-3 PUFA in the prevention of CVD and their biological actions are summarized in Table 5.2.

Table 5.2 Possible mechanisms of favorable effects of n-3 PUFA on cardiovascular health

1. Retard growth of atherosclerotic plaque
Reduce adhesion molecule expression (such as VCAM-1, E-selection, and ICAM-1)
Reduce platelet-derived growth factor
Anti-inflammatory
Suppress production of pro-inflammatory cytokines (such as TNF-alpha and interleukin-1beta and IL-6)
Produce endogenous chemical mediators (such as resolvins and protectins)
Suppress AA cascade
2. Improve vascular function
Promote nitric oxide-induced endothelial relaxation
Increase arterial compliance
3. Mildly hypotensive
4. Lower plasma triglyceride levels
Reduce hepatic synthesis of triglycerides
Increase hepatic fatty acid beta-oxidation
5. Antiarrhythmic
Reduce susceptibility of the heart to arrhythmia
Inhibit fast, voltage-dependent sodium channels
Inhibit L-type calcium channels
Delay rectifier potassium channel
Reduce resting heart rate
6. Prevent heart failure
Increase left ventricular filling capacity
Activate peroxisome proliferator-activator receptor (PPAR)-alpha and PPAR-gamma
7. Antithrombogenic
Decrease platelet aggregation
May enhance fibrinolysis
8. Upregulate adiponectin synthesis
9. Reduce collagen deposition

5.7 N-6 Polyunsaturated Fatty Acids (n-6 PUFA)

Polyunsaturated fatty acids (PUFA) have two or more double bonds. N-6 PUFA is a family of unsaturated fatty acids that have in common a final carbon-carbon double bond in the n-6 position that is the sixth bond from the methyl end. N-6 PUFA is taken in mainly as linoleic acid (LA, 18:2, n-6), the shortest-chained n-6 PUFA, cannot be synthesized by the body, and is therefore an essential fatty acid. The main dietary sources of LA include plant oils such as sunflower, safflower, and corn oils. LA can be desaturated and elongated to form other n-6 PUFA such as γ-linolenic and dihomo-γ-linolenic acids. The latter is converted by the body to the metabolically important n-6 PUFA, arachidonic acid (AA, 20:4). Cyclooxygenases (COX) and lipoxygenases (LOX) can convert AA to the 2-series of prostaglandins (PGE_2, PGI_2, PGD_2, $PGF_{2\alpha}$), the 2-series of thromboxanes (TXA_2, TXB_2), and the 4-series of leukotrienes (LTA_4, LTB_4, LTC_4, LTD_4, LTE_4). These are very important eicosanoids that are involved in various pathological processes mediating pro-inflammatory and prothrombotic conditions such as atherosclerosis (Das 2006). In general, as above, AA-derived eicosanoids are pro-inflammatory, but they have important homeostatic functions in regulating both the promotion and resolution of inflammation in the immune response (Ricciotti and Fitzgerald 2011). Latest studies identified AA-derived eicosanoids to have several beneficial effects including vasodilation, platelet aggregation inhibition, and anti-inflammatory effects (Calder 2009b; Oltman et al. 1998). In contrast, it is known that the n-3 PUFA and their LC derivatives mostly promote anti-inflammatory activities (Tai and Ding 2010). On the other hand, recently, classes of autacoids as lipoxins derived from n-6 PUFA have been identified as specialized mediators that elicit distinct anti-inflammatory and proresolution bioactions and have cytoprotective properties (McMahon and Godson 2004).

Observational studies generally suggest an overall modest benefit of n-6 PUFA intake on CHD risk. In a meta-analysis of 25 case-control studies evaluating blood/tissue n-6 PUFA content and CHD events, LA content was inversely associated with CHD risk, whereas AA was unrelated to CHD risk (Harris et al. 2007).

In the modern Western diet, the n-6:n-3 ratio has increased to be within the range of 10:1–20:1 (Olivier et al. 2011). This increase in the ratio of n-6:n-3 appears to be involved in chronic inflammatory diseases such as CVD, nonalcoholic fatty liver disease, obesity, inflammatory bowel disease, rheumatoid arthritis, and Alzheimer's disease. Diets for appropriate n-6:n-3 ratios should be around 1–4:1.

The role of n-6 PUFA in CVD is much more complex than the role of n-3 PUFA, and the relationship between n-3 and n-6 PUFA is complex, and they are not always in opposition. Pro- and anti-inflammatory effects of n-6 and n-3 PUFA-derived eicosanoids are summarized in Fig. 5.2 (Patterson et al. 2012).

5.8 Trans-Fatty Acids (TFA)

Trans-fatty acids (TFA) are unsaturated fatty acids with at least one carbon-carbon double bond in the trans, rather than the typical cis, configuration. Dietary TFA have two sources: they are formed during the natural bacterial hydrogenation of unsaturated fatty acids in small amounts in the fats of ruminants such as cows, sheep, and goats, or they come from the industrial hydrogenation of unsaturated vegetable oils such as margarines, margarine-based products, shortenings, and fats used for frying in much larger amounts.

Numerous epidemiologic studies and randomized trials have identified higher intake of TFA as an important modifiable risk factor in the development of CVD (Oomen et al. 2001; Lemaitre et al. 2006; Sun et al. 2007). This relation can be explained by several mechanisms. TFA have a wide range of physiologic effect, including both lipid and non-lipid effects. Metabolic studies showed lipid effects of TFA. TFA intakes above the population range of consumption raise LDL

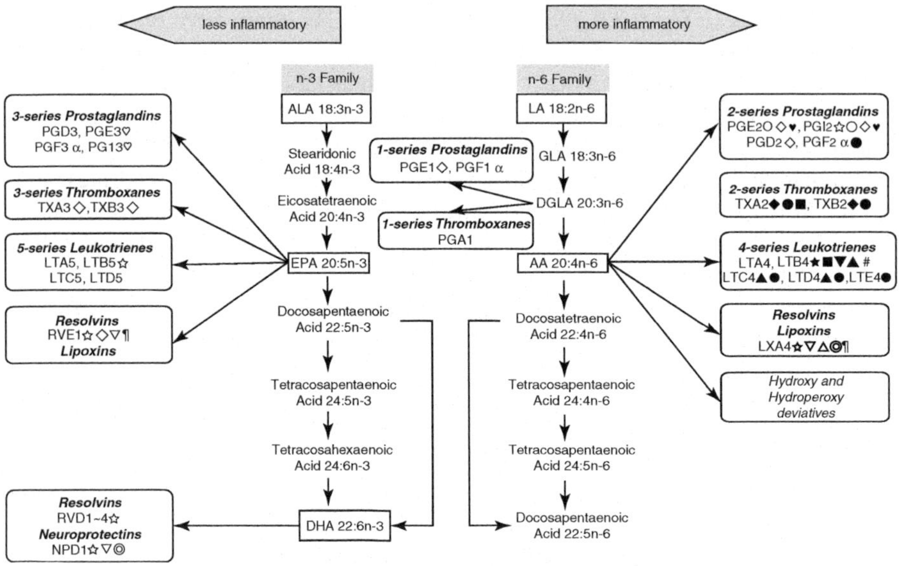

Fig. 5.2 Pro- and anti-inflammatory effects of n-6 and n-3 PUFA-derived eicosanoids. *Abbreviations*: *LA* linoleic acid, *GLA* γ-linolenic acid, *DGLA* dihomo-γ-linolenic acid, *AA* arachidonic acid, *ALA* α-linolenic acid, *EPA* eicosapentaenoic acid, *DHA* docosahexaenoic acid. *Symbols*: ★ pro-inflammatory; ☆ anti-inflammatory; ♥ proarrythmic; ♡ antiarrhythmic; ◆ accelerate platelet aggregation, ◇ inhibit platelet aggregation; ● vasoconstriction; ○ vasodilation; ▲ enhance vascular permeability, △ normalize vascular permeability; ▼ increase oxidative stress; ∇ decrease oxidative stress; # release inflammatory cytokines; ¶ decrease inflammatory cytokines; ■ enhance leucocyte chemotaxis; ⊙ antiapoptotic

cholesterol, lipoprotein (a), and TG levels and reduce HDL cholesterol and LDL particle size (Mozaffarian and Clarke 2009; Mozaffarian et al. 2006). On the other hand, dietary TFA consumption leads to non-lipid risk factors, including systemic inflammation, endothelial dysfunction, and possibly insulin resistance, diabetes, and adiposity.

Recent evidence indicates that TFA intake could also affect biomarkers of inflammation and endothelial dysfunction including C-reactive protein (CRP), interleukin-6 (IL-6), soluble tumor necrosis factor receptor 2 (sTNFR-2), E-selectin, and soluble cell adhesion molecules (sICAM-1 and sVCAM-1) (Mozaffarian et al. 2004; Lopez-Garcia et al. 2005) . In another trial, consumption of TFA impaired endothelial function, as reflected by a reduction in brachial artery flow-mediated vasodilatation (de Roos et al. 2001). RCT in green monkeys indicated that consumption of TFA increased body weight, mainly due to increased deposition of intra-abdominal fat (Kavanagh et al. 2007). In adipocyte, TFA alter the expression of genes for PPAR-γ, resistin, and LPL (Saravanan et al. 2005).

Dietary Guidelines for Americans recommend limiting the amount of TFA to less than 1 % of total daily calories (less than 2 g/day) (Dietary Guidelines Advisory Committee 2005).

5.9 B Vitamins

The water-soluble vitamins of the B class include thiamine (vitamin B1), riboflavin (vitamin B2), niacin (vitamin B3), pantothenic acid (vitamin B5), pyridoxine (vitamin B6), biotin (vitamin B7), folic acid (vitamin B9), and the cobalamins (vitamin B12). B vitamins play a major role in the activities of enzymes, proteins that regulate biochemical reactions in the body, which are important in turning food into energy and other needed substances.

Drug treatment with high-dosed nicotinic acid (vitamin B3, niacin) improves the lipid profile and has been associated with reduction in morbidity and mortality from CVD. These may be due to several beneficial actions of this drug, such as antithrombotic, anti-inflammatory, and antioxidant effects. However, as a side effect, treatment with high-dose nicotinic acid also causes impaired glucose tolerance and elevations in uric acid and homocysteine levels in the blood, which is associated with an increased risk of CVD (Florentin et al. 2011).

The most recognized cardioprotective effects of B vitamins have been estimated with the lowering blood levels of the sulfur-containing amino acid, homocysteine. Several epidemiological studies demonstrated association between dietary intake or concentrations of B vitamins (folic acid, vitamin B6, and vitamin B12), homocysteine, and CVD (Rimm et al. 1998; Ishihara et al. 2008). Elevated homocysteine concentrations have been related to increased oxidant stress, impaired endothelial function, and increased thrombogenicity (Wierzbicki 2007). Each 5 μmol/l of homocysteine has been associated with an approximately 20 % increase in the risk of CHD events, independent of traditional CHD risk factors (Humphrey et al. 2008). Unfortunately, long-term RCTs have not documented benefits of folic acid with or without vitamin B6 and vitamin B12 on CVD outcomes (Bazzano et al. 2006). In some trials, supplemental folic acid was associated with increased risk of CVD (Miller et al. 2010). B vitamins have been shown to play a role in lowering blood levels of homocysteine, but it is not clear whether the lowering levels of homocysteine will reduce risk of heart disease. Further investigation is needed about this discrepancy in effects of B vitamins from epidemiologic studies and clinical trials.

5.10 Vitamin C

Vitamin C (ascorbic acid), a water-soluble vitamin, is widely distributed in fresh fruits and vegetables. Because humans different to rats do not have the ability to biosynthesize their own vitamin C, it must be obtained through their diet. Its deficiency has long been known to cause scurvy that is a lethal condition unless appropriately treated. Vitamin C is available in reduced form (L-ascorbic acid, biologically active form) and oxidized form (L-dehydroascorbic acid).

Results of previous studies on whether vitamin C is helpful for preventing CVD are mixed. But some evidence suggests that it may help protect against atherosclerosis by acting as an antioxidant (Losonczy et al. 1996; Knekt et al. 2004).

Unfortunately, results from most clinical intervention trials have failed to show a beneficial effect of vitamin C supplementation on primary or secondary prevention of CVD (Cook et al. 2007b; Sesso et al. 2008).

Several mechanisms could explain a connection between vitamin C and heart disease. Vitamin C is known to be a powerful antioxidant in the body because, by donating its electrons, it prevents other compounds from being oxidized. As an antioxidant, vitamin C may protect lipids, particularly LDL, from peroxidation, and it keeps LDL ability to bind to LDL receptors (Sakuma et al. 2001). Oxidized LDL can play a key role in the pathogenesis of atherosclerosis, the underlying disorder leading to CVD. Vitamin C *in vitro* also can recycle vitamin E, which can donate electrons to prevent LDL oxidation *in vitro*. As the lipid-phase vitamin E is oxidized, it can be regenerated by aqueous vitamin C. Other possibilities are that vitamin C could improve vasodilatation and vascular reactivity, perhaps by decreasing the interactions of nitric oxide (NO) with oxidants (Vita et al. 1998). Pharmacological doses of vitamin C produce vasodilatation in the coronary arteries (Levine et al. 1996). In healthy subjects, vitamin C administration restored endothelium-dependent vasodilatation that was impaired by acute hyperglycemia (Beckman et al. 2001). Thus, vitamin C may have favorable effects on vascular dilatation, possibly through its antioxidant effects on NO (Jackson et al. 1998). Recent results have shown that endothelial cells recycle and accumulate (May and Qu 2005) and release (Davis et al. 2006) vitamin C by glutathione-dependent mechanisms. These regulation of vitamin C uptake, accumulation, and release by the endothelial cells may maintain the optimal antioxidant capacity of the plasma and the neighboring extracellular space to increase local NO bioavailability and prevent formation of the cytotoxic peroxynitrite ($ONOO^-$).

5.11 Vitamin D

Vitamin D is a fat-soluble vitamin that plays a hormonelike action. Two major forms of vitamin D are vitamin D2 (ergocalciferol) and D3 (cholecalciferol). 1,25-dihydroxyvitamin D [1,25(OH)2D] is the most active vitamin D metabolite, and its receptor (VDR) generates biological responses in the cardiovascular system. Expression of the VDR and vitamin D-metabolizing enzymes in the cardiovascular system including vascular smooth muscle cells, endothelial cells, and cardiomyocytes suggests an important role for vitamin D in the cardiovascular system.

Low-level vitamin D status is a common problem in the general population. The Third National Health and Nutrition Examination Survey (NHANES III) reported the prevalence of vitamin D deficiency in US adults to be 25–57 % (Looker et al. 2002). Risk factors for vitamin D deficiency include advanced age, dark skin color, homebound status, increased distance from the equator, winter season, clothing, the use of sunscreen, air pollution, smoking, obesity, malabsorption, renal disease, liver disease, and different medications (Lee et al. 2008a). Among CVD patients, vitamin D deficiency is a common problem. Vitamin D suppresses the progression of heart failure, hypertension, and myocardial hypertrophy and decreases prevalence of cardiovascular risk factors and related morbidity and mortality.

Several prospective studies indicated that individuals with low level of serum 25-hydroxyvitamin D were at increased risk for CVD (Wang et al. 2008; Giovannucci et al. 2008). Recent meta-analysis of epidemiological studies summarized that high vitamin D concentrations were associated with lower prevalence of CVD (Parker et al. 2010). Although the observational studies were suggestive of increased CVD risks associated with low levels of serum 25-hydroxyvitamin D, recent systematic reviews show that the evidence was inconsistent and insufficient to prove a cause and effect relationship between vitamin D supplementation and prevention of cardiovascular events (Shapses and Manson 2011; Wang et al. 2010).

Vitamin D has a wide variety of the cardioprotective mechanisms including the inhibition of vascular smooth muscle proliferation, the suppression of vascular calcification, the downregulation of pro-inflammatory cytokines, the upregulation of anti-inflammatory cytokines, the suppressive effect on platelet aggregability, the improvement of insulin secretion and sensitivity, and the decreases of renin, angiotensin levels, atrial natriuretic peptide, brain natriuretic peptide, plasminogen activator inhibitor-1, and vascular endothelial growth factor (Zittermann et al. 2005; Darabian et al. 2012; Rammos et al. 2008; Deluca et al. 2001; Gysemans et al. 2005).

5.12 Vitamin E

Vitamin E, a fat-soluble vitamin, is an essential micronutrient. Of the eight naturally occurring forms of vitamin E, only alpha-tocopherol, and to a minor extent gamma-tocopherol, is carried in human blood and is considered to be the active form. Numerous foods provide vitamin E. Nuts, seeds, and vegetable oils are among the best sources of alpha-tocopherol, and significant amounts are available in green leafy vegetables and fortified cereals.

Several prospective cohort studies have associated lower rates of CHD with higher vitamin E consumption or supplementation (Rimm et al. 1993; Stampfer et al. 1993; Knekt et al. 1994). On the other hand, RCTs have failed to show reductions in CVD events with supplemental vitamin E, and two meta-analyses suggest that high-dose vitamin E supplements may increase total mortality (Bjelakovic et al. 2007; Miller et al. 2005). More evidence is needed to determine if there are benefits of vitamin E supplementation.

The cardioprotective effects of vitamin E are attributed to its antioxidant properties that stop the production of reactive oxygen species (ROS) formed when fat undergoes oxidation. Of the antioxidant vitamins, vitamin E may be the most potent inhibitor of lipid oxidation because it is fat-soluble and constitutes part of the LDL molecule (Pryor 2000). It is thought to exert its primary protective effects via the protection of LDL from oxidation. This effect has been demonstrated in laboratory animals *in vivo* (Keaney and Frei 1994), isolated tissues *ex vivo*, and in human populations (Pryor 2000).

Vitamin E may affect the pathogenesis of atherosclerotic vascular disease beyond its direct effects on lipids. Vitamin E could affect CVD morbidity and mortality by reducing platelet adhesion, inhibiting vitamin-K-dependent clotting factors, or stimulating nitric oxide formation by the endothelial cell (Pryor 2000).

5.13 Carotenoids

Carotenoids are a class of more than 600 compounds that are responsible for the red, yellow, and orange fat-soluble pigments found principally in plants. Fruits and vegetables are the main source of carotenoids in the human diet. The most common carotenoids in the human diet are α-carotene, β-carotene, β-cryptoxanthin, lycopene, lutein, zeaxanthin, and astaxanthin.

A number of retrospective and prospective longitudinal studies have identified an inverse association between carotenoid intake or serum concentrations of total or individual carotenoids and CVD risk (Osganian et al. 2003; Sesso et al. 2005; Evans et al. 1998; Buijsse et al. 2008), although the results of these studies are somewhat conflicting.

Known primarily as precursors to vitamin A, carotenoids are also important free radical quenchers and act as potent antioxidants. Two major hypotheses have been proposed to explain the cardioprotective activities of carotenoids, antioxidative and non-antioxidative mechanisms. The facts that LDL is a major transporter of β-carotene and lycopene in the circulation and that these carotenoids have the capacity to trap peroxyl radicals and quench singlet oxygen support the hypothesis that the carotenoids may have a protective role to oxidative stress. Recent studies are also showing that carotenoids may mediate their effects via other mechanisms such as gap junction communication, cell growth regulation, modulating gene expression, immune response, and the action as a hypocholesterolemic agent by inhibiting 3-hydroxy-methylglutaryl-coenzyme A (HMG-CoA) reductase (Paiva and Russell 1999; Bertram 1999; Fuhramn et al. 1997). Unfortunately, RCTs of β-carotene supplements document no benefit in the general population, and in fact, they may even cause an increase in heart disease and total mortality (Bjelakovic et al. 2007).

At the moment, no evidence exists to recommend carotenoids in pharmaceutical form for the treatment or prevention of CHD. Long-term and well-designed future studies to evaluate the role of carotenoids in prevention of CHD are needed.

5.14 Carbohydrates

Carbohydrates (CHO) are an ideal source of energy for the body and are naturally occurring compounds that consist of carbon, hydrogen, and oxygen. CHO encompass a broad range of sugars, starches, and fiber (see Sect. 5.15). CHO that contain only one sugar unit (monosaccharides) or two sugar units (disaccharides) are referred to as simple sugars. Two of the most common monosaccharides are glucose and fructose. Starches are complex CHO without taste or odor. Complex CHO are long chains of simple sugar units bonded together.

It is clear that several aspects of CHO quality influence cardiometabolic health. The glycemic index (GI) is a kinetic parameter that reflects the ability of CHO contained in consumed foods to raise blood glucose *in vivo* (Jenkins et al. 1981). GI is defined as the incremental area under the blood glucose response curve (AUC area under the blood concentration time curve) within a 2-h period elicited by a portion of food containing 50 g of available CHO, relative to the AUC elicited by 50 g glucose. A related metric, glycemic load (GL), is defined as GI × w/100, where w is the grams of available CHO contained in the amount of food consumed (Salmerón et al. 1997).

A recent meta-analysis showed borderline significant associations of increased GI with coronary heart disease (relative risk, 1.25; 95 % CI, 1.00, 1.56) (Barclay et al. 2008). RCTs have shown that low GI/GL diets affect plasma concentrations of LDL cholesterol, HDL cholesterol, total cholesterol, triglycerides, and markers of inflammation and thrombosis, as well as insulin resistance, in ways that would be expected to reduce CVD risk (Rizkalla et al. 2004; McMillan-Price et al. 2006; Maki et al. 2007; Pereira et al. 2004a; Ebbeling et al. 2005; Dickinson et al. 2008). Exaggerated postprandial glycemic spikes generate excess free radicals that can trigger a biochemical cascade resulting in inflammation, endothelial dysfunction, hypercoagulability, and sympathetic hyperactivity (O'Keefe et al. 2008; Weissman et al. 2006). These postprandial changes when repeated multiple times daily eventually lead to atherosclerotic risk factors and CVD. Even hyperglycemic spikes induced artificially using intravenous glucose infusions in lean nondiabetic individuals have been shown to markedly increase free radical generation (Brownlee and Hirsch 2006). At only 80 mg/dl plasma glucose, the cardiovascular risk of postprandial glycemia begins to increase; by 140 mg/dl, the point at which we traditionally only begin to classify patients as glucose intolerant, the risk is already increased by 58 % (Sasso et al. 2004; Mellen et al. 2006).

Additionally, transient hyperglycemia induces long-lasting activating epigenetic changes in the promoter of the nuclear factor κB (NFκB) subunit p65 in aortic endothelial cells both *in vitro* and in nondiabetic mice. NFκB-induced increases in monocyte chemoattractant protein 1 and vascular cell adhesion molecule 1 expression accelerate progression of atherosclerosis (El-Osta et al. 2008).

5.15 Dietary Fiber

Dietary fibers (DF) are highly complex substances described as nondigestible carbohydrates and lignins resistant to digestion by human digestive enzyme and absorption in the small intestine with fermentation in the large intestine. DF is abundant in wholegrain foods, vegetables, fruits, legumes, and nuts. DF have been classified according to solubility in water into two groups: water-soluble fiber (pectin, beta-glucans, inulin, fructans, oligosaccharides, some hemicelluloses, guar, and gums, which were previously classified as viscous fiber) and water-insoluble fiber (hemicellulose, cellulose,

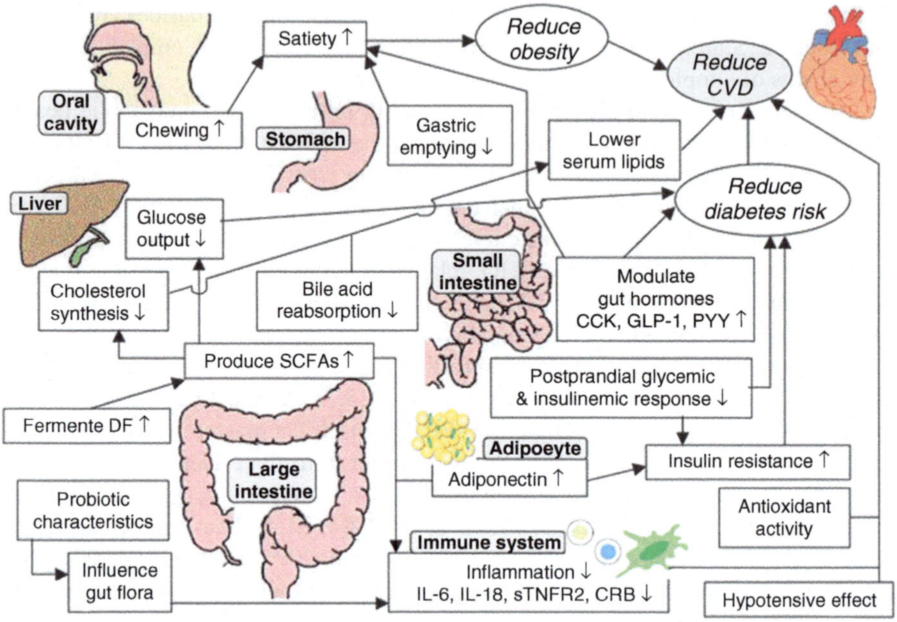

Fig. 5.3 Potential mechanisms of DF in prevention of CVD. *Abbreviations*: *DF* dietary fiber, *CVD* cardiovascular disease, *SCFAs* short-chain fatty acids, *CCK* cholecystokinin, *GLP-1* glucagon-like peptide-1, *PYY* peptide YY, *IL-6* interleukin-6, *IL-18* interleukin-18, *sTNFR2* soluble tumor necrosis factor receptor 2, *CRP* C-reactive protein

resistant starch, and lignin, which were previously classified as nonviscous fiber). Food sources rich in soluble DF include legumes, vegetables, fruits, oat bran, and psyllium seeds. The main source of insoluble DF is whole grain.

Experimental data from both animals and humans, epidemiological studies and feeding trials with supplements support the inverse relationship between dietary fiber and the reduction in developing CVD and major cardiovascular risk factors, such as hypertension, diabetes, obesity, and dyslipidemia (Liu et al. 1999; Liu 2002; Mozaffarian et al. 2003; Rimm et al. 1996; Todd et al. 1999).

A meta-analysis of seven prospective cohort studies including >150,000 persons suggested a 29 % reduction in the risk of CHD on comparing individuals in the highest and lowest quintiles of intake of dietary fiber (Anderson 2003). Recent studies observed that CHD mortality was decreased for every additional 10 g of DF per day by 17–35 % (Pereira et al. 2004b; Streppel et al. 2008).

Although the cardioprotective physiological mechanisms of DF on metabolic health are not fully established, it is speculated to be a result of changes in chewing, plasma lipid profiles, gastric emptying, nutrient absorption, satiety, postprandial glycemic response, insulin sensitivity, blood pressure, production of gut hormones and short-chain fatty acids (SCFAs), gut flora, inflammation, and body weight (Lattimer and Haub 2010). Proposed mechanisms in cardioprotective effects of DF are summarized in Fig. 5.3.

5.16 Summary and Perspectives

Cardiovascular disease can be influenced by lifestyle changes, especially including dietary styles. RCTs of cardiometabolic risk factors and prospective cohort studies of CVD end points provide concordant evidence for cardiovascular effects of several nutrients. Mechanistic studies help to determine the possible molecular and cellular mechanisms and pathways involved in biological functions of nutrients. The major cardioprotective mechanisms of nutrients are classified as follows: (1) activating sirtuins, (2) ameliorating serum lipid profiles, (3) decreasing insulin resistances, (4) anti-inflammatory effects, (5) antioxidants, (6) antithrombogenic actions, (7) improving vascular functions, (8) antiarrhythmic actions, and (9) lowering homocysteine levels. Often, individual nutrients for which epidemiologic studies suggest benefit and mechanistic studies show effects fail to do so in large RCTs. Because many mechanistic studies are done by animal models or *in vitro*, this does not mean that nutrients will necessarily reduce CVD in people. On the other hand, we cannot forget that the reactivity of a cardiovascular system to food is racially or genetically different. For these reasons, further work is needed to better understand the molecular effects of natural dietary compounds in cardiovascular system.

References

Albert CM, Hennekens CH, O'Donnell CJ, Ajani UA, Carey VJ, Willett WC, Ruskin JN, Manson JE (1998) Fish consumption and risk of sudden cardiac death. JAMA 279:23–28

Alderman MH, Cohen HW (2012) Dietary sodium intake and cardiovascular mortality: controversy resolved? Am J Hypertens 25:727–734

Anderson JW (2003) Whole grains protect against atherosclerotic cardiovascular disease. Proc Nutr Soc 62:135–142

Barclay AW, Petocz P, McMillan-Price J, Flood VM, Prvan T, Mitchell P, Brand-Miller JC (2008) Glycemic index, glycemic load, and chronic disease risk – a meta-analysis of observational studies. Am J Clin Nutr 87:627–637

Bazzano LA, Reynolds K, Holder KN, He J (2006) Effect of folic acid supplementation on risk of cardiovascular diseases: a meta-analysis of randomized controlled trials. JAMA 296:2720–2726

Beckman JA, Goldfine AB, Gordon MB, Creager MA (2001) Ascorbate restores endothelium-dependent vasodilation impaired by acute hyperglycemia in humans. Circulation 103:1618–1623

Bertram JS (1999) Carotenoids and gene regulation. Nutr Rev 57:182–191

Bjelakovic G, Nikolova D, Gluud LL, Simonetti RG, Gluud C (2007) Mortality in randomized trials of antioxidant supplements for primary and secondary prevention: systematic review and meta-analysis. JAMA 297:842–857

Bordone L, Guarente L (2005) Calorie restriction, SIRT1 and metabolism: understanding longevity. Nat Rev Mol Cell Biol 6:298–305

Borradaile NM, Pickering JG (2009) NAD+, sirtuins, and cardiovascular disease. Curr Pharm Des 15:110–117

Brownlee M, Hirsch I (2006) Glycemic variability: a hemoglobin A1c-independent risk factor for diabetic complications. JAMA 295:1707–1708

Buijsse B, Feskens EJM, Kwape L, Kok FJ, Kromhout D (2008) Both α-and β-carotene, but Not tocopherols and vitamin C, are inversely related to 15-year cardiovascular mortality in Dutch elderly men. J Nutr 138:344–350

Calder PC (2009a) Polyunsaturated fatty acids and inflammatory processes: new twists in an old tale. Biochimie 91:791–795

Calder PC (2009b) Prostaglandins and inflammation. Arterioscler Biochim 91:791–795

Civitarese AE, Carling S, Heilbronn LK, Hulver MH, Ukropcova B, Deutsch WA, Smith SR, Ravussin E, CALERIE Pennington Team (2007) Calorie restriction increases muscle mitochondrial biogenesis in healthy humans. PLoS Med 3:e76

Cohen HW, Hailpern SM, Fang J, Alderman MH (2006) Sodium intake and mortality in the NHANES II follow-up study. Am J Med 119:275.e7–275.e14

Cohen HY, Miller C, Bitterman KJ, Wall NR, Hekking B, Kessler B, Howitz KT, Gorospe M, de Cabo R, Sinclair DA (2004) Calorie restriction promotes mammalian cell survival by inducing the SIRT1 deacetylase. Science 305:390–392

Cook NR, Cutler JA, Obarzanek E, Buring JE, Rexrode KM, Kumanyika SK, Appel LJ, Whelton PK (2007a) Long term effects of dietary sodium reduction on cardiovascular disease outcomes: observational follow-up of the trials of hypertension prevention (TOHP). BMJ 334:885–888

Cook NR, Albert CM, Gaziano JM, Zaharris E, MacFadyen J, Danielson E, Buring JE, Manson JE (2007b) A randomized factorial trial of vitamins C and E and beta carotene in the secondary prevention of cardiovascular events in women: results from the Women's Antioxidant Cardiovascular Study. Arch Intern Med 167:1610–1618

De Roos NM, Bots ML, Katan MB (2001) Replacement of dietary saturated fatty acids by trans fatty acids lowers serum HDL cholesterol and impairs endothelial function in healthy men and women. Arterioscler Thromb Vasc Biol 21:1233–1237

Darabian S, Rattanasompattikul M, Hatamizadeh P, Bunnapradist S, Budoff MJ, Kovesdy CP, Kalantar-Zadeh K (2012) Cardiorenal syndrome and vitamin D receptor activation in chronic kidney disease. Kidney Res Clin Pract 31(1):12–15

Das UN (2006) Essential fatty acids: biochemistry, physiology and pathology. Biotechnol J 1:420–439

Davis KA, Samson SE, Best K, Mallhi KK, Szewczyk M, Wilson JX, Kwan CY, Grover AK (2006) Ca2+-mediated ascorbate release from coronary artery endothelial cells. Br J Pharmacol 147:131–139

Deluca HF, Cantorna MT (2001) Vitamin D: its role and uses in immunology. FASEB J 15:2579–2585

Dickinson S, Hancock DP, Petocz P, Ceriello A, Brand-Miller J (2008) High-glycemic index carbohydrate increases nuclear factor-kappaB activation in mononuclear cells of young, lean healthy subjects. Am J Clin Nutr 87:1188–1193

Dyerberg J, Bang HO, Stoffersen E, Moncada S, Vane JR (1978) Eicosapentaenoic acid and prevention of thrombosis and atherosclerosis? Lancet 2:117–119

Ebbeling CB, Leidig MM, Sinclair KB, Seger-Shippee LG, Feldman HA, Ludwig DS (2005) Effects of an ad libitum low-glycemic load diet on cardiovascular disease risk factors in obese young adults. Am J Clin Nutr 81:976–982

Elliott P, Stamler J, Nichols R, Dyer AR, Stamler R, Kesteloot H, Marmot M (1996) Intersalt revisited: further analyses of 24 hour sodium excretion and blood pressure within and across populations. Intersalt Cooperative Research Group. BMJ 312:1249–1253

El-Osta A, Brasacchio D, Yao D, Pocai A, Jones PL, Roeder RG, Cooper ME, Brownlee M (2008) Transient high glucose causes persistent epigenetic changes and altered gene expression during subsequent normoglycemia. J Exp Med 205:2409–2417

Evans RW, Shaten BJ, Day BW, Kuller LH (1998) Prospective association between lipid soluble antioxidants and coronary heart disease in men: the Multiple Risk Factor Intervention Trial. Am J Epidemiol 147:180–186

Florentin M, Liberopoulos EN, Kei A, Mikhailidis DP, Elisaf MS (2011) Pleiotropic effects of nicotinic acid: beyond high density lipoprotein cholesterol elevation. Curr Vasc Pharmacol 9:385–400

Fontana L, Meyer TE, Klein S, Holloszy JO (2004) Long-term calorie restriction is highly effective in reducing the risk for atherosclerosis in humans. Proc Natl Acad Sci USA 101:6659–6663

Fuhramn B, Elis A, Aviram M (1997) Hypocholesterolemic effect of lycopene and β-carotene is related to suppression of cholesterol synthesis and augmentation of LDL receptor activity in macrophage. Biochem Biophys Res Commun 233:658–662

Geleijnse JM, Witteman JC, Stijnen T, Kloos MW, Hofman A, Grobbee DE (2007) Sodium and potassium intake and risk of cardiovascular events and all-cause mortality. Eur J Epidemiol 22:763–770

Giovannucci E, Liu Y, Hollis BW, Rimm EB (2008) 25-hydroxyvitamin D and risk of myocardial infarction in men: a prospective study. Arch Intern Med 168:1174–1180

Gordon T, Kannel WB (1982) Multiple risk functions for predicting coronary heart disease: the concept, accuracy, and application. Am Heart J 103:1031–1039

Graudal NA, Hubeck-Graudal T, Juergens G (2012) Effects of low-sodium diet vs. high-sodium diet on blood pressure, renin, aldosterone, catecholamines, cholesterol, and triglyceride (Cochrane review). Am J Hypertens 25:1–15

Grassi G, Dell'Oro R, Seravalle G, Foglia G, Trevano FQ, Mancia G (2002) Short- and long-term neuroadrenergic effects of moderate dietary sodium restriction in essential hypertension. Circulation 106:1957–1961

Gysemans CA, Cardozo AK, Callewaert H, Giulietti A, Hulshagen L, Bouillon R, Eizirik DL, Mathieu C (2005) 1,25-Dihydroxyvitamin D 3 modulates expression of chemokines and cytokines in pancreatic islets: implications for prevention of diabetes in nonobese diabetic mice. Endocrinology 146:1956–1964

Harris WS, Miller M, Tighe AP, Davidson MH, Schaefer EJ (2008) Omega-3 fatty acids and coronary heart disease risk: clinical and mechanistic perspectives. Atherosclerosis 197:12–24

Harris WS, Poston WC, Haddock CK (2007) Tissue n-3 and n-6 fatty acids and risk for coronary heart disease events. Atherosclerosis 193:1–10

Humphrey LL, Fu R, Rogers K, Freeman M, Helfand M (2008) Homocysteine level and coronary heart disease incidence: a systematic review and meta-analysis. Mayo Clin Proc 83:1203–1212

Ishihara J, Iso H, Inoue M, Iwasaki M, Okada K, Kita Y, Kokubo Y, Okayama A, Tsugane S, JPHC Study Group (2008) Intake of folate, vitamin B6 and vitamin B12 and the risk of CHD: the Japan Public Health Center-Based Prospective Study Cohort I. J Am Coll Nutr 27:127–136

Jackson TS, Xu A, Vita JA, Keaney JF Jr (1998) Ascorbate prevents the interaction of superoxide and nitric oxide only at very high physiological concentrations. Circ Res 83:916–922

Jenkins DJ, Wolever TM, Taylor RH, Barker H, Fielden H, Baldwin JM, Bowling AC, Newman HC, Jenkins AL, Goff DV (1981) Glycemic index of foods: a physiological basis for carbohydrate exchange. Am J Clin Nutr 34:362–366

Jialal I, Devaraj S, Huet BA, Traber M (1999) Dietary supplementation with n-3 polyunsaturated fatty acids and vitamin E after myocardial infarction: results of the GISSI-Prevenzione trial. Gruppo Italiano per lo Studio della Sopravvivenza nell'Infarto miocardico. Lancet 354:447–455

Kanbay M, Chen Y, Solak Y, Sanders PW (2011) Mechanisms and consequences of salt sensitivity and dietary salt intake. Curr Opin Nephrol Hypertens 20:37–43

Kavanagh K, Jones KL, Sawyer J, Kelley K, Carr JJ, Wagner JD, Rudel LL (2007) Trans fat diet induces abdominal obesity and changes in insulin sensitivity in monkeys. Obesity (Silver Spring) 15:1675–1684

Keaney JF, Frei B (1994) Antioxidant protection of low density lipoprotein and its role in the prevention of atherosclerotic vascular disease. In: Frei B (ed) Natural antioxidants in human health and disease. Academic, San Diego, pp 303–351

Kemi M, Keenan KP, McCoy C, Hoe CM, Soper KA, Ballam GC, van Zwieten MJ (2000) The relative protective effects of moderate dietary restriction versus dietary modification on spontaneous cardiomyopathy in male Sprague-Dawley rats. Toxicol Pathol 28:285–296

Keys AB, Aravanis C (1980) Seven countries: a multivariate analysis of death and coronary heart disease. Harvard University Press, Cambridge

Knekt P, Reunanen A, Jarvinen R, Seppanen R, Heliovaara M, Aromaa A (1994) Antioxidant vitamin intake and coronary mortality in a longitudinal population study. Am J Epidemiol 139:1180–1189

Knekt P, Ritz J, Pereira MA, O'Reilly EJ, Augustsson K, Fraser GE, Goldbourt U, Heitmann BL, Hallmans G, Liu S, Pietinen P, Spiegelman D, Stevens J, Virtamo J, Willett WC, Rimm EB, Ascherio A (2004) Antioxidant vitamins and coronary heart disease risk: a pooled analysis of 9 cohorts. Am J Clin Nutr 80:1508–1520

Knoops KT, de Groot LC, Kromhout D, Perrin AE, Moreiras-Varela O, Menotti A, van Staveren WA (2004) Mediterranean diet, lifestyle factors, and 10-year mortality in elderly European men and women: the HALE project. JAMA 292:1433–1439

Kris-Etherton PM (1999) Monounsaturated fatty acids and risk of cardiovascular disease. Circulation 100:1253–1258

Kromhout D, Bosschieter EB, de Lezenne CC (1985) The inverse relation between fish consumption and 20-year mortality from coronary heart disease. N Engl J Med 312:1205–1209

Lai CQ, Parnell LD, Arnett DK, García-Bailo B, Tsai MY, Kabagambe EK, Straka RJ, Province MA, An P, Borecki IB, Tucker KL, Ordovás JM (2009) WDTC1, the ortholog of Drosophila adipose gene, associates with human obesity, modulated by MUFA intake. Obesity 17:593–600

Lattimer JM, Haub MD (2010) Effects of dietary fiber and its components on metabolic health. Nutrients 2:1266–1289

Lee JH, O'Keefe JH, Bell D, Hensrud DD, Holick MF (2008a) Vitamin D deficiency an important, common, and easily treatable cardiovascular risk factor? J Am Coll Cardiol 52:1949–1956

Lee JH, O'Keefe JH, Lavie CJ, Marchioli R, Harris WS (2008b) Omega-3 fatty acids for cardioprotection. Mayo Clin Proc 83:324–332

Lemaitre RN, King IB, Mozaffarian D, Sotoodehnia N, Rea TD, Kuller LH, Tracy RP, Siscovick DS (2006) Plasma phospholipid trans fatty acids, fatal ischemic heart disease, and sudden cardiac death in older adults: the cardiovascular health study. Circulation 114:209–215

Levine GN, Frei B, Koulouris SN, Gerhard MD, Keaney JF Jr, Vita JA (1996) Ascorbic acid reverses endothelial vasomotor dysfunction in patients with coronary artery disease. Circulation 93:1107–1113

Lin SJ, Ford E, Haigis M, Liszt G, Guarente L (2004) Calorie restriction extends yeast life span by lowering the level of NADH. Genes Dev 18:12–16

Liu S (2002) Intake of refined carbohydrates and whole grain foods in relation to risk of type 2 diabetes mellitus and coronary heart disease. J Am Coll Nutr 21:298–306

Liu S, Stampfer MJ, Hu FB, Giovannucci E, Rimm E, Manson JE, Hennekens CH, Willett WC (1999) Whole-grain consumption and risk of coronary heart disease: results from the Nurses' Health study. Am J Clin Nutr 70:412–419

Looker AC, Dawson-Hughes B, Calvo MS, Gunter EW, Sahyoun NR (2002) Serum 25-hydroxyvitamin D status of adolescents and adults in two seasonal subpopulations from NHANES III. Bone 30:771–777

Lopez-Garcia E, Schulze MB, Meigs JB, Manson JE, Rifai N, Stampfer MJ, Willett WC, Hu FB (2005) Consumption of trans fatty acids is related to plasma biomarkers of inflammation and endothelial dysfunction. J Nutr 135:562–566

Lopez-Segura F, Velasco F, Lopez-Miranda J, Castro P, Lopez-Pedrera R, Blanco A, Jimenez-Pereperez J, Torres A, Trujillo J, Ordovas JM, Perez-Jimenez F (1996) Monounsaturated fatty acid-enriched diet decreases plasma plasminogen activator inhibitor type 1. Arterioscler Thromb Vasc Biol 16:82–88

Losonczy KG, Harris TB, Havlik RJ (1996) Vitamin E and vitamin C supplement use and risk of all-cause and coronary heart disease mortality in older persons: the Established Populations for Epidemiologic Studies of the Elderly. Am J Clin Nutr 64:190–196

MaCay CM, Crowell MF, Maynard LA (1935) The effect of retarded growth upon the length of life span and upon the ultimate body size. J Nutr 10:63–79

Maki KC, Rains TM, Kaden VN, Raneri KR, Davidson MH (2007) Effects of a reduced-glycemic-load diet on body weight, body composition, and cardiovascular disease risk markers in overweight and obese adults. Am J Clin Nutr 85:724–734

Martinez-Gonzalez MA, Fernandez-Jarne E, Serrano-Martinez M, Marti A, Martinez JA, Martin-Moreno JM (2002) Mediterranean diet and reduction in the risk of a first acute myocardial infarction: an operational healthy dietary score. Eur J Nutr 41:153–160

Marzetti E, Wohlgemuth SE, Anton SD, Bernabei R, Carter CS, Leeuwenburgh C (2009) Cellular mechanisms of cardioprotection by calorie restriction: state of the science and future perspectives. Clin Geriatr Med 25:715–732

May JM, Qu ZC (2005) Transport and intracellular accumulation of vitamin C in endothelial cells: relevance to collagen synthesis. Arch Biochem Biophys 434:178–186

McDonald BE, Gerrard JM, Bruce VM, Corner EJ (1989) Comparison of the effect of canola oil and sunflower oil on plasma lipids and lipoproteins and on in vivo thromboxane A2 and prostacyclin production in healthy young men. Am J Clin Nutr 50:1382–1388

McMahon B, Godson C (2004) Lipoxins: endogenous regulators of inflammation. Am J Physiol Renal Physiol 286:F189–F201

McMillan-Price J, Petocz P, Atkinson F, O'neill K, Samman S, Steinbeck K, Caterson I, Brand-Miller J (2006) Comparison of 4 diets of varying glycemic load on weight loss and cardiovascular risk reduction in overweight and obese young adults: a randomized controlled trial. Arch Intern Med 166:1466–1475

Mellen P, Cefalu W, Herrington D (2006) Diabetes, the metabolic syndrome, and angiographic progression of coronary artery disease in postmenopausal women. Arterioscler Thromb Vasc Biol 26:189–193

Meyer TE, Kovács SJ, Ehsani AA, Klein S, Holloszy JO, Fontana L (2006) Long-term caloric restriction ameliorates the decline in diastolic function in humans. J Am Coll Cardiol 47:398–402

Miller ER 3rd, Pastor-Barriuso R, Dalal D, Riemersma RA, Appel LJ, Guallar E (2005) Meta-analysis: high-dosage vitamin E supplementation may increase all-cause mortality. Ann Intern Med 142:37–46

Miller ER 3rd, Juraschek S, Pastor-Barriuso R, Bazzano LA, Appel LJ, Guallar E (2010) Meta-analysis of folic acid supplementation trials on risk of cardiovascular disease and risk interaction with baseline homocysteine levels. Am J Cardiol 106:517–527

Mozaffarian D, Clarke R (2009) Quantitative effects on cardiovascular risk factors and coronary heart disease risk of replacing partially hydrogenated vegetable oils with other fats and oils. Eur J Clin Nutr 63(Suppl 2):S22–S33

Mozaffarian D, Kumanyika SK, Lemaitre RN, Olson JL, Burke GL, Siscovick DS (2003) Cereal, fruit, and vegetable fiber intake and the risk of cardiovascular disease in elderly individuals. JAMA 289:1659–1666

Mozaffarian D, Pischon T, Hankinson SE, Rifai N, Joshipura K, Willett WC, Rimm EB (2004) Dietary intake of trans fatty acids and systemic inflammation in women. Am J Clin Nutr 79:606–612

Mozaffarian D, Katan MB, Ascherio A, Stampfer MJ, Willett WC (2006) Trans fatty acids and cardiovascular disease. N Engl J Med 354:1601–1613

Nisoli E, Tonello C, Cardile A, Cozzi V, Bracale R, Tedesco L, Falcone S, Valerio A, Cantoni O, Clementi E, Moncada S, Carruba MO (2005) Calorie restriction promotes mitochondrial biogenesis by inducing the expression of eNOS. Science 310:314–317

O'Donnell MJ, Yusuf S, Mente A, Gao P, Mann JF, Teo K, McQueen M, Sleight P, Sharma AM, Dans A, Probstfield J, Schmieder RE (2011) Urinary sodium and potassium excretion and risk of cardiovascular events. JAMA 306:2229–2238

O'Keefe JH Jr, Abuissa H, Sastre A, Steinhaus DM, Harris WS (2006) Effects of omega-3 fatty acids on resting heart rate, heart rate recovery after exercise, and heart rate viability in men with healed myocardial infarctions and depressed ejection fractions. Am J Cardiol 97:1127–1130

O'Keefe JH, Gheewala NM, O'Keefe JO (2008) Dietary strategies for improving post-prandial glucose, lipids, inflammation, and cardiovascular health FREE. J Am Coll Cardiol 51:249–255

Olivier MC, Vanessa L, Isabelle A (2011) Why and how meet n-3 PUFA dietary recommendations? Gastroenterol Res Pract 2011:1–11

Oltman CL, Weintraub NL, VanRollins M, Dellsperger KC (1998) Epoxyeicosatrienoic acids and dihydroxyeicosatrienoic acids are potent vasodilators in the canine coronary microcirculation. Circ Res 83:932–939

Oomen CM, Ocké MC, Feskens EJ, van Erp-Baart MA, Kok FJ, Kromhout D (2001) Association between trans fatty acid intake and 10-year risk of coronary heart disease in the Zutphen Elderly Study: a prospective population-based study. Lancet 357:746–751

Osganian SK, Stampfer MJ, Rimm E, Spiegelman D, Manson JAE, Willett WC (2003) Dietary carotenoids and risk of coronary artery disease in women. Am J Clin Nutr 77:1390–1399

Paiva S, Russell R (1999) β-carotene and other carotenoids as antioxidants. J Am Coll Nutr 18:426–433

Parker J, Hashmi O, Dutton D, Mavrodaris A, Stranges S, Kandala NB, Clarke A, Franco OH (2010) Levels of vitamin D and cardiometabolic disorders: systematic review and meta-analysis. Maturitas 65:225–236

Patterson E, Wall R, Fitzgerald GF, Ross RP, Stanton C (2012) Health implications of high dietary omega-6 polyunsaturated fatty acids. J Nutr Metab 2012:1–16

Pereira MA, Swain J, Goldfine AB, Rifai N, Ludwig DS (2004a) Effects of a low-glycemic load diet on resting energy expenditure and heart disease risk factors during weight loss. JAMA 292:2482–2490

Pereira MA, O'Reilly E, Augustsson K, Fraser GE, Goldbourt U, Heitmann BL, Hallmans G, Knekt P, Liu SM, Pietinen P, Spiegelman D, Stevens J, Virtamo J, Willett WC, Ascherio A (2004b) Dietary fiber and risk of coronary heart disease – a pooled analysis of cohort studies. Arch Intern Med 164:370–376

Petrie JR, Morris AD, Minamisawa K, Hilditch TE, Elliott HL, Small M, McConnell J (1998) Dietary sodium restriction impairs insulin sensitivity in noninsulin-dependent diabetes mellitus. J Clin Endocrinol Metab 83:1552–1557

Pryor W (2000) Vitamin E and heart disease: basic science to clinical intervention trials. Free Radic Biol Med 28:141–164

Rammos G, Tseke P, Ziakka S (2008) Vitamin D, the renin-angiotensin system, and insulin resistance. Int Urol Nephrol 40:419–426

Reiffel JA, McDonald A (2006) Antiarrhythmic effects of omega-3 fatty acids. Am J Cardiol 98:50i–60i

Ricciotti E, Fitzgerald GA (2011) Prostaglandins and inflammation. Arterioscler Thromb Vasc Biol 31:986–1000

Rimm EB, Stampfer MJ, Ascherio A, Giovannucci E, Colditz GA, Willett WC (1993) Vitamin E consumption and the risk of coronary heart disease in men. N Engl J Med 328:1450–1456

Rimm EB, Ascherio A, Giovannucci E, Spiegelman D, Stampfer MJ, Willett WC (1996) Vegetable, fruit, and cereal fiber intake and risk of coronary heart disease among men. JAMA 275:447–451

Rimm EB, Willett WC, Hu FB, Sampson L, Colditz GA, Manson JE, Hennekens C, Stampfer MJ (1998) Folate and vitamin B6 from diet and supplements in relation to risk of coronary heart disease among women. JAMA 279:359–364

Rizkalla SW, Taghrid L, Laromiguiere M, Huet D, Boillot J, Rigoir A, Elgrably F, Slama G (2004) Improved plasma glucose control, whole-body glucose utilization, and lipid profile on a low-glycemic index diet in type 2 diabetic men: a randomized controlled trial. Diabetes Care 27:1866–1872

Sacks FM, Svetkey LP, Vollmer WM, Appel LJ, Bray GA, Harsha D, Obarzanek E, Conlin PR, Miller ER 3rd, Simons-Morton DG, Karanja N, Lin PH (2001) DASH-Sodium Collaborative Research Group. Effects on blood pressure of reduced dietary sodium and the Dietary Approaches to Stop Hypertension (DASH) diet. N Engl J Med 344:3–10

Sakuma N, Yoshikawa M, Hibino A, Sato A, Kamiya Y, Ohte N, Tamai N, Kijnimatsu M, Kimura G, Inoue M (2001) Ascorbic acid protects against peroxidative modification of low-density lipoprotein, maintaining its recognition by LDL receptors. J Nutr Sci Vitaminol (Tokyo) 47:28–31

Salmerón J, Manson JE, Stampfer MJ, Colditz GA, Wing AL, Willett WC (1997) Dietary fiber, glycemic load, and risk of non-insulin-dependent diabetes mellitus in women. JAMA 277:472–477

Sanada H, Jones JE, Jose PA (2011) Genetics of salt-sensitive hypertension. Curr Hypertens Rep 13:55–66

Saravanan N, Haseeb A, Ehtesham NZ, Ghafoorunissa (2005) Differential effects of dietary saturated and trans-fatty acids on expression of genes associated with insulin sensitivity in rat adipose tissue. Eur J Endocrinol 153:159–165

Sasso FC, Carbonara O, Nasti R, Campana B, Marfella R, Torella M, Nappi G, Torella R, Cozzolino D (2004) Glucose metabolism and coronary heart disease in patients with normal glucose tolerance. JAMA 291:1857–1863

Serhan CN (2007) Resolution phase of inflammation: novel endogenous anti-inflammatory and proresolving lipid mediators and pathways. Annu Rev Immunol 25:101–137

Sesso HD, Buring JE, Christen WG, Kurth T, Belanger C, MacFadyen J, Bubes V, Manson JE, Glynn RJ, Gaziano JM (2008) Vitamins E and C in the prevention of cardiovascular disease in men: the Physicians' Health Study II randomized controlled trial. JAMA 300:2123–2133

Sesso HD, Buring JE, Norkus EP, Gaziano JM (2005) Plasma lycopene, other carotenoids, retinol and the risk of cardiovascular disease in men. Am J Clin Nutr 81:990–997

Shapses SA, Manson JE (2011) Vitamin D and prevention of cardiovascular disease and diabetes: why the evidence falls short. JAMA 305:2565–2566

Sirtori CR, Tremoli E, Gatti E, Montanari G, Sirtori M, Colli S, Gianfranceschi G, Maderna P, Dentone CZ, Testolin G, Galli C (1986) Controlled evaluation of fat intake in the Mediterranean diet: comparative activities of olive oil and corn oil on plasma lipids and platelets in high-risk patients. Am J Clin Nutr 44:635–642

Stampfer MJ, Hennekens CH, Manson JE, Colditz GA, Rosner B, Willett WC (1993) Vitamin E consumption and the risk of coronary disease in women. N Engl J Med 328:1444–1449

Stolarz-Skrzypek K, Kuznetsova T, Thijs L, Tikhonoff V, Seidlerová J, Richart T, Jin Y, Olszanecka A, Malyutina S, Casiglia E, Filipovský J, Kawecka-Jaszcz K, Nikitin Y, Staessen JA, European Project on Genes in Hypertension (EPOGH) Investigators (2011) Fatal and nonfatal outcomes, incidence of hypertension, and blood pressure changes in relation to urinary sodium excretion. JAMA 305:1777–1785

Strazzullo P, D'Elia L, Kandala NB, Cappuccio FP (2009) Salt intake, stroke, and cardiovascular disease: meta-analysis of prospective studies. BMJ 339:b4567

Streppel MT, Ocke MC, Boshuizen HC, Kok FJ, Kromhout D (2008) Dietary fiber intake in relation to coronary heart disease and all-cause mortality over 40 y: the Zutphen Study. Am J Clin Nutr 88:1119–1125

Sun Q, Ma J, Campos H, Hankinson SE, Manson JE, Stampfer MJ, Rexrode KM, Willett WC, Hu FB (2007) A prospective study of trans fatty acids in erythrocytes and risk of coronary heart disease. Circulation 115:1858–1865

Tai CC, Ding ST (2010) N-3 polyunsaturated fatty acids regulate lipid metabolism through several inflammation mediators: mechanisms and implications for obesity prevention. J Nutr Biochem 21:357–363

Tavazzi L, Maggioni AP, Marchioli R, Barlera S, Franzosi MG, Latini R, Lucci D, Nicolosi GL, Porcu M, Tognoni G, Gissi-HF Investigators (2008) Effect of n-3 polyunsaturated fatty acids in patients with chronic heart failure (the GISSI-HF trial): a randomised, double-blind, placebo-controlled trial. Lancet 372:1223–1230

Taylor RS, Ashton KE, Moxham T, Hooper L, Ebrahim S (2011) Reduced dietary salt for the prevention of cardiovascular disease: a meta-analysis of randomized controlled trials (Cochrane review). Am J Hypertens 24:843–853

The Report of the Dietary Guidelines Advisory Committee on Dietary Guidelines for Americans, 2005 Dietary Guidelines Advisory Committee (2005) Available from: http://www.health.gov/dietaryguidelines/dga2005/report/

Todd S, Woodward M, Tunstall-Pedoe H, Bolton-Smith C (1999) Dietary antioxidant vitamins and fiber in the etiology of cardiovascular disease and all-causes mortality: results from the Scottish Heart Health Study. Am J Epidemiol 150:1073–1080

Umesawa M, Iso H, Date C, Yamamoto A, Toyoshima H, Watanabe Y, Kikuchi S, Koizumi A, Kondo T, Inaba Y, Tanabe N, Tamakoshi A, JACC Study Group (2008) Relations between dietary sodium and potassium intakes and mortality from cardiovascular disease. Am J Clin Nutr 88:195–202

Vessby B, Uusitupa M, Hermansen K, Riccardi G, Rivellese AA, Tapsell LC, Nälsén C, Berglund L, Louheranta A, Rasmussen BM, Calvert GD, Maffetone A, Pedersen E, Gustafsson IB, Storlien LH, KANWU Study (2001) Substituting dietary saturated for monounsaturated fat impairs insulin sensitivity in healthy men and women: the KANWU study. Diabetologia 44:312–319

Vita JA, Keaney JF, Raby KE (1998) Low plasma ascorbic acid independently predicts the presence of an unstable coronary syndrome. J Am Coll Cardiol 120:1480–1485

Wall R, Ross RP, Fitzgerald GF, Stanton C (2010) Fatty acids from fish: the anti-inflammatory potential of long-chain omega-3 fatty acids. Nutr Rev 68:280–289

Wang L, Manson JE, Song Y, Sesso HD (2010) Systematic review: vitamin D and calcium supplementation in prevention of cardiovascular events. Ann Intern Med 152:315–323

Wang TJ, Pencina MJ, Booth SL, Jacques PF, Ingelsson E, Lanier K, Benjamin EJ, D'Agostino RB, Wolf M, Vasan RS (2008) Vitamin D deficiency and risk of cardiovascular disease. Circulation 117:503–511

Warodomwichit D, Shen J, Arnett DK, Tsai MY, Kabagambe EK, Peacock JM, Hixson JE, Straka RJ, Province MA, An P, Lai CQ, Parnell LD, Borecki IB, Ordovas JM (2009) ADIPOQ polymorphisms, monounsaturated fatty acids, and obesity risk: the GOLDN study. Obesity 17:510–517

Weinberger MH, Miller JZ, Luft FC, Grim CE, Fineberg NS (1986) Definitions and characteristics of sodium sensitivity and blood pressure resistance. Hypertension 6(2):II127–II134

Weinberger MH, Fineberg NS, Fineberg SE, Weinberger M (2001) Salt sensitivity, pulse pressure, and death in normal and hypertensive humans. Hypertension 2(2):429–432

Weissman A, Lowenstein L, Peleg A, Thaler I, Zimmer E (2006) Power spectral analysis of heart rate variability during the 100-g oral glucose tolerance test in pregnant women. Diabetes Care 29:571–574

Wierzbicki AS (2007) Homocysteine and cardiovascular disease: a review of the evidence. Diab Vasc Dis Res 4:143–150

Yang Q, Liu T, Kuklina EV, Flanders WD, Hong Y, Gillespie C, Chang MH, Gwinn M, Dowling N, Khoury MJ, Hu FB (2011) Sodium and potassium intake and mortality among US adults. Arch Intern Med 171:1183–1191

Yokoyama M, Origasa H, Matsuzaki M, Matsuzawa Y, Saito Y, Ishikawa Y, Oikawa S, Sasaki J, Hishida H, Itakura H, Kita T, Kitabatake A, Nakaya N, Sakata T, Shimada K, Shirato K, Japan EPA lipid intervention study (JELIS) Investigators (2007) Effects of eicosapentaenoic acid on major coronary events in hypercholesterolaemic patients (JELIS): a randomised open-label, blinded endpoint analysis. Lancet 369:1090–1098

Zittermann A, Schleithoff SS, Koerfer R (2005) Putting cardiovascular disease and vitamin D insufficiency into perspective. Br J Nutr 94:483–492

Physical Activity and Cardiovascular Diseases Epidemiology and Primary Preventive and Therapeutic Targets

6

Martin Burtscher and Erich Gnaiger

Abstract

Following the publication of the landmark study of Morris and colleagues, a plethora of evidences confirmed the inverse and independent relationship between physical activity and cardiovascular as well as overall mortality. It has been established that regular physical activity elicits its beneficial health effects by reducing especially those cardiovascular risk factors which are associated with metabolic disorders, e.g., hyperlipidemia, glucose intolerance, or systemic hypertension, but physical activity has also been shown directly to inhibit the development of atherosclerosis and associated cardiovascular diseases, e.g., by preventing or correcting endothelial dysfunction or due to cardiovascular remodeling. Nowadays, fascinating experimental studies more and more discover cellular and molecular mechanisms as primary risk factors and explain how physical activity fights the development of cardiovascular diseases. Oxidative stress, low NO bioavailability, and inflammation are considered as primary targets for modification of risk factors by regular exercise training. All these factors are closely interrelated and may play important roles in the development of atherosclerosis.

Keywords

Exercise • Training • Risk factors • Disease • Cardiovascular • Metabolism • Atherosclerosis • Oxidative stress • Inflammation

M. Burtscher, MD, PhD (✉)
Department of Sport Science, Medical Section,
University of Innsbruck, Fürstenweg 186, Innsbruck A-6020, Austria
e-mail: martin.burtscher@uibk.ac.at

E. Gnaiger, PhD
Daniel Swarovski Research Laboratory,
Medical University of Innsbruck,
Innrain 66/6, Innsbruck A-6020, Austria

I. Wakabayashi, K. Groschner (eds.), *Interdisciplinary Concepts in Cardiovascular Health*,
DOI 10.1007/978-3-7091-1334-9_6, © Springer-Verlag Wien 2013

6.1 Introduction

Survival of our earliest ancestors was largely dependent on the availability of food and food supply was closely related to physical fitness and activity. Because food was not constantly available, their genes were adapted to store fuel for periods of hunger and maintain aerobic capacity during moderate caloric restriction. Whereas the human genetic constitutions remained almost constant since the appearance of Homo sapiens about 40,000 years ago, lifestyle changed dramatically within a short evolutionary period of time. Consequently, the human genome has not adapted adequately to the sedentary lifestyle and constant food availability in today's developed countries (Eaton and Konner 1985; Szostak and Laurant 2011) likely resulting in increased food storages associated with their insufficient utilization. Metabolic disorders and the increasing prevalence of cardiovascular diseases are considered as direct consequences of this imbalance between food surplus and physical activity (Wallace 2005). About six decades ago the landmark study of Morris and colleagues convincingly demonstrated reduced mortality rates in a large cohort of civil servants with physically demanding occupations when compared to desk clerks (Morris et al. 1953). Subsequently, a plethora of evidences confirmed the inverse and independent relationship between physical activity and cardiovascular as well as overall mortality (Paffenbarger and Hale 1975; Paffenbarger et al. 1986; Blair et al. 1989, 1995; Myers et al. 2002; Kokkinos et al. 2011). It has been established that regular physical activity elicits its beneficial health effects by reducing especially those cardiovascular risk factors which are associated with metabolic disorders, e.g., hyperlipidemia, glucose intolerance, or systemic hypertension (Stamler et al. 1993; Wei et al. 1999; Thompson et al. 2003). On the other hand physical activity has been shown directly to inhibit the development of atherosclerosis and associated cardiovascular diseases, e.g., by preventing or correcting endothelial dysfunction or due to cardiovascular remodeling (Ellison et al. 2012). Nowadays, fascinating experimental studies more and more discover cellular and molecular mechanisms as primary risk factors and explaining how physical activity fights the development of cardiovascular diseases (CVD). The control of mitochondrial biogenesis and the capacity of oxidative phosphorylation (OXPHOS) emerge as integrative elements of health and disease not only in skeletal and cardiac muscle but also in other tissues, particularly the brain.

6.2 The Dynamics of Degenerative Diseases

Dysfunction and disease are a consequence of damaged or defective structures which fail to support corresponding functions. From an evolutionary and dynamic perspective, however, any complex system requires energetically coupled processes to build and maintain structures, and loss of function is then a cause rather than consequence of structural degeneration. Deficient structure results from dysfunction, and disease is the inability to counteract spontaneous or enforced decay. Cause and effect cannot be separated in complex systems where structure and function are embedded in a network of circular loops developing over time.

In the extreme, the static or deterministic perspective presents structure without time scale. At the strike of a mutation, the molecular function of the originally encoded enzyme is lost, which then provides the definite cause of a disease, like a bullet shot through the heart. Aging and degenerative diseases are phenomena which are dominated by the time scale, where the dynamics of stochastic and chaotic patterns eliminates the theoretical basis for distinguishing cause and consequence, and effects turn instantaneously into cause. Deterministic concepts fail in most instances to understand and treat degenerative diseases. When feedback loops are disturbed, stabilizing feedback may turn into feedforward control as the basis of a self-amplifying vicious circle, marking the transition from latent to manifest, from aging to aged, from proinflammatory to disease. Simple "static" interventions, such as various antioxidant treatments, may elicit no or paradoxical results. Lifestyle interventions are as complex as the organismic system itself and thus remain partly or largely elusive to simple factorial analysis. These rather abstract considerations provide the basis of a working hypothesis that dynamic intervention on an integrated system's level is necessary to combat degradation over time. Degradation is the sum of processes that put the aging organism at risk of degeneration, including CVD. Integrative dynamic intervention imposes periodic perturbations at different hierarchical levels simultaneously. An accessible and economic intervention of this kind is exercise.

Mitochondria are the evolutionary ancient symbionts in our cells which provide most of the vital energy by burning carbohydrate and lipid to produce ATP and heat. Lifestyle sets the cellular and mitochondrial environment in terms of energy, food, and metabolites. Lifestyle in turn sets the relative value of energy and time in terms of velocity, metabolic flows, power, and efficiency (Gnaiger 1993). Striking changes of our cellular environment were brought about by the industrial revolution. The shift from energy limitation to time limitation increased power and the "speed of life." For the first time in human history, the distance of transport and velocity of locomotion became dissociated completely from physical exercise, muscular power, and cellular metabolism. External energy flows and artificial structures (heat engines) replaced the cellular metabolic machinery. As a consequence, industrial societies generated a lifestyle with food supply topping metabolic demands. This energy imbalance disturbs the cellular and mitochondrial environment, causes metabolic shifts, and manifests itself as an increasing body mass index (BMI). Although average life span increased, the sedentary overfed lifestyle reduces the quality of life and induces a proinflammatory state with an increasing risk of acquiring degenerative age-associated diseases including various forms of cancer (Wallace 2005).

6.3 Physical Activity and Longevity

The unique work by Morris and colleagues (1953) initiated numerous large studies investigating the association between physical activity and longevity. Paffenbarger and coworkers (1986, 1993) demonstrated a strong negative association between physical activity and morbidity and mortality from cardiovascular and other causes.

Lee and Skerrett (2001) reviewed 44 observational studies from 1966 to 2000 again convincingly confirming the negative relationship between physical activity and all-cause mortality. More recent studies started to assess the individual fitness level as an objective measure of physical activity consistently showing a linear dose-response relation between the fitness level and mortality. Kokkinos and colleagues (2008) demonstrated an approximately 20 % lower mortality risk for subjects with an exercise capacity of 5–7 metabolic equivalents (METs; 1 MET = 1 ml $O_2 \cdot min^{-1} \cdot kg^{-1}$) compared to those achieving below 5 METs. This risk was reduced by 50 % with an exercise capacity of 7–10 METs and even by 70 % in those achieving more than 10 METs. The observed mortality reduction was similar in subjects with and without cardiovascular disease. Exercise capacity turned out to be a more powerful predictor of mortality than well-established risk factors like systemic hypertension, dyslipidemia, diabetes, or smoking (Myers et al. 2002; Kokkinos et al. 2008). Many other studies confirmed these results. Generally, the reduction of the mortality risk ranges between 10 and 25 % for each MET increase in exercise capacity (Kokkinos et al. 2011).

6.4 Type and Dose of Physical Activity

Many of the aforementioned studies assessed the amount of physical activity as energy expenditure related to such activities. Paffenbarger and colleagues (1986) found a 25–33 % lower mortality risk in subjects with a weekly energy expenditure of 2,000 kcal or more from leisure time activities compared to those below 2,000 kcal. Noteworthy the mortality risk tended to increase when the activity related energy expenditure exceeded 3,500 kcal/week. On the other hand the mortality rates still decreased when the individual exercise capacity exceeded 10 METs. When various types of exercise (walking, stair climbing, sports playing) were compared a threshold exercise intensity of about 5–6 METs was found with only minimal additional benefits when performing at higher intensities (Paffenbarger et al. 1986). Regarding exercise effects on the risk factor profile, several clinical trials report greater improvements after vigorous (>6 METs) than moderate intensity exercise for diastolic blood pressure, glucose control, and aerobic capacity. However, these studies did not see such intensity effects on systolic blood pressure, lipid profile, or body fat loss (Swain and Franklin 2006). Our studies on risk factors for sudden cardiac death during mountain sport activities indicate that sport-specific training intensities help best to prevent cardiac deaths (Burtscher and Ponchia 2010). Mitochondrial biogenesis in skeletal muscle is induced by both strength and endurance training (Pesta et al. 2011), but in cardiac tissue only high-intensity interval training increases citrate synthase (CS) activity and OXPHOS capacity per unit CS activity (Hafstad et al. 2011). Taken together, exercise at a vigorous intensity elicits greater cardioprotective benefits than exercise of a moderate intensity provided that total energy expenditure of exercise is held constant. General training prescriptions for patients with CVD recommend endurance exercise at a minimum of 3 times a week for at least 30 min, including 5 min of warm-up and cool-down calisthenics and at least 20 min of exercise at an intensity requiring 70–85 % of the predetermined peak heart rate (Thompson 2005).

6.5 Physical Activity as a Modifier of Risk Factors

There is ample evidence that physical activity is an independent and important factor for the assessment of the CVD risk and that increasing levels of physical fitness protect against elevations in most risk factors in subjects with and without CVD. Despite these facts physical activity is often ignored in large epidemiologic studies (Burtscher 2012).

Regular physical activity may be considered the most important condition counteracting metabolic and cardiovascular diseases (Szostak and Laurant 2011). Pedersen (2009) suggested that physical inactivity leads to accumulation of visceral fat which subsequently activates several inflammatory pathways. Consequences include insulin resistance and atherosclerosis. The resulting cluster of diseases has been termed "diseasome" (Pedersen 2009). When rats were selected to be low- or high-capacity runners, a connection between endurance capacity and the development of risk factors for metabolic and cardiovascular diseases has been demonstrated (Kivelä et al. 2010; Szostak and Laurant 2011). In these rats 239 genes in skeletal muscle were differently expressed between low- and high-capacity runners indicating phenotypic differences (Kivelä et al. 2010).

Nowadays it is well established that physical activity elicits its beneficial effects by modifying classical risk factors for CVD and direct effects on the cardiovascular system. Regular physical activity reduces all traditional risk factors, i.e., systemic arterial hypertension, elevated low-density (LDL) and reduced high-density lipoproteins (HDL), increased triglyceride levels, insulin resistance, and glucose intolerance (Thompson et al. 2003). As mentioned above the modification of risk factors depends at least partly on the type, the duration, and the intensity of exercise.

6.5.1 Hypercholesterolemia

Hypercholesterolemia (total cholesterol level >200 mg/dl) is part of the traditional risk factor set for CVD. Whereas higher levels of HDL cholesterol are associated with a reduced risk for cardiovascular events, elevated LDL cholesterol and triglycerides are considered to be detrimental with regard to cardiovascular outcomes (Stapleton et al. 2010). Hypercholesterolemia leads to endothelial dysfunction and is very likely related to the reduced nitric oxide (NO) bioavailability, elevated oxidative stress, and the development of proinflammatory conditions (Goodwill et al. 2008; Stapleton et al. 2010). Of course, hypercholesterolemia is often associated with certain genetic risk factors, and pharmacological treatments focus on lowering of cholesterol levels by preventing its formation in the liver or its absorption in the intestine (Stapleton et al. 2010). Exercise training is well established to increase HDL cholesterol levels and improving cardiovascular function. In addition, regular physical activity decreases oxidative stress and chronic systemic inflammation and increases NO bioactivity thus improving hypercholesterolemia-related endothelial dysfunction (Goto et al. 2003; Szostak and Laurant 2011).

6.5.2 Hypertension

As hyperlipidemia systemic hypertension is part of the set of traditional risk factors for CVD. Blood pressure values below 120 mmHg systolic and 80 mmHg diastolic are considered optimal and values above 140 mmHg systolic and 90 mmHg indicate hypertension. Hypertension has been termed as a "silent killer" because it may be responsible for 13.5 % of all deaths worldwide (Arima et al. 2011). Fifty-four percent of strokes and 47 % of coronary heart diseases are attributed to elevated blood pressure values (Arima et al. 2011). Beneficial effects of regular exercise on hypertension have been repeatedly demonstrated (Arakawa 1996). Besides improvements in sympathetic regulation and volume-depleting actions, the expression of endothelial NO synthase (eNOS) and NO bioactivity contribute to the antihypertensive effects of exercise (Arakawa 1996; Ohta et al. 2005). However, increased NO production is not necessarily associated with improvement of physiological function likely due to the fact that NO is highly reactive with superoxide resulting in powerful oxidant products (Ohta et al. 2005). Thus, the ratio between NO and the amount of superoxide determines NO bioavailability. Longtime exercise training is suggested to increase NO and antioxidant defense thereby enhancing NO bioavailability and favorable effects on systemic blood pressure (Suzuki et al. 2000).

6.5.3 Type 2 Diabetes

The worldwide increase in the prevalence of type 2 diabetes mellitus (T2DM) poses an enormous health burden in both developed and developing countries (Burtscher et al. 2009). The importance of the healthcare problem of T2DM results not just from the disease itself but also from its association with obesity and cardiovascular risk factors, particularly dyslipidemia and hypertension (Zimmet et al. 2001). Epidemiologic data suggest that lifestyle changes especially involving increased regular physical activity would prevent T2DM to a large degree (Tuomilehto et al. 2001). Exercise training has been demonstrated to increase whole-body insulin-mediated glucose disposal in T2DM patients due to an increase in glucose transporters type 4 (GLUT4) protein and independent of changes in the insulin-signaling cascade (O'Gorman et al. 2006). Additionally, sufficient levels of NO are essential for normal skeletal muscle and endothelial function, but they are also necessary for optimal insulin secretion from pancreatic β-cells thereby counteracting the development of T2DM (Newsholme et al. 2010). As discussed above, low levels of oxidative stress will enhance NO bioavailability and the related beneficial metabolic effects (Ohta et al. 2005). Besides, exercise activates AMP-activated protein kinase (AMPK). AMPK then stimulates energy-generating processes like the oxidation of fatty acids and glucose (Richter and Ruderman 2009) and contributes to mechanisms leading to improved insulin secretion and actions resulting in improved metabolic function (Newsholme et al. 2010). Finally, increasing anti-inflammatory effects by regular exercise training, at least partly mediated by skeletal muscle derived interleukin-6 (IL-6), have also been suggested to exert positive effects on metabolic disorders as T2DM (Pedersen and Steensberg 2002).

6.5.4 Low Mitochondrial Power: Link from Obesity to Cardiovascular Risk

BMI is an integrator of energy balance in active versus sedentary lifestyles. Since energetically coupled processes are required for physiological function and maintenance of structure and physical exercise is a major sink for food-derived energy, it is reasonable to focus on the energetically dominating metabolic pathway in the living cell, which is oxidative phosphorylation (OXPHOS, Fig. 6.1). An abnormal expression of genes regulating oxidative phosphorylation in skeletal muscle has been shown in patients suffering from T2DM (Mootha et al. 2003). OXPHOS capacity of vastus lateralis in overweight and T2DM patients is low (Boushel et al. 2007; Larsen et al. 2009; Rabøl et al. 2009), and exercise training increases insulin sensitivity and OXPHOS capacity in obese patients (BMI of 30–33 kg/m^2) with and without T2DM (Hey-Mogensen et al. 2010; Phielix et al. 2010). Comparable to humans (Gnaiger 2009), OXPHOS capacity in skeletal muscle declines in obese horses and increases by about 2.5-fold from untrained to competitive race horses (Votion et al. 2012). Functional OXPHOS analysis in various tissues of mice suggests that obesity alters mitochondrial density and respiratory performance according to tissue-specific control mechanisms (Holmström et al. 2012).

As physical activity is diminished despite sufficient energy supply, the electron transport system becomes more reduced (2[H]), the electrical potential across the inner mitochondrial membrane is increased, and electron leak to oxygen increases the generation of reactive oxygen species (ROS; Fig. 6.1). Under these conditions of diminished energy utilization, reduction of mitochondrial density may be considered adaptive (Fig. 6.2). At low mitochondrial density and tissue mass-specific OXPHOS capacity, the work to be done per mitochondrial unit in the tissue is increased, keeping hyperpolarization in check. Simultaneously, the density is decreased of mitochondrial ROS production sites (Complexes I and III). The respiratory LEAK/OXPHOS flux control ratio is diminished in untrained healthy and obese subjects (Pesta et al. 2011; Hey-Mogensen et al. 2010), thus increasing the proton leak and maintaining the proton-motive force lower at rest (Fig. 6.1). Importantly, however, healthy mitochondria act not only as ROS producers but also as ROS scavengers (particularly CoQ and cytochrome c); mitochondria are important redox regulators, and these cell-signaling functions are progressively lost as mitochondrial density declines. Below a threshold level of low mitochondrial density, the adaptive range for downregulation of mitochondrial biogenesis is exceeded.

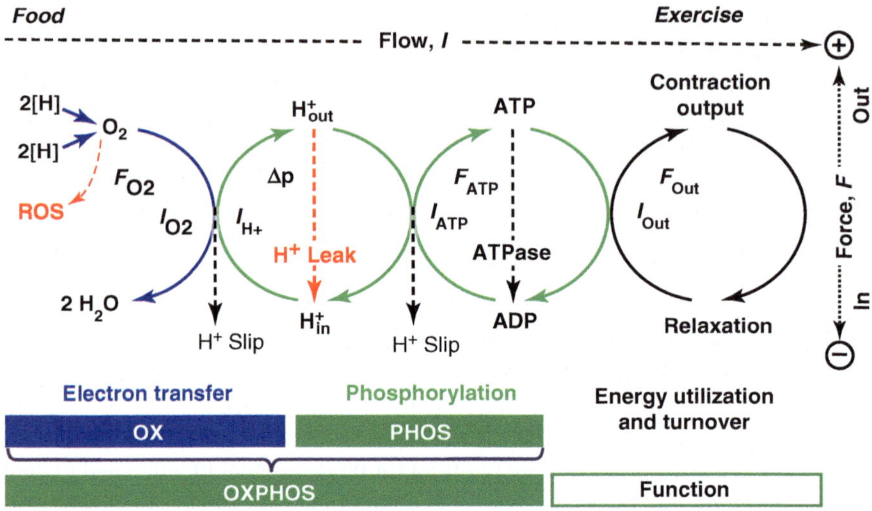

Fig. 6.1 Dynamics of energy transformation in coupled flows, I, and forces, F, of aerobic metabolism and energy utilization. 2[H] indicates the reduced hydrogen equivalents of carbon substrates with electron transfer to molecular oxygen. F_{O2} is the Gibbs force or molar Gibbs energy of oxygen consumption (typically –450 kJ/mol O_2). H^+_{in} to H^+_{out} is the process of proton translocation across the inner mitochondrial membrane, generating a proton motive force, $FH^+ = \Delta p$ which has a chemical and electrical component (proton gradient and membrane potential). A short circuit (H^+ leak) dissipates the energy of the translocated protons, which otherwise is used to phosphorylate ADP to ATP, with an output force of 52–66 kJ/mol ATP under intracellular conditions (Gnaiger 1993). The efficiency of coupling is diminished by potential proton slips of the proton-energy conserving pumps (Complexes CI, CIII, and CIV) and the proton-slip of proton-energy utilizing pump (ATP synthase or CV). Production of reactive oxygen species (*ROS*) is another component reducing the coupling between oxygen flux and ATP turnover. ATPase activity may short-circuit the ATP cycle and dissipate the energy stored in ATP, which otherwise is used to drive various output flows, from biosynthetic pathways, ion pumping to muscle contraction

This is the transition to a proinflammatory state ("mitochondrial fever"), putting the cell at risk of further degeneration and apoptosis, with all sequelae of age-related diseases.

In rats artificially selected for low aerobic capacity, impaired regulation of oxidative pathways (PGC-1α; Fig. 6.2) in skeletal muscle precedes the increase in body weight and is linked to a high incidence of cardiovascular disease (Wisløff et al. 2005). In this model of the metabolic syndrome, high-intensity exercise training is superior to moderate-intensity exercise for ameliorating cardiovascular risks (Haram et al. 2009). Selective breeding for high running capacity, in turn, increases OXPHOS capacity and enhances ROS production, whereas oxidative DNA damage is decreased due to a compensatory increase in antioxidant capacity (Tweedie et al. 2011). A high-fat diet increases mitochondrial H_2O_2 production and decreases redox-buffering capacity (Anderson et al. 2009). While mitochondrial oxidative stress mediates heart failure (Dai et al. 2011), endurance exercise increases antioxidant enzyme activity in cardiac mitochondria, attenuates ROS-induced cytochrome c release, and reduces the

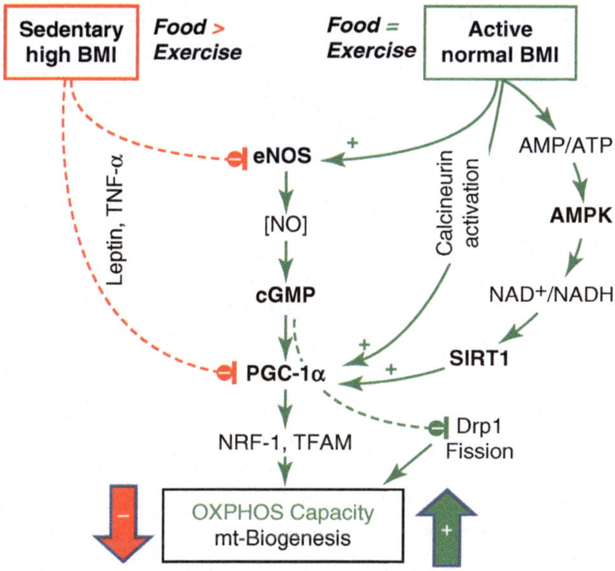

Fig. 6.2 Central role of peroxisome proliferator-activated receptor γ coactivator-1α (*PGC-1α*) in the control of tissue OXPHOS capacity and mitochondrial (*mt*) biogenesis. Calcineurin (Lin et al. 2002), eNOS (Nisoli et al. 2007), and AMP-activated protein kinase (*AMPK*) (Cantó et al. 2009) increase PGC-1α expression, which in turn increases coordinated expression of nuclear DNA and mtDNA through downstream transcription factors, nuclear respiratory factor 1 (*NRF-1*), and mt transcription factor A (*TFAM*). PGC-1α expression is inhibited by the proinflammatory cytokine tumor necrosis factor α (*TNF-α*) and high levels of leptin. At a given mt volume density as a net result of mt biogenesis and mt degradation, NO exerts an additional role on OXPHOS and muscular repair by elongation of tubular mitochondria through a reduction of fission by inhibition of dynamin-related protein (*Drp1*) (De Palma et al. 2010). In normal-weight animals, moderate caloric restriction enhances OXPHOS capacity by increasing eNOS expression (Nisoli et al. 2005)

threshold for ROS-induced apoptosis (Kavazis et al. 2008). Chronic heart failure (CHF) in humans is associated with a decreased OXPHOS capacity particularly for fatty acid oxidation (Lemieux et al. 2011). In vastus lateralis of CHF patients, OXPHOS capacity (with glutamate+malate as substrates) and the positive response to training are preserved with respect to sedentary controls (Zoll et al. 2003; Garnier et al. 2005). Stimulation of mitochondrial biogenesis entails beneficial effects on various levels, including impaired carcinogenesis (Wang and Moraes 2011).

6.6 Direct Effects of Physical Activity on the Cardiovascular System

Whereas several decades ago physicians recommended prolonged rest for patients with heart diseases, nowadays regular physical activity or even vigorous exercise is an evidence-based part of the prevention and therapy of various heart diseases

(Shephard and Balady 1999). "To date, the only practical and sustainable countermeasure capable of promoting cardioprotection is regular bouts of endurance exercise" (Kavazis et al. 2008), reducing the risk of death due to ischemia-reperfusion in humans (Ignarro et al. 2007) and animal models (Powers et al. 2002). The favorable effects of exercise are partly due to the modification of cardiovascular risk factors but also to direct effects on the heart, i.e., cellular and molecular remodeling (Ellison et al. 2012). Exercise-induced cardiac hypertrophy, a balanced increase of left ventricular mass accompanied by neoangiogenesis, represents the most widely known structural adaptation of the heart to regular endurance exercise (Weiner and Baggish 2012). In contrast to the old paradigm that the heart represents a post-mitotic organ without regenerative capacity (Ellison et al. 2012), it has recently been demonstrated that 50 % of cardiomyocytes are exchanged during a normal human life (Bergmann et al. 2009). Waring et al. (2010) reported an elevated proliferation, number, and cardiogenic differentiation of endogenous cardiac stem and progenitor cells following controlled endurance training. Animal studies indicated that exercise-induced cardiomyocyte hypertrophy and renewal depended on a reduced expression of the transcription factor C/EBPb and an increase in the expression of CBP/p300-interacting transactivator (Boström et al. 2010). Also functional benefits like improved contractility have been shown in unloaded cardiomyocytes after endurance training, probably due to a modification of intracellular handling of Ca^{2+} (Ellison et al. 2012). Additionally, regular endurance exercise increases the number and size of coronary vessels, endothelial function, coronary blood flow, oxygen transport capacity, and oxygen extraction (Duncker and Bache 2008; Ellison et al. 2012). These adaptations are triggered by signals including hemodynamic forces like flow and stretch, tissue ischemia, and growth factors like vascular endothelial growth factor (VEGF) and fibroblast growth factor-2 (FGF-2) (Ellison et al. 2012). Exercise is also an important regulator of eNOS-mediated vascular release of NO (Balligand et al. 2009) which is closely associated with cardiovascular remodeling, improved Ca^{2+} handling and endothelial function, and mitochondrial biogenesis (Fig. 6.2).

6.6.1 Endothelial Dysfunction

Blunted endothelium-dependent vasodilatation is an early sign in patients with risk factors for CVD (McVeigh et al. 1992). This dysfunction of the endothelium is typically caused by a reduction of NO bioactivity. NO inactivation is caused by excessive ROS production and has been demonstrated in patients with risk factors for CVD (Cai and Harrison 2000). There are several lines of evidence supporting the potential of regular physical activity to prevent or reverse endothelial dysfunction, e.g., in patients with coronary artery disease or T2DM (Szostak and Laurant 2011). This becomes evident since exercise training represents one of the most important regulators of eNOS-mediated release of NO from the endothelium (Indolfi et al. 2002). NO release results at least partly from the increased vascular shear stress during exercise (Tuteja et al. 2004). Elevated NO production not only favors vasorelaxation but also reduces platelet adhesion and aggregation thereby eliciting its beneficial effects in patients with coronary artery disease (Tuteja et al. 2004). Exercise training represents an antioxidant therapy thereby increasing NO bioavailability and improving endothelial function (Di Francescomarino et al. 2009). Both reduced NO bioactivity and oxidative stress seem to contribute to chronic systemic inflammation and endothelial dysfunction (Szostak and Laurant 2011; Rizvi 2009) which seems to be improved by regular exercise training (Petersen and Pedersen 2005).

6.7 Primary Preventive and Therapeutic Targets of Physical Activity

Oxidative stress, low NO bioavailability, and inflammation are considered as primary targets for modification of risk factors by regular exercise training. All these factors are closely interrelated and may play important roles in the development of traditional risk factors and atherosclerosis.

ROS inactivate NO resulting in endothelial dysfunction, promote LDL oxidation, and the progression of atherosclerosis (Davignon and Ganz 2004). During acute exercise the mitochondrial electron transfer system contributes to an elevation of ROS production (Bloomer 2008). Concomitantly, there is an increase of the prooxidant enzymes which is associated with ROS markers like F2-isoprostanes (Brennan et al. 2003). NADPH oxidase activity is closely related to ROS generation and has been shown to be associated to physical inactivity (Szostak and Laurant 2011). During regular and chronic exercise training, antioxidant enzymes increase, e.g., superoxide dismutase (SOD) and glutathione peroxidase, resulting in a more powerful oxidant handling (Stapleton et al. 2010). Consequently, oxidative stress markers like F2-isoprostane decrease (Seals et al. 2008).

Whereas NO production by vasoactive agonists is related to the increase of calcium concentration in endothelial cells, fluid shear stress during exercise activates eNOS by phosphorylation of the serine/threonine protein kinase Akt/protein kinase B in a phosphoinositide 3-kinase-dependent manner (Wyatt et al. 2004). Oxidative stress, caused by an imbalance between oxidant and anti-oxidant enzymes, inactivates NO and favors low NO bioavailability (Szostak and Laurant 2011). The consequences of decreased NO bioavailability include upregulation of endothelin-1 (ET-1) and activation of platelets and coagulation pathways and thus endothelial dysfunction. On the other hand, regular exercise increases NO bioavailability directly by eNOS activation and indirectly by diminishing oxidative stress thereby improving endothelial function (Goto et al. 2003; Szostak and Laurant 2011).

Thus, exercise increases antiatherogenic markers like eNOS and SODs and decreases the expression of ET-1, adhesion molecules like VCAM-1, and ROS as markers of a proatherogenic endothelial cell phenotype (Laughlin et al. 2008) (Fig. 6.3).

Both elevated ROS production and decreased NO bioactivity seem to be causally linked to chronic inflammation representing a pathogenic feature of atherosclerosis (Ross 1999). For example, ROS and ROS products activate the nuclear factor κB

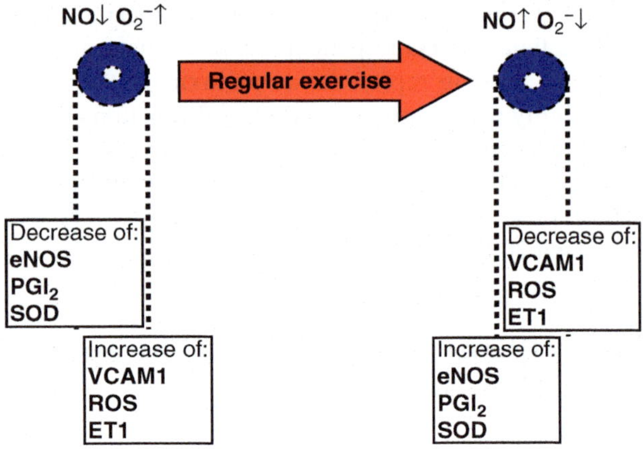

Fig. 6.3 Regular exercise increases antiatherogenic markers like endothelial nitric oxide synthase (*eNOS*); prostacyclin (*PGI2*), and superoxide dismutase (*SOD*) and decreases the expression of endothelin-1 (*ET-1*), adhesion molecules like VCAM-1, and reactive oxygen species (*ROS*) as markers of a proatherogenic endothelial cell phenotype (Laughlin et al. 2008)

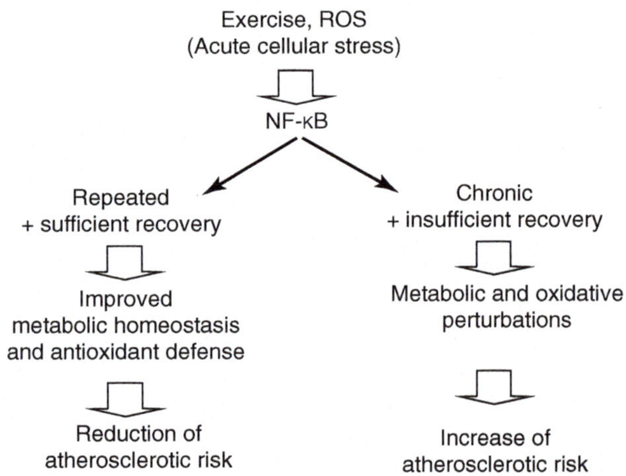

Fig. 6.4 Nuclear factor κB (*NFκB*) is activated by acute cellular stress. When NFκB is chronically activated in diseases like diabetes, the related metabolic and oxidative perturbations increase the atherosclerotic risk. However, repeated NFκB activation with exercise (and sufficient recovery) improves metabolic homeostasis and antioxidant defense resulting in reduction of the atherosclerotic risk (Kramer and Goodyear 2007)

(NFκB) which contributes to the regulation of proinflammatory genes. NFκB seems to be activated during healthy and diseased conditions. It seems reasonable that the severity and the frequency of stress determine whether responses are favorable or, with inadequate recovery, rather detrimental. Thus, regular exercise with sufficient recovery will promote positive adaptations making the organism more resistant to future perturbations (Kramer and Goodyear 2007) (Fig. 6.4).

NFκB has also been suggested to be involved in alterations of fuel metabolism during and following exercise by increasing transcription of the IL-6 gene (Kramer and Goodyear 2007). IL-6 is produced by the contracting skeletal muscle and favors glucose transport and lipid oxidation (Pedersen and Steensberg 2002) and stimulates satellite cell proliferation (Toth et al. 2011). On the other hand IL-6 diminishes tumor necrosis factor-alpha (TNF-α) release. TNF-α is known to promote vascular inflammation, oxidative stress, and reduced NO bioactivity resulting in endothelial

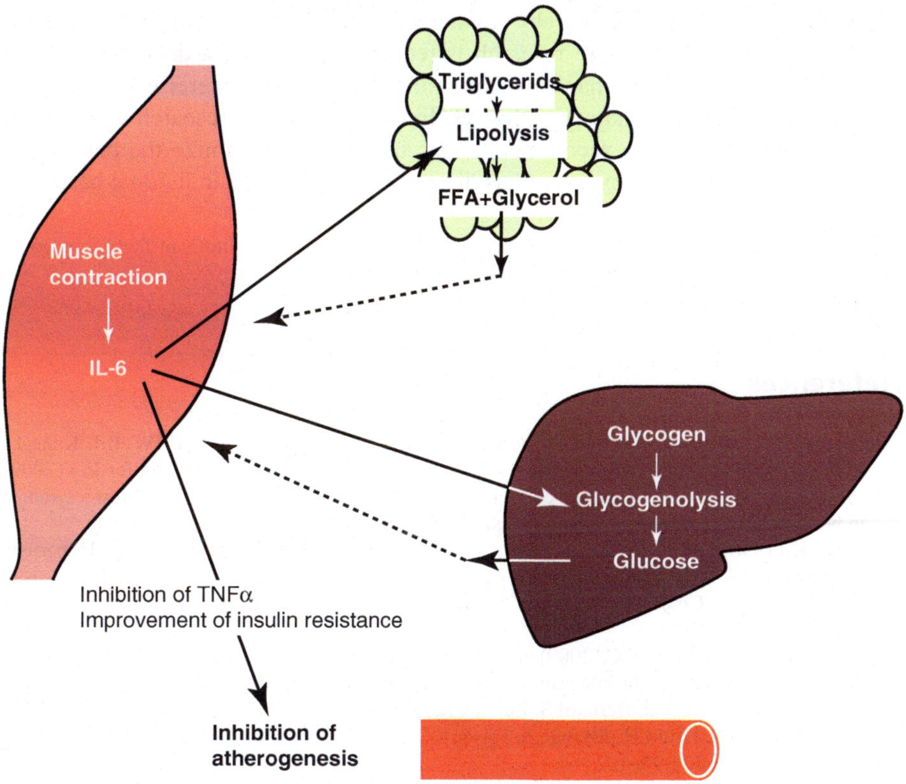

Fig. 6.5 Interleukin-6 (*IL-6*) is produced by the contracting skeletal muscle and favors glucose transport and lipid oxidation. Regular physical activity is also associated with decreasing concentrations of proinflammatory markers like tumor necrosis factor-alpha (*TNF-α*) and improved insulin sensitivity thereby inhibiting atherogenesis (Pedersen et al. 2007)

dysfunction (Szostak and Laurant 2011) and arrest of mitochondrial biosynthesis (Fig. 6.2). TNF-α also contributes to insulin resistance by increasing lipolysis and the related release of nonesterified fatty acid (NEFA, Plomgaard et al. 2005). Regular physical activity, however, is associated with decreasing concentrations of proinflammatory markers like TNF-α and improved insulin sensitivity thereby inhibiting atherogenesis (Pedersen et al. 2007) (Fig. 6.5).

6.8 Summary and Perspectives

In recent years, many studies have focused on the pathophysiological pathways associated with the involvement of oxidative stress and inflammation in the development of metabolic disorders and cardiovascular diseases and the preventive impact of regular exercise training. Ample evidence has been provided that regular exercise increases anti-inflammatory environment, improves anti-oxidant defense, and reduces oxidative stress thereby counteracting the development of atherosclerosis. Nevertheless, much more research is required to provide a deeper insight in cellular and molecular mechanisms explaining the effects of exercise on oxidative stress effects and inflammatory pathways. Findings from this research will help to establish optimal training recommendations in order to maximize the benefits of exercise training for healthy as well as diseased subjects on an individual basis.

Acknowledgments This chapter is a contribution to K-Regio project MitoCom Tyrol, funded by the Tyrolian Government and the European Regional Development Fund.

References

Anderson EJ, Lustig ME, Boyle KE, Woodlief TL, Kane DA, Lin CT, Price JW 3rd, Kang L, Rabinovitch PS, Szeto HH, Houmard JA, Cortright RN, Wasserman DH, Neufer PD (2009) Mitochondrial H_2O_2 emission and cellular redox state link excess fat intake to insulin resistance in both rodents and humans. J Clin Invest 119:573–581

Arakawa K (1996) Effect of exercise on hypertension and associated complications. Hypertens Res 19(Suppl 1):S87–S91

Arima H, Barzi F, Chalmers J (2011) Mortality patterns in hypertension. J Hyperterns 29 Suppl 1:S3–7

Balligand JL, Feron O, Dessy C (2009) eNOS activation by physical forces: from short-term regulation of contraction to chronic remodeling of cardiovascular tissues. Physiol Rev 89:481–534

Bergmann O, Bhardwaj RD, Bernard S, Zdunek S, Barnabé-Heider F, Walsh S, Zupicich J, Alkass K, Buchholz BA, Druid H, Jovinge S, Frisén J (2009) Evidence for cardiomyocyte renewal in humans. Science 324:98–102

Blair SN, Kohl HW, Paffenbarger RS, Clark DG, Cooper KH, Gibbons LW (1989) Physical fitness and all-cause mortality: a prospective study of healthy men and women. JAMA 262: 2395–2401

Blair SN, Kohl HW, Barlow CE, Paffenbarger RS, Gibbons LW, Macera CA (1995) Changes in physical fitness and all-cause mortality: a prospective study in healthy and unhealthy men. JAMA 273:1093–1098

Bloomer RJ (2008) Effect of exercise on oxidative stress biomarkers. Adv Clin Chem 46:1–50

Boström P, Mann N, Wu J, Quintero PA, Plovie ER, Panáková D, Gupta RK, Xiao C, MacRae CA, Rosenzweig A, Spiegelman BM (2010) C/EBPb controls exercise-induced cardiac growthand protects against pathological cardiac remodeling. Cell 143:1072–1083

Boushel R, Gnaiger E, Schjerling P, Skovbro M, Kraunsøe R, Dela F (2007) Patients with type 2 diabetes have normal mitochondrial function in skeletal muscle. Diabetologia 50:790–796

Brennan ML, Penn MS, Van Lente F, Nambi V, Shishehbor MH, Aviles RJ, Goormastic M, Pepoy ML, McErlean ES, Topol EJ, Nissen SE, Hazen SL (2003) Prognostic value of myeloperoxidase in patients with chest pain. N Engl J Med 349:1595–1604

Burtscher M (2012) Lifetime risks of cardiovascular disease. N Engl J Med 366:1642

Burtscher M, Ponchia A (2010) The risk of cardiovascular events during leisure time activities at altitude. Prog Cardiovasc Dis 52:507–511

Burtscher M, Gatterer H, Kunczicky H, Brandstätter E, Ulmer H (2009) Supervised exercise in patients with impaired fasting glucose: impact on exercise capacity. Clin J Sport Med 19:394–398

Cai H, Harrison DG (2000) Endothelial dysfunction in cardiovascular diseases: the role of oxidant stress. Circ Res 87:840–844

Cantó C, Gerhart-Hines Z, Feige JN, Lagouge M, Noriega L, Milne JC, Elliott PJ, Puigserver P, Auwerx J (2009) AMPK regulates energy expenditure by modulating NAD+ metabolism and SIRT1 activity. Nature 458:1056–1060

Dai DF, Johnson SC, Villarin JJ, Chin MT, Nieves-Cintrón M, Chen T, Marcinek DJ, Dorn GW 2nd, Kang YJ, Prolla TA, Santana LF, Rabinovitch PS (2011) Mitochondrial oxidative stress mediates angiotensin II-induced cardiac hypertrophy and Galphaq overexpression-induced heart failure. Circ Res 108:837–846

Davignon J, Ganz P (2004) Role of endothelial dysfunction in atherosclerosis. Circulation 109(supplIII):27–32

De Palma C, Falcone S, Pisoni S, Cipolat S, Panzeri C, Pambianco S, Pisconti A, Allevi R, Bassi MT, Cossu G, Pozzan T, Moncada S, Scorrano L, Brunelli S, Clementi E (2010) Nitric oxide inhibition of Drp1-mediated mitochondrial fission is critical for myogenic differentiation. Cell Death Differ 17:1684–1696

Di Francescomarino S, Sciartilli A, Di Valerio V, Di Baldassarre A, Gallina S (2009) The effect of exercise on endothelial function. Sports Med 39:797–812

Duncker DJ, Bache RJ (2008) Regulation of coronary blood flow during exercise. Physiol Rev 88:1009–1086

Eaton SB, Konner M (1985) Paleolithic nutrition: a consideration of its nature and current implications. N Engl J Med 312:283–289

Ellison GM, Waring CD, Vicinanza C, Torella D (2012) Physiological cardiac remodelling in response to endurance exercise training: cellular and molecular mechanisms. Heart 98:5–10

Garnier A, Fortin D, Zoll J, N'Guessan B, Mettauer B, Lampert E, Veksler V, Ventura-Clapier R (2005) Coordinated changes in mitochondrial function and biogenesis in healthy and diseased human skeletal muscle. FASEB J 19:43–52

Gnaiger E (1993) Efficiency and power strategies under hypoxia. Is low efficiency at high glycolytic ATP production a paradox? In: Hochachka PW, Lutz PL, Sick T, Rosenthal M, Van den Thillart G (eds) Surviving hypoxia: mechanisms of control and adaptation. CRC Press, Boca Raton/Ann Arbor/London, pp 77–109

Gnaiger E (2009) Capacity of oxidative phosphorylation in human skeletal muscle. New perspectives of mitochondrial physiology. Int J Biochem Cell Biol 41:1837–1845

Goodwill AG, Stapleton PA, James ME, D'Audiffret AC, Frisbee JC (2008) Increased arachidonic acid-induced thromboxane generation impairs skeletal muscle arteriolar dilation with genetic dyslipidemia. Microcirculation 15:621–631

Goto C, Higashi Y, Kimura M, Noma K, Hara K, Nakagawa K, Kawamura M, Chayama K, Yoshizumi M, Nara I (2003) Effect of different intensities of exercise on endothelium-dependent vasodilation in human: role of endothelium-dependent nitric oxide and oxidative stress. Circulation 108:530–535

Hafstad AD, Boardman NT, Lund J, Hagve M, Khalid AM, Wisløff U, Larsen TS, Aasum E (2011) High intensity interval training alters substrate utilization and reduces oxygen consumption in the heart. J Appl Physiol 111:1235–1241

Haram PM, Kemi OJ, Lee SJ, Bendheim MØ, Al-Share QY, Waldum HL, Gilligan LJ, Koch LG, Britton SL, Najjar SM, Wisløff U (2009) Aerobic interval training vs. continuous moderate exercise in the metabolic syndrome of rats artificially selected for low aerobic capacity. Cardiovasc Res 81:723–732

Hey-Mogensen M, Højlund K, Vind BF, Wang L, Dela F, Beck-Nielsen H, Fernström M, Sahlin K (2010) Effect of physical training on mitochondrial respiration and reactive oxygen species release in skeletal muscle in patients with obesity and type 2 diabetes. Diabetologia 53:1976–1985

Holmström MH, Iglesias-Gutierrez E, Zierath JR, Garcia-Roves PM (2012) Tissue-specific control of mitochondrial respiration in obesity-related insulin resistance and diabetes. Am J Physiol Endocrinol Metab 302:E731–E739

Ignarro LJ, Balestrieri ML, Napoli C (2007) Nutrition, physical activity, and cardiovascular disease: an update. Cardiovasc Res 73:326–340

Indolfi C, Torella D, Coppola C, Curcio A, Rodriguez F, Bilancio A, Leccia A, Arcucci O, Falco M, Leosco D, Chiariello M (2002) Physical training increases eNOS vascular expression and activity and reduces restenosis after balloon angioplasty or arterial stenting in rats. Circ Res 91:1190–1197

Kavazis AN, McClung JM, Hood DA, Powers SK (2008) Exercise induces a cardiac mitochondrial phenotype that resists apoptotic stimuli. Am J Physiol Heart Circ Physiol 294:H928–H935

Kivelä R, Silvennoinen M, Lehti M, Rinnankoski-Tuikka R, Purhonen T, Ketola T, Pullinen K, Vuento M, Mutanen N, Sartor MA, Reunanen H, Koch LG, Britton SL, Kainulainen H (2010) Gene expression centroids that link with low intrinsic aerobic capacity and complex disease risk. FASEB J 24:4565–4574

Kokkinos P, Myers J, Kokkinos JP, Pittaras A, Narayan P, Manolis A, Karasik P, Greenberg M, Papademetriou V, Singh S (2008) Exercise capacity and mortality in black and white men. Circulation 117:614–622

Kokkinos P, Sheriff H, Kheirbek R (2011) Physical inactivity and mortality risk. Cardiol Res Pract 2011:924945

Kramer HF, Goodyear LJ (2007) Exercise, MAPK, and NF-κB signaling in skeletal muscle. J Appl Physiol 103:388–395

Larsen S, Ara I, Rabøl R, Andersen JL, Boushel R, Dela F, Helge JW (2009) Are substrate use during exercise and mitochondrial respiratory capacity decreased in arm and leg muscle in type 2 diabetes? Diabetologia 52:1400–1408

Laughlin MH, Newcomer SC, Bender SB (2008) Importance of hemodynamic forces as signals for exercise-induced changes in endothelial cell phenotype. J Appl Physiol 104:588–600

Lee IM, Skerrett PJ (2001) Physical activity and all-cause mortality: what is the dose-response relation? Med Sci Sports Exerc 33:S459–S471

Lemieux H, Semsroth S, Antretter H, Hoefer D, Gnaiger E (2011) Mitochondrial respiratory control and early defects of oxidative phosphorylation in the failing human heart. Int J Biochem Cell Biol 43:1729–1738

Lin J, Wu H, Tarr PT, Zhang CY, Wu Z, Boss O, Michael LF, Puigserver P, Isotani E, Olson EN, Lowell BB, Bassel-Duby R, Spiegelman BM (2002) Transcriptional co-activator PGC-1 alpha drives the formation of slow-twitch muscle fibres. Nature 418:797–801

McVeigh GE, Brennan GM, Johnston GD, McDermott BJ, McGrath LT, Henry WR, Andrews JW, Hayes JR (1992) Impaired endothelium-dependent and independent vasodilation in patients with type 2 (non-insulin dependent) diabetes mellitus. Diabetologia 35:771–776

Mootha VK, Lindgren CM, Eriksson KF, Subramanian A, Sihag S, Lehar J, Puigserver P, Carlsson E, Ridderstråle M, Laurila E, Houstis N, Daly MJ, Patterson N, Mesirov JP, Golub TR, Tamayo P, Spiegelman B, Lander ES, Hirschhorn JN, Altshuler D, Groop LC (2003) PGC-1α-responsive genes involved in oxidative phosphorylation are coordinately downregulated in human diabetes. Nat Genet 34:267–273

Morris JN, Heady JA, Raffle PAB, Roberts CG, Parks JW (1953) Coronary heart-disease and physical activity of work. Lancet 265:1111–1120

Myers J, Prakash M, Froelicher V, Do D, Partington S, Atwood JE (2002) Exercise capacity and mortality among men referred for exercise testing. N Engl J Med 346:793–801

Newsholme P, Homem De Bittencourt PI, O' Hagan C, De Vito G, Murphy C, Krause MS (2010) Exercise and possible molecular mechanisms of protection from vascular disease and diabetes: the central role of ROS and nitric oxide. Clin Sci 118:341–349

Nisoli E, Tonello C, Cardile A, Cozzi V, Bracale R, Tedesco L, Falcone S, Valerio A, Cantoni O, Clementi E, Moncada S, Carruba MO (2005) Calorie restriction promotes mitochondrial biogenesis by inducing the expression of eNOS. Science 310:314–317

Nisoli E, Clementi E, Carruba MO, Moncada S (2007) Defective mitochondrial biogenesis. A hallmark of the high cardiovascular risk in the metabolic syndrome? Circ Res 100:795–806

O'Gorman DJ, Karlsson HK, McQuaid S, Yousif O, Rahman Y, Gasparro D, Glund S, Chibalin AV, Zierath JR, Nolan JJ (2006) Exercise training increases insulin-stimulated glucose disposal and GLUT4 (SLC2A4) protein content in patients with type 2 diabetes. Diabetologia 49:2983–2992

Ohta M, Nanri H, Matsushima Y, Sato Y, Ikeda M (2005) Blood pressure-lowering effects of lifestyle modification: possible involvement of nitric oxide bioavailability. Hypertens Res 28:779–786

Paffenbarger RS, Hale WE (1975) Work activity and coronary heart mortality. N Engl J Med 292:545–550

Paffenbarger RS, Hyde RT, Wing AL, Hsieh CC (1986) Physical activity, all-cause mortality, and longevity of college alumni. N Engl J Med 314:605–613

Paffenbarger RS, Hyde RT, Wing AL, Lee IM, Jung DL, Kampert JB (1993) The association of changes in physical-activity level and other life-style characteristics with mortality among men. N Engl J Med 328:538–545

Pedersen BK (2009) The diseasome of physical inactivity: and the role of myokines in muscle-fat-cross talk. J Physiol 587:5559–5568

Pedersen BK, Steensberg A (2002) Exercise and hypoxia: effects on leukocytes and interleukin-6-shared mechanisms? Med Sci Sports Exerc 34:2004–2013

Pedersen BK, Akerstrom TC, Nielsen AR, Fischer CP (2007) Role of myokines in exercise and metabolism. J Appl Physiol 103:1093–1098

Pesta D, Hoppel F, Macek C, Messner H, Faulhaber M, Kobel C, Parson W, Burtscher M, Schocke M, Gnaiger E (2011) Similar qualitative and quantitative changes of mitochondrial respiration following strength and endurance training in normoxia and hypoxia in sedentary humans. Am J Physiol Regul Integr Comp Physiol 301:R1078–1087

Petersen AMW, Pedersen BK (2005) The anti-inflammatory effect of exercise. J Appl Physiol 98:1154–1162

Phielix E, Meex R, Moonen-Kornips E, Hesselink MK, Schrauwen P (2010) Exercise training increases mitochondrial content and ex vivo mitochondrial function similarly in patients with type 2 diabetes and in control individuals. Diabetologia 53:1714–1721

Plomgaard P, Bouzakri K, Krogh-Madsen R, Mittendorfer B, Zierath JR, Pedersen BK (2005) Tumor necrosis factor-α induces skeletal muscle insulin resistance in healthy human subjects via inhibition of Akt substrate 160 phosphorylation. Diabetes 54:2939–2945

Powers SK, Lennon SL, Quindry J, Mehta JL (2002) Exercise and cardioprotection. Curr Opin Cardiol 17:495–502

Rabøl R, Højberg PM, Almdal T, Boushel R, Haugaard SB, Madsbad S, Dela F (2009) Effect of hyperglycemia on mitochondrial respiration in type 2 diabetes. J Clin Endocrinol Metabol 94:1372–1378

Richter EA, Ruderman NB (2009) AMPK and the biochemistry of exercise: implications for human health and disease. Biochem J 418:261–275

Rizvi AA (2009) Cytokine biomarkers, endothelial inflammation, and atherosclerosis in the metabolic syndrome: emerging concepts. Am J Med Sci 338:310–318

Ross R (1999) Atherosclerosis: an inflammatory disease. N Engl J Med 340:115–116

Seals DR, Desouza CA, Donato AJ, Tanaka H (2008) Habitual exercise and arterial aging. J Appl Physiol 105:1323–1332

Shephard RJ, Balady GJ (1999) Exercise as cardiovascular therapy. Circulation 99:963e72

Stamler J, Stamler R, Neaton JD (1993) Blood pressure, systolic and diastolic, and cardiovascular risks: US population data. Arch Intern Med 153:598–615

Stapleton PA, Goodwill AG, James ME, Brock RW, Frisbee JC (2010) Hypercholesterolemia and microvascular dysfunction: interventional strategies. J Inflamm 7:54

Suzuki K, Ohno H, Oh-ishi S, Kizaki T, Ookawara T, Fujii J, Radak Z, Taniguchi N (2000) Superoxide dismutases in exercise and disease. In: Sen CK, Packer L, Häninen O (eds) Handbook of oxidants and antioxidants in exercise. Elsevier, Amsterdam, pp 243–295

Swain DP, Franklin BA (2006) Comparison of cardioprotective benefits of vigorous versus moderate intensity aerobic exercise. Am J Cardiol 97:141–147

Szostak J, Laurant P (2011) The forgotten face of regular physical exercise: a natural anti-atherogenic activity. Clin Sci 121:91–106

Thompson PD (2005) Exercise prescription and proscription for patients with coronary artery disease. Circulation 112:2354–2363

Thompson PD, Buchner D, Pina IL, Balady GJ, Williams MA, Marcus BH, Berra K, Blair SN, Costa F, Franklin B, Fletcher GF, Gordon NF, Pate RR, Rodriguez BL, Yancey AK, Wenger NK, American Heart Association Council on Clinical Cardiology Subcommittee on Exercise, Rehabilitation, and Prevention, American Heart Association Council on Nutrition, Physical Activity, and Metabolism Subcommittee on Physical Activity (2003) Exercise and physical activity in the prevention and treatment of atherosclerotic cardiovascular disease. Circulation 107:3109–3116

Toth KG, McKay BR, De Lisio M, Little JP, Tarnopolsky MA, Parise G (2011) IL-6 induced STAT3 signalling is associated with the proliferation of human muscle satellite cells following acute muscle damage. PLoS One 6:e17392

Tuomilehto J, Lindström J, Eriksson JG, Valle TT, Hämäläinen H, Ilanne-Parikka P, Keinänen-Kiukaanniemi S, Laakso M, Louheranta A, Rastas M, Salminen V, Uusitupa M, Finnish Diabetes Prevention Study Group (2001) Prevention of type 2 diabetes mellitus by changes in lifestyle among subjects with impaired glucose tolerance. N Engl J Med 344:134–150

Tuteja N, Chandra M, Tuteja R, Misra MK (2004) Nitric oxide as a unique bioactive signaling messenger in physiology and pathophysiology. J Biomed Biotechnol 4:227–237

Tweedie C, Romestaing C, Burelle Y, Safdar A, Tarnopolsky MA, Seadon S, Britton SL, Koch LG, Hepple RT (2011) Lower oxidative DNA damage despite greater ROS production in muscles from rats selectively bred for high running capacity. Am J Physiol Regul Integr Comp Physiol 300:R544–R553

Votion DM, Gnaiger E, Lemieux H, Mouithys-Mickalad A, Serteyn D (2012) Physical fitness and mitochondrial respiratory capacity in horse skeletal muscle. PLoS One 7:e34890

Wallace DC (2005) A mitochondrial paradigm of metabolic and degenerative diseases, aging, and cancer: a dawn for evolutionary medicine. Annu Rev Genet 39:359–407

Wang X, Moraes CT (2011) Increases in mitochondrial biogenesis impair carcinogenesis at multiple levels. Mol Oncol 5:399–409

Waring CD, Papalambrou A, Sharp L, Smith AJ, Purushothaman S, Vicinanza C, Goldspink D, Torella D, Nadal-Ginard B, Ellison GM (2010) Cardiac stem cell activation and ensuing myogenesis and angiogenesis contribute to cardiac adaptation following intensity-controlled exercise training. Circulation 122:A19155

Wei M, Gibbons LW, Mitchell TL, Kampert JB, Lee CD, Blair SN (1999) The association between cardiorespiratory fitness and impaired fasting glucose and type 2 diabetes mellitus in men. Ann Intern Med 130:89–96

Weiner RB, Baggish AL (2012) Exercise-induced cardiac remodeling. Prog Cardiovasc Dis 54:380–386

Wisløff U, Najjar SM, Ellingsen O, Haram PM, Swoap S, Al-Share Q, Fernström M, Rezaei K, Lee SJ, Koch LG, Britton SL (2005) Cardiovascular risk factors emerge after artificial selection for low aerobic capacity. Science 307:418–420

Wyatt AW, Steinert JR, Mann GE (2004) Modulation of the L-arginine/nitric oxide signalling pathway in vascular endothelial cells. Biochem Soc Symp 71:143–156

Zimmet P, Alberti K, Shaw J (2001) Global and societal implications of the diabetes epidemic. Nature 414:782–787

Zoll J, N'Guessan B, Ribera F, Lampert E, Fortin D, Veksler V, Bigard X, Geny B, Lonsdorfer J, Ventura-Clapier R, Mettauer B (2003) Preserved response of mitochondrial function to short-term endurance training in skeletal muscle of heart transplant recipients. J Am Coll Cardiol 42:126–132

Alcohol

7

Ichiro Wakabayashi

Abstract

Alcohol drinking is a general modifiable risk factor for various diseases including cardiovascular disease. A J-shaped relationship is known to exist between alcohol intake and risk of cardiovascular disease as well as between alcohol intake and total mortality. Risks of ischemic arterial diseases, such as coronary artery disease, ischemic stroke, and peripheral arterial disease, are lower in light-to-moderate drinkers (not more than one drink per day for women and not more than two drinks per day for men) than in abstainers. Beneficial effects of alcohol drinking on the risks of ischemic arterial diseases are mainly explained by the actions of alcohol on cholesterol metabolism and the blood coagulation-fibrinolysis system. Light-to-moderate alcohol consumption has also been shown to be associated with decrease in production of proinflammatory cytokines and increase in insulin sensitivity. Additional benefits for cardiovascular protection are expected in polyphenols, contained in red wine, which exert cardiovascular protective actions represented by antioxidative action. On the other hand, heavy drinking and binge drinking increase the risks of hemorrhagic types of stroke, such as cerebral hemorrhage and subarachnoid hemorrhage; cardiomyopathy; and cardiac arrhythmia, particularly atrial fibrillation. The association between alcohol and hemorrhagic stroke is speculated to be due to alcohol-induced hypertension in combination with impaired blood coagulation-fibrinolysis balance. Drinking alcohol beverages, particularly beer, causes increase in blood uric acid level, an independent risk factor for cardiovascular disease. A variety of molecular pathophysiological mechanisms, e.g., oxidative stress, protein phosphorylation, and alterations in Ca^{2+} homeostasis and gene expression, explained in this

I. Wakabayashi, MD, PhD
Department of Environmental and Preventive Medicine,
Hyogo College of Medicine, Mukogawa-cho 1-1,
Nishinomiya, Hyogo 663-8501, Japan
e-mail: wakabaya@hyo-med.ac.jp

I. Wakabayashi, K. Groschner (eds.), *Interdisciplinary Concepts in Cardiovascular Health*,
DOI 10.1007/978-3-7091-1334-9_7, © Springer-Verlag Wien 2013

chapter are involved in the beneficial and detrimental effects of alcohol on cardiovascular disease and its risk factors. Furthermore, the relationships between alcohol intake and cardiovascular risk may be modified by individual genetic backgrounds represented by polymorphisms of alcohol-metabolizing enzymes. Although light-to-moderate drinking is associated with reduced risk of cardiovascular disease, the use of alcohol as a strategy for cardiovascular protection is not recommended for abstainers because of the lack of randomized outcome data and the future possibility of alcohol abuse.

Keywords
Alcohol • Arrhythmia • Cardiomyopathy • Cholesterol • Coagulation-fibrinolysis balance • Hypertension • Inflammation • Insulin sensitivity • Polyphenol • Uric acid

7.1 Introduction

Alcohol drinking, a common habit, often influences individual health. According to a meta-analysis study, consumption of one drink (corresponding to 12 g ethanol) daily for women and one or two drinks daily for men was associated with a reduction of total mortality by 18 %, but more alcohol consumption was associated dose dependently with increased mortality, thus showing a J-shaped relationship between alcohol and total mortality (Di Castelnuovo et al. 2006). A J-shaped relationship has also been demonstrated between alcohol consumption and the risk of cardiovascular disease. There are inverse associations between light-to-moderate alcohol consumption and risks for coronary artery disease (Corrao et al. 2000), ischemic stroke (Reynolds et al. 2003), and peripheral arterial disease (Camargo et al. 1997), while alcohol consumption, particularly heavy drinking and binge drinking, is associated with increased risks for hemorrhagic types of stroke such as cerebral hemorrhage and subarachnoid hemorrhage (Reynolds et al. 2003).

Beneficial effects of alcohol on the risk for ischemic arterial disease are mainly explained by alcohol-induced increase in HDL cholesterol and decrease in blood coagulability through inhibiting platelet aggregation and blood coagulation and augmenting blood fibrinolysis (Rimm et al. 1999). A linear positive association between hemorrhagic stroke and alcohol consumption has been revealed by a systemic review (Mazzaglia et al. 2001). Chronic alcohol consumption is a risk factor for hypertension (Klatsky 1996). Therefore, the association between alcohol and hemorrhagic stroke is speculated to be due to alcohol-induced hypertension in combination with impaired blood coagulation-fibrinolysis balance, which weakens the small deep cerebral arteries (Puddey and Beilin 2006). Heavy alcohol drinking is also associated with several cardiovascular adverse effects on cardiomyopathy (Urbano-Marquez et al. 1989) and cardiac arrhythmia, particularly atrial fibrillation (Balbão et al. 2009). Beneficial and detrimental effects of alcohol drinking on cardiovascular risk factors and cardiovascular events are summarized in Fig. 7.1.

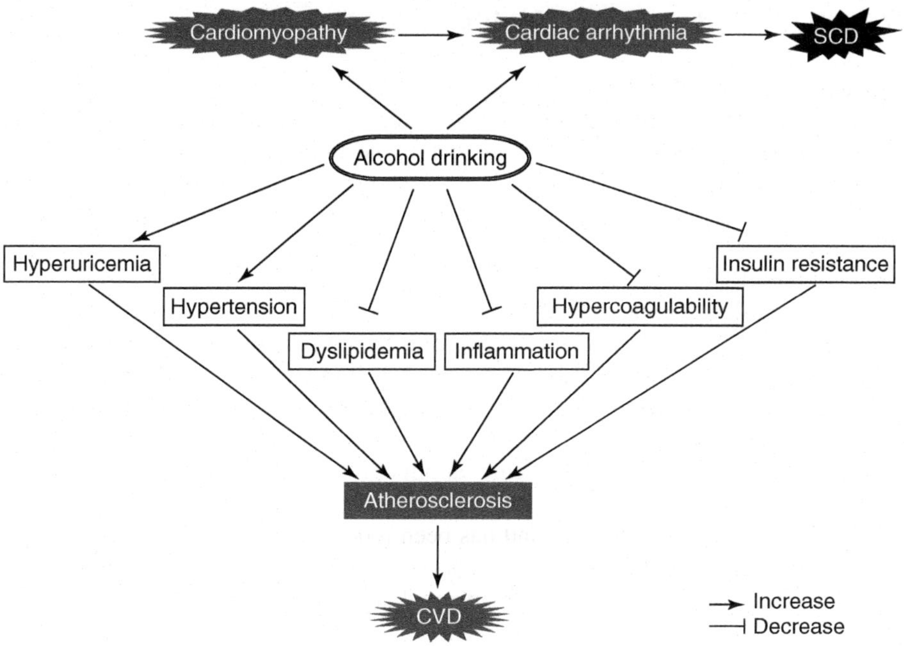

Fig. 7.1 Beneficial and detrimental effects of alcohol drinking on cardiovascular risk factors and cardiovascular events. Each effect of alcohol depends on the degree of alcohol consumption. *CVD* cardiovascular diseases, *SCD* sudden cardiac death

Among alcoholic beverages, additional benefits for cardiovascular protection are expected in red wine because it contains polyphenols, which exert cardiovascular protective actions represented by antioxidative action (Rodrigo et al. 2011). However, whether red wine is superior to other alcoholic drinks or not is still debatable because there is a possibility of confounding by socioeconomic factors such as a healthier lifestyle of wine drinkers compared to that of beer and spirit drinkers (van de Wiel and de Lange 2008). Incidental heavy or binge drinking (generally defined as five or more drinks on an occasion) is known to increase cardiovascular events, and thus the pattern of drinking is suggested to be more important than the kind of alcoholic beverage (van de Wiel and de Lange 2008). Wine consumption with a meal has been shown to suppress blood level of postprandial lipid hydroperoxides (Ursini et al. 1998), and alcohol intake with a meal may have beneficial effects on metabolism of alcohol by its retarded absorption. Therefore, regular light-to-moderate drinking (no more than one drink per day for women and two drinks per day for men) with a meal is the way to go for drinkers. On the other hand, the use of alcohol as a cardioprotective strategy is not recommended for abstainers because of no proof of causality and no way to predict individual possibility of alcohol abuse in the future.

7.2 Alcohol and Blood Lipids

An abnormal blood cholesterol profile, such as high LDL cholesterol and low HDL cholesterol, is the most important risk factor for coronary arterial disease. Habitual alcohol drinking results in elevation of HDL cholesterol and reduction of LDL cholesterol (Castelli et al. 1977; Langer et al. 1992; Gaziano et al. 1993). About half of the protection against coronary heart disease afforded by moderate alcohol consumption has been reported to be mediated by an increase in HDL cholesterol, and an additional 18 % of this protection has been shown to be attributable to a decrease in LDL cholesterol (Langer et al. 1992). Alcohol induces elevation of HDL cholesterol, especially its subfractions of HDL2 and HDL3 (Gaziano et al. 1993). In fact, HDL2 sampled from heavy drinkers has been demonstrated to enhance cholesterol efflux from cultured macrophages *in vitro* (Mäkelä et al. 2008). Non-HDL cholesterol, calculated as the difference between total cholesterol and HDL cholesterol, contains particles of all atherogenic apolipoprotein B-containing lipoproteins, including very-low-density lipoprotein (VLDL), intermediate-density lipoprotein (IDL), LDL, and lipoprotein(a), and has been reported to be superior to LDL cholesterol in predicting cardiovascular events (Cui et al. 2001). Non-HDL cholesterol has also been shown to be lower in drinkers than in nondrinkers (Wakabayashi and Groschner 2009).

As mechanisms for the effects of alcohol on lipid metabolism, habitual alcohol drinking has been shown to increase lipoprotein lipase (LPL) activity and decrease cholesteryl ester transfer protein (CETP) activity, resulting in elevation of HDL cholesterol and reduction of LDL cholesterol (Savolainen et al. 1990; Hannuksela et al. 2004). CETP mediates transfer of cholesteryl esters from HDL to LDL and VLDL particles. Alcohol itself or its metabolite, acetaldehyde, has been suggested to cause defective sialylation and inhibit glycosylation of CETP, which is necessary for binding of CETP to lipoproteins, in alcohol drinkers (Liinamaa et al. 2006). Polymorphism of the CETP gene (TaqIB) is known to influence HDL cholesterol level and risk of coronary artery disease (Boekholdt et al. 2005). The TaqIB SNP has been reported to modulate effects of alcohol on level of HDL cholesterol and risk of coronary artery disease: Individuals carrying the TaqIB2 allele showed stronger associations of alcohol intake with elevation of HDL cholesterol and reduction of the risk of coronary artery disease than did individuals carrying the TaqIB1 allele (Fumeron et al. 1995). Regarding polymorphism of LPL, S447X polymorphism has been reported to modulate the association between alcohol drinking and plasma HDL cholesterol level in women (Lee et al. 2004).

A meta-analysis study using 35 data records has shown that triglyceride concentrations increased by 0.19 mg/dl/g of alcohol consumed a day and that this association was stronger in men than in women and slightly stronger in beer drinkers than in wine drinkers (Rimm et al. 1999). Excessive alcohol drinking causes hypertriglyceridemia in both postprandial and fasting states, which may be mainly due to an increase in the synthesis of large VLDL particles in the liver, and an acute inhibitory effect of alcohol on LPL activity is also involved in postprandial hypertriglyceridemia in drinkers (Van de Wiel 2012). On the other hand, a well-controlled intervention study has demonstrated that consumption of alcohol (15 and 30 g/day) dose dependently reduced triglyceride levels in healthy postmenopausal women (Davies et al. 2002). Therefore, the relationship between alcohol and triglycerides may be greatly modified by diet and other lifestyle factors.

7.3 Alcohol and Blood Coagulation-Fibrinolysis Balance

Thrombus formation on the arterial wall is an important etiology for progression of atherosclerosis. Therefore, a disorder of blood coagulation-fibrinolysis balance determined by activities of platelets, coagulation factors, and fibrinolysis factors is a major risk factor for atherosclerotic disease. Inhibitory effects of alcohol on platelet function and blood coagulation factors and augmenting effects of alcohol on the blood fibrinolysis pathway partly explain the inverse association between light-to-moderate alcohol consumption and the risk for ischemic types of atherosclerotic diseases such as ischemic heart disease, cerebral infarction, and peripheral arterial disease. On the other hand, hemorrhagic stroke is induced by heavy drinking through potentiation of fibrinolytic function by alcohol.

Alcohol inhibits platelet activation *in vivo* and *in vitro* in response to diverse agonists such as collagen, adenosine diphosphate (ADP), thrombin, and platelet-activating factor (PAF). Ethanol inhibits the formation of arachidonic acid and thromboxane A_2, while free arachidonic acid-induced platelet aggregation was not affected by ethanol (Rubin 1999). Direct inhibition of membrane-associated phospholipase A_2 by ethanol was also demonstrated (Stubbs and Rubin 1992). Therefore, the major inhibitory focus of ethanol is thought to be phospholipase A_2. A physiological agonist, such as thrombin, induces platelet aggregation via phospholipase C-mediated phosphoinositide turnover, resulting in production of inositol trisphosphate and diacylglycerol. The former induces release of Ca^{2+} from its intraplatelet storage pools and triggers transmembraneous Ca^{2+} entry, called capacitative Ca^{2+} entry (CCE) or store-operated Ca^{2+} entry (SOCE), in which canonical transient receptor potential channel (TRPC) 1/4/5 is involved. The latter product induces transmembraneous Ca^{2+} entry through activating TRPC 3/6/7. Ethanol has been demonstrated to inhibit agonist-independent CCE induced by a Ca^{2+}-ATPase inhibitor, thapsigargin (Wakabayashi and Marumo 2002), while Ca^{2+} entry induced by a diacylglycerol mimic, 1-oleoyl-2-acetyl-sn-glycerol (OAG), was not inhibited by ethanol (Marumo and Wakabayashi 2010). Therefore, ethanol is thought to selectively inhibit TRPC 1/4/5. As another inhibitory focus of ethanol, ADP- and thrombin-induced p-selectin expression in platelets has been shown to be attenuated by ethanol, and this inhibitory effect of ethanol is suggested to be independent of the effect of ethanol on arachidonate metabolism, because aspirin treatment did not attenuate this effect (Nguyen et al. 1998; Mukamal et al. 2005).

Elevated fibrinogen level in blood is known to be an independent cardiovascular risk factor, although it is still under debate whether fibrinogen is just a marker of cardiovascular risk or whether lowering circulating levels of fibrinogen will result in a significant decrease in clinically relevant endpoints (Kakafika et al. 2007).

An inverse association between alcohol consumption and fibrinogen level has been shown in many epidemiological studies. A 0.78 % decrease in fibrinogen concentration with each 10 g/day increase in alcohol intake has been indicated by results of a prospective study (Meade et al. 1987). An experimental study demonstrated that blood fibrinogen concentration was lowered by daily ethanol administration in rats (5 % ethanol v/v in drinking water for 4 weeks) and that fibrinogen production and mRNA expression of fibrinogen in lined hepatoma cells were diminished by exposure of the cells to ethanol at 10 and 20 mM (Wang et al. 1999). In addition to fibrinogen level, levels of von Willebrand factor and factor VII have been shown to be lower in light-to-moderate alcohol drinkers than in nondrinkers (Mukamal et al. 2001).

The results of previous epidemiological studies concerning relationships of alcohol drinking with fibrinolytic factors are controversial. Positive associations of alcohol intake with tissue plasminogen activator (tPA) antigen and activity and with urine plasminogen activator (uPA) activity were demonstrated (Sumi et al. 1988; Ridker et al. 1994; Mukamal et al. 2001), while plasminogen activator inhibitor-1 (PAI-1) antigen and activity were reportedly increased by acute alcohol drinking (Mukamal et al. 2001; van de Wiel et al. 2001). There were reports showing no relationships of regular alcohol drinking with tPA antigen and activity levels and PAI-1 antigen and activity levels (van Golde et al. 2002; Volpato et al. 2004). In cultured human monocytes, fibrinolytic activity and mRNA expression of tPA and uPA were increased by incubation with ethanol (0.1 % v/v) (Tabengwa et al. 2002). Ethanol *in vitro* also increased tPA and uPA antigen and their mRNA expression in cultured human vascular endothelial cells, while PAI-1 antigen and its mRNA expression were decreased by ethanol *in vitro* (Booyse et al. 1999).

7.4 Alcohol-Induced Hypertension

Since the 1970s, many longitudinal and cross-sectional studies have demonstrated that chronic alcohol consumption is a risk factor for hypertension independent of the type of alcoholic beverage, adiposity, education, smoking, salt intake, and several other traits (Klatsky 1996). A study analyzing WHO Global Burden of Disease has shown that 16 % of all hypertensive diseases were attributed to alcohol (Rehm et al. 2003). A variety of mechanisms for the relationship between alcohol and blood pressure have been suggested by previous studies using acute and chronic experiments. The mechanisms include increased levels of catecholamines, renin, and cortisol; sympathetic nervous system effect; increased peripheral vascular muscle tone; abnormality in calcium transport; erythrocyte membrane changes; insulin resistance; and magnesium depletion (Klatsky 2004). However, no convincing conclusion with respect to the exact mechanism for alcohol-induced hypertension has been obtained. According to a systemic review on the relationship between alcohol and blood pressure including studies using ambulatory or home blood pressure monitoring, alcohol shows two opposite effects, an early effect (in the hours after drinking) leading to a reduction of blood pressure and a later effect (next day) leading to an elevation of blood pressure (McFadden et al. 2005). Because habitual drinkers experience repeated alcohol withdrawal, adrenergic discharge following withdrawal from alcohol intake has been suggested to be an important mechanism for alcohol-induced hypertension (Potter et al. 1983; Kawano 2010). As the mechanism for the acute depressing effect of alcohol on blood pressure, several actions of acetaldehyde, such as direct vasodilating action (Altura and Altura 1982), release of prostacyclin (Guivernau et al. 1987), and activation of glandular kallikreins (Hatake et al. 1990), have been proposed. Since changes in average 24-h blood pressure after drinking alcohol are smaller than alcohol-induced changes in blood pressure at casual office- or clinic-based measurements, a possibility of overestimation of the hypertensive effect of alcohol has been pointed out (Kawano 2010).

7.5 Alcohol and Cardiac Arrhythmias

High risk for cardiac rhythm disorders, particularly atrial fibrillation, by heavy alcohol drinking on weekends or holidays is known as the "holiday heart syndrome" (Ettinger et al. 1978). Meta-analysis studies have demonstrated a dose-dependent relationship between alcohol drinking and risk of atrial fibrillation: The risk was estimated to be increasing 1.5-fold by the highest alcohol intake compared to that by the lowest alcohol intake or no alcohol intake, while there may be no difference in the risk between abstainers and moderate drinkers (Samokhvalov et al. 2010; Kodama et al. 2011). Decrease in calcium currents and shortening of action potential duration by alcohol were demonstrated in *in vitro* studies but were not reproduced in studies using intact animal models, and results regarding electrophysiological effects of alcohol in humans were controversial (Balbão et al. 2009). Therefore, at present, alcohol seems not to exert direct myocardial toxicity inducing atrial fibrillation and is speculated to cause susceptibility to induction of atrial fibrillation through increasing levels of circulating catecholamines, inducing oxidative stress, and releasing plasma free fatty acids in individuals with structural heart diseases or in long-term alcohol drinkers with subclinical cardiac abnormalities (Balbão et al. 2009).

Heavy drinking is also associated with increased risk of ventricular tachyarrhythmias and sudden cardiac death (SCD). Potential mechanisms that could account for these associations are related to multiple physiological aberrancies in chronic alcohol abuse, including increased QT interval leading to ventricular tachyarrhythmias, proarrhythmic electrolyte abnormalities such as hypomagnesemia and hypokalemia, sympathoadrenal stimulation, and reduced baroreflex sensitivity (Lucas et al. 2005; George and Figueredo 2010). In contrast, a prospective study has shown reduced risk of SCD in light alcohol drinkers (two to six drinks per week) compared with that in subjects who rarely or never drank alcohol and compared with heavy drinkers (six or more drinks per day) and binge drinkers (Albert et al. 1999). This inverse association between modest alcohol consumption and risk of SCD may be attributable to reduced risk of coronary artery disease in light-to-moderate alcohol drinkers.

7.6 Alcohol and Cardiomyopathy

Alcoholic cardiomyopathy, an important etiology of mortality in chronic alcoholics, is observed in those excessively consuming alcohol such as greater than 100 and 80 g of ethanol per day in men and women, respectively, for longer than 10 years in the absence of coronary artery disease and nutritional disorders (Urbano-Marquez et al. 1989). Thus, women are more sensitive to toxic effects of alcohol on the myocardium than are men. There may also be a genetic involvement in vulnerability to alcoholic cardiomyopathy: Alcoholics with DD polymorphism of angiotensin-converting enzyme (ACE) showed a 16-fold higher risk of cardiomyopathy than did those with other ACE genotypes (Fernández-Solà et al. 2002). Left ventricular diastolic dysfunction has been proposed to be the earliest sign of subclinical alcoholic cardiomyopathy. After subclinical alcoholic cardiomyopathy, patients continuing to drink alcohol heavily suffer from congestive heart failure and have a higher incidence of arrhythmias and sudden death.

Macro- and microscopic findings of alcoholic cardiomyopathy are indistinguishable from those of other dilated cardiomyopathy. *In vitro* and *ex vivo* experiments revealed that long-term alcohol consumption induced histological and cellular abnormalities such as myocyte loss, intracellular organelle dysfunction, reduction of contractile proteins, and impairment of calcium homeostasis (Laonigro et al. 2009). Alcohol has been shown to have apoptotic effects on cardiac myocytes and increase expression of pro-apoptotic proteins such as Bax and Bcl-2 with degrees similar to the effects of long-standing hypertension (Fernández-Solà et al. 2006). Increase in myocardial lipid peroxidation and protein oxidation and decrease in glutathione content have been demonstrated in alcohol-intoxicated animal models, and thus reactive oxygen species (ROS), which are produced from ethanol through cytochrome P450 2E1 after heavy drinking, play an important role in alcohol-induced cardiomyopathy. The renin-angiotensin system (RAS) is activated by prolonged alcohol consumption (Wright et al. 1986). Alcoholic cardiomyopathy has also been proposed to be caused by alcohol-generated nitrative stress through activation of protein kinase C (PKC)-β1-dependent expression of NOX (nicotinamide adenine dinucleotide phosphate oxidase) by interaction of stimulation of angiotensin II with its type 1 receptor (Tan et al. 2012). A mass spectrometry-based proteomic analysis demonstrated that chronic ethanol administration caused reduction of myofibrillar, sarcoplasmic, membrane-associated, and mitochondrial proteins due to fundamental alteration of cardiac muscle (Fogle et al. 2010). Chronic alcohol consumption has been shown to impair myocardial protein synthesis through RNA translation following reduction of active eIF4E (eukaryotic translation initiation factor 4E) and p70 ribosomal S6 phosphorylation (Vary et al. 2001). Ethanol is known to impair cardiac excitation-contraction coupling, and abnormalities in calcium homeostasis, such as impairment of cytosolic Ca^{2+} transients, decrease in myofilament Ca^{2+} sensitivity, and upregulation of L-type Ca^{2+} channels, have been demonstrated in alcohol-intoxicated rat hearts (Laonigro et al. 2009). The above-proposed mechanisms for alcohol-induced cardiomyopathy are summarized in Fig. 7.2.

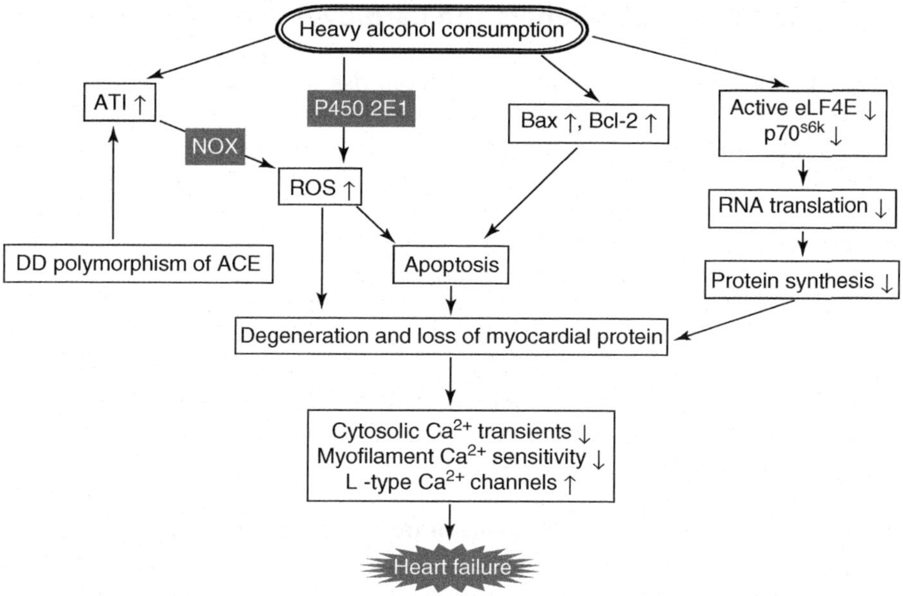

Fig. 7.2 Proposed mechanisms for alcohol-induced cardiomyopathy. *AT1* angiotensin type 1 receptor, *ACE* angiotensin-converting enzyme, *NOX* nicotinamide adenine dinucleotide phosphate oxidase, *p70^{S6K}* p70 ribosomal S6 kinase, *ROS* reactive oxygen species

7.7 Alcohol and Hyperuricemia

In addition to relations to established cardiovascular risk factors such as gender, age, obesity, and blood pressure, elevated serum uric acid level is an independent risk factor for cardiovascular disease (Feig et al. 2008). However, it has not yet been determined whether lowering uric acid levels would be of clinical benefit for the prevention of cardiovascular disease (Feig et al. 2008). *In vivo* and *in vitro* findings suggest that uric acid contributes to endothelial dysfunction and has proinflammatory and proliferative effects on vascular smooth muscle cells (Kanellis and Kang 2005). In a US prospective study (NHANES III), serum uric acid level tended to increase with an increase in frequency of drinking and was highest in habitual beer drinkers among various alcohol beverage drinkers, while there was an inverse relationship between drinking frequency and serum uric acid level in habitual wine drinkers (Choi and Curhan 2004). In addition to intake of purine body contained in alcohol beverages, particularly in beer, alcohol drinking increases uric acid level by reducing renal excretion of uric acid due to elevation of blood lactate and by increasing uric acid production due to enhanced turnover of adenine nucleotides following consumption of ATP in the liver (Yamamoto et al. 2005). On the other hand, acute elevation of uric acid has recently been reported to increase antioxidant capacity and to be a protective factor when drinking red wine (Modun et al. 2008).

7.8 Red Wine Polyphenols and Cardiovascular Disease

Polyphenols are mainly responsible for reduced risk of cardiovascular diseases by red wine, which is known as French paradox and Mediterranean diet. Polyphenols in red wine, most of which are derived from the skin and seeds of grapes, possess antioxidant and free radical-scavenging properties and show beneficial effects through acting on signaling pathways of platelets and vascular cells such as endothelial and smooth muscle cells. In addition, lipid- and lipoprotein-lowering actions of polyphenols are also involved in reduction of cardiovascular disease risk by red wine and its polyphenol constituents. Major wine-containing polyphenols are stilbenes, flavan-3-ols, flavonols, anthocyanins, hydroxybenzoic acid, procyanidins, and hydroxycinnamic acids, and they are largely grouped as flavonoids and nonflavonoids (Smirnoff 2005). Well-known flavonoids in red wine are flavonols (e.g., quercetin) and flavan-3-ols (e.g., tannins and catechin), while resveratrol is the best known nonflavonoid.

Phenolic compounds in red wine, but not its alcohol component, have antioxidative characteristics including scavenging of free radicals, chelation of metals, sparing of vitamin E and carotenoids, and increasing serum paraoxonase activity (Fuhrman and Aviram 2001). Increase in blood antioxidant activity following red wine consumption reduces LDL sensitivity to peroxidation and thereby suppresses foam cell formation and early atherosclerotic plaque development. Figure 7.3 shows various mechanisms explained below for suppression of atherosclerosis by polyphenols in alcohol beverages.

Red wine has been demonstrated to inhibit agonist-induced platelet aggregation and thromboxane A_2 formation by *in vitro* and *ex vivo* experiments (Pace-Asciak et al. 1996). Raising levels of cAMP and cGMP in platelets due to decreased activities of cAMP and cGMP phosphodiesterases and subsequent decrease in cytosolic calcium levels have been shown to be responsible for polyphenol-induced inhibition of platelet function (Soleas et al. 1997). Red wine polyphenols also decreased platelet production of hydrogen peroxide and inhibited the phospholipase C pathway (Pignatelli et al. 2000). Nitric oxide (NO) inhibits platelet aggregation through augmenting cGMP guanylate cyclase, resulting in an increase in cGMP level in platelets. Grape juice has been shown to attenuate platelet aggregation, increase platelet-derived NO release, and decrease superoxide production in platelets (Freedman et al. 2001).

The vascular endothelium plays obligatory roles in preserving blood flow and keeping a good coagulation-fibrinolysis balance. For these purposes, NO is a key substance produced in and released from endothelial cells. Red wine polyphenols are known to enhance endothelial NO production and NO-mediated vasodilation (Fitzpatrick et al. 1993). Oxidative stress reduces vascular bioavailability of NO, causing an increase in vascular tone. Polyphenols have been shown to reduce vascular oxidative stress through reducing expression and activity of NADPH oxidase and increasing antioxidant enzymes such as catalase (Ying et al. 2003). Several mechanisms for enhancement of endothelial NO synthase (eNOS) by polyphenols have been proposed: Src-dependent activation of the PI3-kinase pathways in endothelial cells is induced by prooxidant response to polyphenols, resulting in Akt phosphorylation, which increases activity of eNOS via phosphorylation of Ser1177,

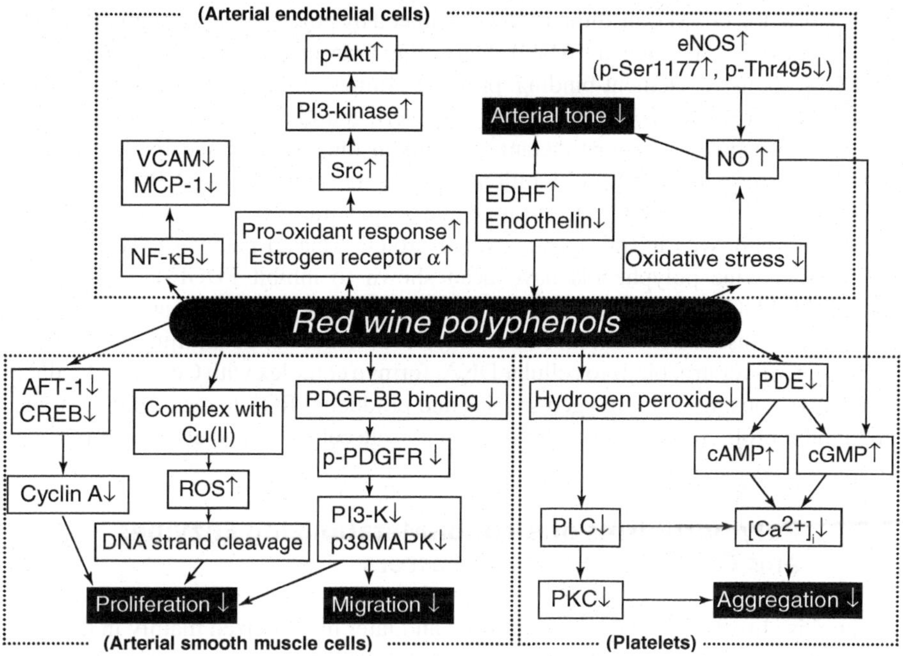

Fig. 7.3 Proposed mechanisms for suppression of atherosclerosis by polyphenols contained in alcohol beverages. *ATF-1* activating transcription factor-1, *CREB* cAMP-responsive element-binding protein, *eNOS* endothelial nitric oxide synthase, *EDHF* endothelium-derived hyperpolarizing factor, *MCP-1* monocyte chemotactic protein-1, *NFκB* nuclear factor κB, *NO* nitric oxide, *PDE* phosphodiesterase, *PDGF* platelet-derived growth factor, *p-PDGFR* phosphorylated platelet-derived growth factor receptor, *PI3-K* phosphoinositide 3-kinase, *PLC* phospholipase C, *PKC* protein kinase C, *ROS* reactive oxygen species, *VCAM* vascular cell adhesion molecule

an activator site of eNOS, while Thr495, an inhibitor site of eNOS, is dephosphorylated by red wine polyphenols (Ndiaye et al. 2005; Anselm et al. 2007). Estrogen receptor α has been suggested to be a pivotal receptor of red wine polyphenols, particularly delphinidin, involved in Src-dependent activation of eNOS expression (Chalopin et al. 2010). As other mechanisms for the vasodilating activity of red wine polyphenols, inhibition of endothelin-1 synthesis (Corder et al. 2001) and induction of EDHF (endothelium-derived hyperpolarizing factor)-mediated vasorelaxation (Ndiaye et al. 2003) have been reported in several vessels.

Endothelial expression of an adhesion molecule, VCAM-1 (vascular cell adhesion molecule-1), which is activated through the NFκB (nuclear factor κB) pathway, is increased at atherosclerotic lesions. Red wine inhibits NFκB activation, resulting in downregulation of VCAM-1 (Carluccio et al. 2003). Red wine also inhibits endothelial expression of MCP-1 (monocyte chemotactic protein-1), resulting in attenuated monocyte migration into the subendothelial lesion (Feng et al. 1999).

Vascular smooth muscle cell proliferation and migration, which are important components of atherogenesis, are inhibited by polyphenols. Several mechanisms for the antiproliferative action of red wine have been proposed: Gene expression of cyclin A, a cell-cycle regulator functioning at two stages of the cell cycle, at the G1-S transition and again at the G2-M transition, is downregulated by polyphenols through decreasing transcription factors such as AFT-1 (activating transcription factor-1) and CREB (cAMP-responsive element-binding protein) (Iijima et al. 2000). Platelet-derived growth factor (PDGF)-BB, released from platelets, endothelial cells, and smooth muscle cells, is a potent chemoattractant for smooth muscle cells and induces smooth muscle cell migration and proliferation through phosphorylation of PDGF receptor (PDGFR) and subsequent activation of the signal transduction cascade including PI3-K and p38 MAPK (mitogen-activated protein kinase) pathways (Heldin et al. 1998). Red wine polyphenols have been shown to inhibit PDGF-BB binding to βPDGFR, PDGF-BB-induced PI3-K and p38 MAPK activation, and subsequent migration of vascular smooth muscle cells (Rosenkranz et al. 2002; Iijima et al. 2002). Moreover, resveratrol binds to cellular DNA, forms a complex with Cu(II), and reduces it to Cu(I), causing redox oxygen species and producing ROS to cleave DNA strands and inhibit proliferation of vascular smooth muscle cells (Ahmad et al. 2000).

7.9 Polymorphisms of Alcohol-Metabolizing Enzymes and Cardiovascular Risk Factors

Ethanol is absorbed by the small intestine and later metabolized mainly in the liver. Alcohol dehydrogenases (ADHs) are cytosolic, dimeric, zinc-containing NAD-dependent enzymes that oxidize ethanol into acetaldehyde. When alcohol consumption is high, cytochrome P450 2E1 (CYP2E1, a member of the cytochrome P450 superfamily) can also catalyze ethanol into acetaldehyde while producing reactive oxygen species (ROS). Subsequently, acetaldehyde is converted into acetate by aldehyde dehydrogenases (ALDHs). Although there are multiple forms of ALDH in the liver, the enzyme encoded by mitochondrial ALDH2 on chromosome 12 has a very low Michaelis constant for acetaldehyde (about 1 μmol/l) and is thought to oxidize most of the acetaldehyde generated during alcohol metabolism.

Functional relevant polymorphism is known in the genes encoding ADH1 (ADH1B, Arg48His; ADH1C, Arg272Gln and Ile350Val) and ALDH2 (Glu504Lys). Exchange of amino acid in each polymorphism introduced here is based on the information obtained from the Single Nucleotide Polymorphism Database (dbSNP, http://www.ncbi.nlm.nih.gov/snp/) (build 137). The ADH1B 48Arg and homozygous ADH1C 350Val have been shown to be associated with slow metabolism of ethanol (Osier et al. 1999). ALDH2 504Lys is an inactive variant that is common among Asians but nearby absent among white populations and protects against alcoholism. Homozygous 504Lys persons are unable to metabolize acetaldehyde and drink considerably less alcohol than do homozygous 504Glu persons due to Oriental flushing syndrome characterized by facial flushing, palpitation, nausea, and headache (Brooks et al. 2009). Persons with heterozygous ALDH2 drink intermediate amounts of alcohol. The ALDH1B1 gene also shows polymorphism at two different residues of Ala86Val and Arg107Leu, though functional relevance of this polymorphism is unknown (Sherman et al. 1993). Polymorphisms of ADH1 and ALDH2 are related to outcomes of various diseases including cancer both in Asians and white populations (Seitz and Meier 2007).

Since amount of alcohol consumption is strongly related to ALDH2 polymorphism in Asians and habitual alcohol drinking is an important risk factor for hypertension, it is understandable that blood pressure is associated with ALDH2 polymorphism in Asians. According to a recent meta-analysis study including ten studies using Asian subjects, overall odds ratios for hypertension in the wild-type homozygotes and the heterozygotes vs. the null variant have been reported to be 2.42 and 1.72, respectively, and systolic blood pressure levels were 7.44 and 4.24 mmHg greater in the wild-type homozygotes and in the heterozygotes, respectively, than in the null variant (Chen et al. 2008). ADH1B variant and ALDH1B1 variant (Ala86Val) were reportedly associated with systolic blood pressure in Japanese (Hashimoto et al. 2002) and with diastolic blood pressure in Danish (Husemoen et al. 2008), respectively, though there have been only a limited number of studies on the relationships between these polymorphisms and blood pressure. In addition, a fast-metabolizing variant of ADH1B has also been shown to be associated with high triglycerides and high uric acid (Hashimoto et al. 2002). On the other hand, it has been shown that there were no significant relationships of the ADH1B variant with LDL and HDL cholesterols (Hashimoto et al. 2002; Whitfield et al. 2003). Amount of alcohol consumption was not significantly different in the three ADH1B variant groups (Hashimoto et al. 2002), and the above ADH1B-related increases in prevalence of cardiovascular risk factors might be caused by high velocity of alcohol oxidation, resulting in accelerated production of ROS. ADH1C (rs1693482) fast-metabolizing variant has been shown to be associated with an increased risk of impaired glucose tolerance/diabetes, though no clear dose-response relationship between the risk and ADH1C polymorphism was observed (Husemoen et al. 2010).

As aforementioned, flushing is a typical symptom of high sensitivity to alcohol drinking often found in Asians due to accumulation of acetaldehyde resulting from low activity of ALDH. Persons with high sensitivity to alcohol evaluated by the symptom of flushing showed a stronger association between alcohol intake and increased risk for hypertension in Japanese (Itoh et al. 1997; Wakabayashi 2005) and Chinese (Zhang et al. 2009). Thus, there is an interaction between alcohol and flushing with respect to blood pressure, suggesting that ALDH2 variant (rs671) is a risk factor for hypertension in drinkers. Similarly, HDL cholesterol and insulin resistance have been shown to be more prone to be affected by alcohol in flushers than in non-flushers by Japanese and Korean groups (Wakabayashi and Masuda 2006; Jung et al. 2010). However, a recent study has shown the opposite finding that systolic and diastolic blood pressure in Japanese heavy drinkers was higher in drinkers with the active variant of ALDH2 rs671 than in drinkers with its inactive variant (Tsuchihashi-Makaya et al. 2009). This discrepancy might be due to differences in the amount of alcohol consumption between subjects with active and inactive variants of ALDH2 rs671 in each drinker subgroup. Moderate drinkers with the homozygote of slow-oxidizing ADH1C allele reportedly had higher HDL cholesterol levels and a substantially decreased risk of myocardial infarction, suggesting a significant interaction regarding HDL cholesterol and myocardial infarction between alcohol and ADH1C (Hines et al. 2001). Interaction between alcohol and ADH1C (rs698) has also been shown regarding adiposity indices such as body mass index and waist circumference in men (Latella et al. 2009). Significant interactions

Table 7.1 Major polymorphisms of alcohol-metabolizing enzymes related to cardiovascular risk factors

Gene	dbSNP number	Chromosomal location (GRCh37/hg19)	Alleles	Exchange of amino acid or nucleotide	Cardiovascular risk factors possibly influenced by polymorphism
ADH1B	rs1229984	4q23 (100239319)	A/G	p.His48Arg	Blood pressure, triglycerides, uric acid
ADH1C	rs1693482	4q23 (100263965)	C/T	p.Arg272Gln	Glucose tolerance
	rs698	4q23 (100260789)	A/G/T	p.Ile350Val/ p.Ile350Phe	HDL chol, adiposity
ALDH1B1	rs2228093	9p13.2 (38396002)	C/T	p.Ala86Val	Blood pressure
	rs2073478	9p13.2 (38396065)	G/T	p.Arg107Leu	Unknown
ALDH2	rs671	12q24.12 (112241766)	A/G	p.Glu504Lys	Blood pressure, HDL chol, insulin resistance
	rs886205	12q24.12 (112204427)	C/T	p.Glu457Lys c.-96-264A>G	Glucose tolerance

The information on each gene and its polymorphism was obtained from the Single Nucleotide Polymorphism Database (dbSNP, http://www.ncbi.nlm.nih.gov/snp/) (build 137). *Chol* cholesterol

have been reported between alcohol and ADH1B (rs1229984) with respect to LDL cholesterol and between alcohol and ALDH2 (rs886205) with respect to glucose tolerance (Husemoen et al. 2010).

Table 7.1 summarizes major polymorphisms obtained by web search (http://www.ncbi.nlm.nih.gov/snp/) of alcohol-metabolizing enzymes related to cardiovascular disease and its risk factors. Polymorphisms of alcohol-metabolizing enzymes such as ADH1B, ADH1C, ALDH1B1, and ALDH2 are associated with cardiovascular risk factors including blood pressure, blood lipids, blood uric acid, and glucose tolerance. However, some of the results of previous studies are inconsistent, and there are clear ethnic differences in frequencies of the polymorphisms. There has been limited knowledge on the relationships among alcohol, cardiovascular risks, and polymorphism of CYP2E1: Results of previous studies using human and animal models suggested no contribution of CYP2E1 and its polymorphism to alcohol-induced blood pressure elevation (Yamada et al. 2002; Cowpland et al. 2006). Further studies using larger populations as well as longitudinal studies are needed to clarify the relationships among alcohol, polymorphisms of its metabolizing enzymes, and cardiovascular risks.

7.10 Alcohol and Inflammation

Inflammatory mechanisms couple dyslipidemia to atheroma formation and promote thrombosis, a late and dreaded complication of atherosclerosis responsible for cardiovascular events such as myocardial infarctions and most strokes

(Libby 2002). In addition to a role as an immunosuppressant (Nelson and Kolls 2002), alcohol shows anti-inflammatory actions, which are mediated through changing cytokine profiles. Moderate alcohol intake has been shown to reduce LPS-stimulated early production of proinflammatory cytokines, such as TNF-α and IL-1, through inhibiting NFκB activation and to augment late production of IL-10, an anti-inflammatory cytokine in monocytes (Mandrekar et al. 2006). An *in vitro* study using human monocytes has also revealed that alcohol inhibited NFκB translocation by decreasing IκB kinase activity and subsequent p65 phosphorylation as well as by promoting proteolytic degradation of IκBα (Mandrekar et al. 2007). Ethanol has been suggested to inhibit LPS-induced STAT1 activation and subsequent overproduction of NO through the inducible NO synthase pathway in a murine RAW 264.7 monocyte/macrophage line (Wakabayashi and Negoro 2002). Moderate alcohol consumption has been shown to be inversely associated with inflammatory markers, such as TNF, IL-6, and CRP, in men and women (Pai et al. 2006).

7.11 Alcohol and Insulin Sensitivity

Meta-analysis studies have shown a 30 % decreased risk of type 2 diabetes in moderate alcohol drinkers (Koppes et al. 2005; Carlsson et al. 2005), although further investigations are needed to determine if the risk is increased by high alcohol consumption. Decreased risk of diabetes in moderate drinkers may be due to increased insulin sensitivity by alcohol (Hulthe and Fagerberg 2005). Alcohol consumption is thought to promote insulin sensitivity independent of body weight and body fat (Hong et al. 2009; Paulson et al. 2010). Since the association between chronic alcohol intake and insulin sensitivity is supported mainly by cross-sectional studies, further large-scale longitudinal studies as well as intervention studies are needed.

Adiponectin promotes insulin sensitivity and possesses anti-inflammatory properties (Aldhahi and Hamdy 2003). A consistent significant increase in blood adiponectin levels after alcohol consumption has been shown by a meta-analysis study (Brien et al. 2011), and alcohol-induced increase in adiponectin is attributable, at least in part, to transcriptional alterations in adipose tissue (Joosten et al. 2008). Leptin, another adipocyte-secreted hormone, increases glucose catabolism and insulin sensitivity (Mantzoros et al. 2011). By a gene array analysis using mouse white adipose tissue, mRNA expression of leptin and anti-inflammatory factors, such as IL-10 and adrenergic β receptor kinase 1 (Adbk1), as well as mRNA expression of adiponectin, has been demonstrated to be increased by chronic alcohol administration (Paulson et al. 2010). There is also a hypothesis to explain the inverse association between alcohol and insulin sensitivity: Metabolism of acetate in peripheral tissues generates sufficient levels of AMP to temporarily stimulate the AMP-activated protein kinase, which in turn induces the synthesis of certain long-lived proteins that act to boost insulin sensitivity and possibly aid the efficiency of fat oxidation as well (McCarty 2001).

7.12 Summary

Alcohol has both beneficial and detrimental effects on the risk of cardiovascular disease. The former is thought to be mediated by alcohol actions such as improvement of cholesterol profile, inhibition of blood coagulation, and promotion of insulin sensitivity, while the latter is explained by alcohol-induced hypertension, hyperuricemia, and hypertriglyceridemia. In addition, contents other than ethanol in alcoholic beverage, particularly polyphenols in red wine, work preventively for cardiovascular disease mainly through their antioxidative actions. In general, no more than one drink per day for women and two drinks per day for men are recommended. The relationships between alcohol consumption and risks of cardiovascular disease are modified by alcohol-metabolizing enzymes such as ADH and ALDH, and there may also be other individual factors, e.g., smoking and medication, influencing the relationships. The positive and inverse associations of alcohol intake with blood pressure and LDL cholesterol, respectively, have been shown to be stronger in smokers than in nonsmokers (Wakabayashi 2011). Alcohol intake has been shown to be associated with blood pressure and pulse pressure in older men not receiving therapy for hypertension but not in those receiving antihypertensive therapy (Wakabayashi 2010), indicating a possibility that alcohol drinking does not have a substantial impact on blood pressure among treated hypertensive patients. Therefore, in order to determine the precise limit of the amount of alcohol consumption for each person, individual backgrounds, including age, gender, smoking, medication, body composition, and genetic polymorphism of alcohol-metabolizing enzymes, should be further taken into account (Fig. 7.4). Knowledge of the relation of alcohol to cardiovascular disease has been strengthened by a numerous epidemiological studies. However, its causality has not

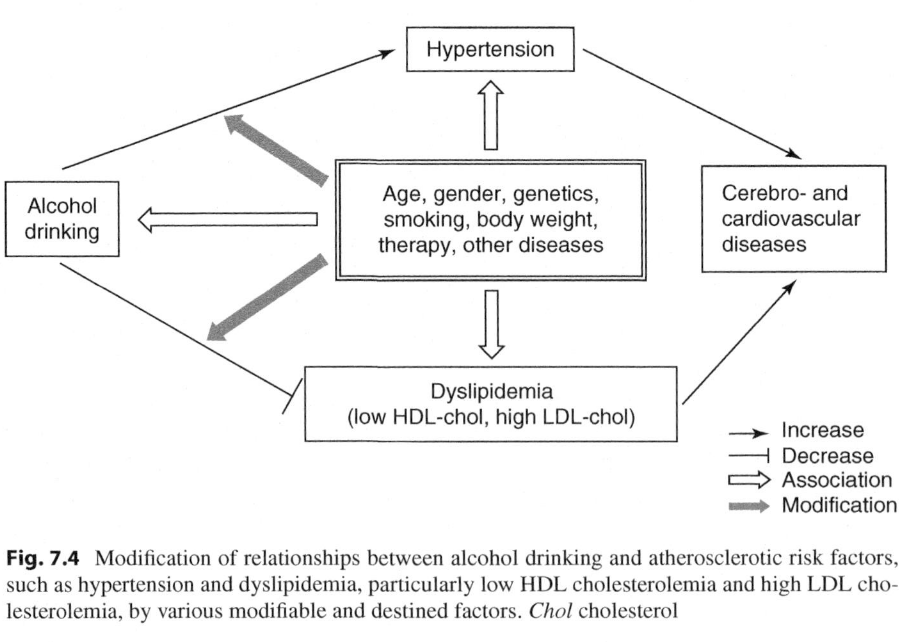

Fig. 7.4 Modification of relationships between alcohol drinking and atherosclerotic risk factors, such as hypertension and dyslipidemia, particularly low HDL cholesterolemia and high LDL cholesterolemia, by various modifiable and destined factors. *Chol* cholesterol

been established, and subcellular mechanisms of various alcohol actions remain unknown. Therefore, fundamental *in vitro* and *in vivo* studies on the effects of alcohol on the cardiovascular system are needed in the future.

References

Ahmad A, Farhan Asad S, Singh S, Hadi SM (2000) DNA breakage by resveratrol and Cu(II): reaction mechanism and bacteriophage inactivation. Cancer Lett 154:29–37

Albert CM, Manson JE, Cook NR, Ajani UA, Gaziano JM, Hennekens CH (1999) Moderate alcohol consumption and the risk of sudden cardiac death among US male physicians. Circulation 100:944–950

Aldhahi W, Hamdy O (2003) Adipokines, inflammation, and the endothelium in diabetes. Curr Diab Rep 3:293–298

Altura BM, Altura BT (1982) Microvascular and vascular smooth muscle actions of ethanol, acetaldehyde, and acetate. Fed Proc 41:2447–2451

Anselm E, Chataigneau M, Ndiaye M, Chataigneau T, Schini-Kerth VB (2007) Grape juice causes endothelium-dependent relaxation via a redox-sensitive Src- and Akt-dependent activation of eNOS. Cardiovasc Res 73:404–413

Balbão CE, de Paola AA, Fenelon G (2009) Effects of alcohol on atrial fibrillation: myths and truths. Ther Adv Cardiovasc Dis 3:53–63

Boekholdt SM, Sacks FM, Jukema JW, Shepherd J, Freeman DJ, McMahon AD, Cambien F, Nicaud V, de Grooth GJ, Talmud PJ, Humphries SE, Miller GJ, Eiriksdottir G, Gudnason V, Kauma H, Kakko S, Savolainen MJ, Arca M, Montali A, Liu S, Lanz HJ, Zwinderman AH, Kuivenhoven JA, Kastelein JJ (2005) Cholesteryl ester transfer protein TaqIB variant, high-density lipoprotein cholesterol levels, cardiovascular risk, and efficacy of pravastatin treatment: individual patient meta-analysis of 13,677 subjects. Circulation 111:278–287

Booyse FM, Aikens ML, Grenett HE (1999) Endothelial cell fibrinolysis: transcriptional regulation of fibrinolytic protein gene expression (t-PA, u-PA, and PAI-1) by low alcohol. Alcohol Clin Exp Res 23:1119–1124

Brien SE, Ronksley PE, Turner BJ, Mukamal KJ, Ghali WA (2011) Effect of alcohol consumption on biological markers associated with risk of coronary heart disease: systematic review and meta-analysis of interventional studies. BMJ 342:d636

Brooks PJ, Enoch MA, Goldman D, Li TK, Yokoyama A (2009) The alcohol flushing response: an unrecognized risk factor for esophageal cancer from alcohol consumption. PLoS Med 6:e50

Camargo CA Jr, Stampfer MJ, Glynn RJ, Gaziano JM, Manson JE, Goldhaber SZ, Hennekens CH (1997) Prospective study of moderate alcohol consumption and risk of peripheral arterial disease in US male physicians. Circulation 95:577–580

Carlsson S, Hammar N, Grill V (2005) Alcohol consumption and type 2 diabetes meta-analysis of epidemiological studies indicates a U-shaped relationship. Diabetologia 48:1051–1054

Carluccio MA, Siculella L, Ancora MA, Massaro M, Scoditti E, Storelli C, Visioli F, Distante A, De Caterina R (2003) Olive oil and red wine antioxidant polyphenols inhibit endothelial activation: antiatherogenic properties of Mediterranean diet phytochemicals. Arterioscler Thromb Vasc Biol 23:622–629

Castelli WP, Doyle JT, Gordon T, Hames CG, Hjortland MC, Hulley SB, Kagan A, Zukel WJ (1977) Alcohol and blood lipids. The cooperative lipoprotein phenotyping study. Lancet 2:153–155

Chalopin M, Tesse A, Martínez MC, Rognan D, Arnal JF, Andriantsitohaina R (2010) Estrogen receptor alpha as a key target of red wine polyphenols action on the endothelium. PLoS One 5:e8554

Chen L, Davey Smith G, Harbord RM, Lewis SJ (2008) Alcohol intake and blood pressure: a systematic review implementing a Mendelian randomization approach. PLoS Med 5:e52

Choi HK, Curhan G (2004) Beer, liquor, and wine consumption and serum uric acid level: the Third National Health and Nutrition Examination Survey. Arthritis Rheum 51:1023–1029

Corder R, Douthwaite JA, Lees DM, Khan NQ, Viseu Dos Santos AC, Wood EG, Carrier MJ (2001) Endothelin-1 synthesis reduced by red wine. Nature 414:863–864
Corrao G, Rubbiati L, Bagnardi V, Zambon A, Poikolainen K (2000) Alcohol and coronary heart disease: a meta-analysis. Addiction 95:1505–1523
Cowpland C, Su GM, Murray M, Puddey IB, Croft KD (2006) Effect of alcohol on cytochrome p450 arachidonic acid metabolism and blood pressure in rats and its modulation by red wine polyphenolics. Clin Exp Pharmacol Physiol 33:183–188
Cui Y, Blumenthal RS, Flaws JA, Whiteman MK, Langenberg P, Bachorik PS, Bush TL (2001) Non-high-density lipoprotein cholesterol level as a predictor of cardiovascular disease mortality. Arch Intern Med 161:1413–1419
Davies MJ, Baer DJ, Judd JT, Brown ED, Campbell WS, Taylor PR (2002) Effects of moderate alcohol intake on fasting insulin and glucose concentrations and insulin sensitivity in postmenopausal women: a randomized controlled trial. JAMA 287:2559–2562
Di Castelnuovo A, Costanzo S, Bagnardi V, Donati MB, Iacoviello L, de Gaetano G (2006) Alcohol dosing and total mortality in men and women: an updated meta-analysis of 34 prospective studies. Arch Intern Med 166:2437–2445
Ettinger PO, Wu CF, De La Cruz C Jr, Weisse AB, Ahmed SS, Regan TJ (1978) Arrhythmias and the "Holiday Heart": alcohol-associated cardiac rhythm disorders. Am Heart J 95:555–562
Feig DI, Kang DH, Johnson RJ (2008) Uric acid and cardiovascular risk. N Engl J Med 359:1811–1821
Feng AN, Chen YL, Chen YT, Ding YZ, Lin SJ (1999) Red wine inhibits monocyte chemotactic protein-1 expression and modestly reduces neointimal hyperplasia after balloon injury in cholesterol-fed rabbits. Circulation 100:2254–2259
Fernández-Solà J, Nicolás JM, Oriola J, Sacanella E, Estruch R, Rubin E, Urbano-Márquez A (2002) Angiotensin-converting enzyme gene polymorphism is associated with vulnerability to alcoholic cardiomyopathy. Ann Intern Med 137:321–326
Fernández-Solà J, Fatjó F, Sacanella E, Estruch R, Bosch X, Urbano-Márquez A, Nicolás JM (2006) Evidence of apoptosis in alcoholic cardiomyopathy. Hum Pathol 37:1100–1110
Fitzpatrick DF, Hirschfield SL, Coffey RG (1993) Endothelium-dependent vasorelaxing activity of wine and other grape products. Am J Physiol 265:H774–H778
Fogle RL, Lynch CJ, Palopoli M, Deiter G, Stanley BA, Vary TC (2010) Impact of chronic alcohol ingestion on cardiac muscle protein expression. Alcohol Clin Exp Res 34:1226–1234
Freedman JE, Parker C 3rd, Li L, Perlman JA, Frei B, Ivanov V, Deak LR, Iafrati MD, Folts JD (2001) Select flavonoids and whole juice from purple grapes inhibit platelet function and enhance nitric oxide release. Circulation 103:2792–2798
Fuhrman B, Aviram M (2001) Flavonoids protect LDL from oxidation and attenuate atherosclerosis. Curr Opin Lipidol 12:41–48
Fumeron F, Betoulle D, Luc G, Behague I, Ricard S, Poirier O, Jemaa R, Evans A, Arveiler D, Marques-Vidal P (1995) Alcohol intake modulates the effect of a polymorphism of the cholesteryl ester transfer protein gene on plasma high density lipoprotein and the risk of myocardial infarction. J Clin Invest 96:1664–1671
Gaziano JM, Buring JE, Breslow JL, Goldhaber SZ, Rosner B, VanDenburgh M, Willett W, Hennekens CH (1993) Moderate alcohol intake, increased levels of high-density lipoprotein and its subfractions, and decreased risk of myocardial infarction. N Engl J Med 329:1829–1834
George A, Figueredo VM (2010) Alcohol and arrhythmias: a comprehensive review. J Cardiovasc Med (Hagerstown) 11:221–228
Guivernau M, Baraona E, Lieber CS (1987) Acute and chronic effects of ethanol and its metabolites on vascular production of prostacyclin in rats. J Pharmacol Exp Ther 240:59–64
Hannuksela ML, Rämet ME, Nissinen AE, Liisanantti MK, Savolainen MJ (2004) Effects of ethanol on lipids and atherosclerosis. Pathophysiology 10:93–103
Hashimoto Y, Nakayama T, Futamura A, Omura M, Nakarai H, Nakahara K (2002) Relationship between genetic polymorphisms of alcohol-metabolizing enzymes and changes in risk factors for coronary heart disease associated with alcohol consumption. Clin Chem 48:1043–1048
Hatake K, Taniguchi T, Ouchi H, Sakaki N, Hishida S, Ijiri I (1990) Possible involvement of kinins in cardiovascular changes after alcohol intake. Pharmacol Biochem Behav 35:437–442

Heldin CH, Ostman A, Rönnstrand L (1998) Signal transduction via platelet-derived growth factor receptors. Biochim Biophys Acta 1378:F79–F113
Hines LM, Stampfer MJ, Ma J, Gaziano JM, Ridker PM, Hankinson SE, Sacks F, Rimm EB, Hunter DJ (2001) Genetic variation in alcohol dehydrogenase and the beneficial effect of moderate alcohol consumption on myocardial infarction. N Engl J Med 344:549–555
Hong J, Smith RR, Harvey AE, Núñez NP (2009) Alcohol consumption promotes insulin sensitivity without affecting body fat levels. Int J Obes (Lond) 33:197–203
Hulthe J, Fagerberg B (2005) Alcohol consumption and insulin sensitivity: A review. Metab Syndr Relat Disord 3:45–50
Husemoen LL, Fenger M, Friedrich N, Tolstrup JS, Beenfeldt Fredriksen S, Linneberg A (2008) The association of ADH and ALDH gene variants with alcohol drinking habits and cardiovascular disease risk factors. Alcohol Clin Exp Res 32:1984–1991
Husemoen LL, Jørgensen T, Borch-Johnsen K, Hansen T, Pedersen O, Linneberg A (2010) The association of alcohol and alcohol metabolizing gene variants with diabetes and coronary heart disease risk factors in a white population. PLoS One 5:e11735
Iijima K, Yoshizumi M, Hashimoto M, Kim S, Eto M, Ako J, Liang YQ, Sudoh N, Hosoda K, Nakahara K, Toba K, Ouchi Y (2000) Red wine polyphenols inhibit proliferation of vascular smooth muscle cells and downregulate expression of cyclin A gene. Circulation 101:805–811
Iijima K, Yoshizumi M, Hashimoto M, Akishita M, Kozaki K, Ako J, Watanabe T, Ohike Y, Son B, Yu J, Nakahara K, Ouchi Y (2002) Red wine polyphenols inhibit vascular smooth muscle cell migration through two distinct signaling pathways. Circulation 105:2404–2410
Itoh T, Matsumoto M, Nakamura M, Okada A, Shirahashi N, Hougaku H, Hashimoto H, Sakaguchi M, Handa N, Takeshita T, Morimoto K, Hori M (1997) Effects of daily alcohol intake on the blood pressure differ depending on an individual's sensitivity to alcohol: oriental flushing as a sign to stop drinking for health reasons. J Hypertens 15:1211–1217
Joosten MM, Beulens JW, Kersten S, Hendriks HF (2008) Moderate alcohol consumption increases insulin sensitivity and ADIPOQ expression in postmenopausal women: a randomised, crossover trial. Diabetologia 51:1375–1381
Jung JG, Kim JS, Oh MK (2010) The role of the flushing response in the relationship between alcohol consumption and insulin resistance. Alcohol Clin Exp Res 34:1699–1704
Kakafika AI, Liberopoulos EN, Mikhailidis DP (2007) Fibrinogen: a predictor of vascular disease. Curr Pharm Des 13:1647–1659
Kanellis J, Kang DH (2005) Uric acid as a mediator of endothelial dysfunction, inflammation, and vascular disease. Semin Nephrol 25:39–42
Kawano Y (2010) Physio-pathological effects of alcohol on the cardiovascular system: its role in hypertension and cardiovascular disease. Hypertens Res 33:181–191
Klatsky AL (1996) Alcohol and hypertension. Clin Chim Acta 246:91–105
Klatsky AL (2004) Alcohol-associated hypertension: when one drinks makes a difference. Hypertension 44:805–806
Kodama S, Saito K, Tanaka S, Horikawa C, Saito A, Heianza Y, Anasako Y, Nishigaki Y, Yachi Y, Iida KT, Ohashi Y, Yamada N, Sone H (2011) Alcohol consumption and risk of atrial fibrillation: a meta-analysis. J Am Coll Cardiol 57:427–436
Koppes LL, Dekker JM, Hendriks HF, Bouter LM, Heine RJ (2005) Moderate alcohol consumption lowers the risk of type 2 diabetes: a meta-analysis of prospective observational studies. Diabetes Care 28:719–725
Langer RD, Criqui MH, Reed DM (1992) Lipoproteins and blood pressure as biological pathways for effect of moderate alcohol consumption on coronary heart disease. Circulation 85:910–915
Laonigro I, Correale M, Di Biase M, Altomare E (2009) Alcohol abuse and heart failure. Eur J Heart Fail 11:453–462
Latella MC, Di Castelnuovo A, de Lorgeril M, Arnout J, Cappuccio FP, Krogh V, Siani A, van Dongen M, Donati MB, de Gaetano G, Iacoviello L, European Collaborative Group of the IMMIDIET Project (2009) Genetic variation of alcohol dehydrogenase type 1C (ADH1C), alcohol consumption, and metabolic cardiovascular risk factors: results from the IMMIDIET study. Atherosclerosis 207:284–290

Lee J, Tan CS, Chia KS, Tan CE, Chew SK, Ordovas JM, Tai ES (2004) The lipoprotein lipase S447X polymorphism and plasma lipids: interactions with APOE polymorphisms, smoking, and alcohol consumption. J Lipid Res 45:1132–1139

Libby P (2002) Inflammation in atherosclerosis. Nature 420:868–874

Liinamaa MJ, Hannuksela ML, Rämet ME, Savolainen MJ (2006) Defective glycosylation of cholesteryl ester transfer protein in plasma from alcohol abusers. Alcohol Alcohol 41:18–23

Lucas DL, Brown RA, Wassef M, Giles TD (2005) Alcohol and the cardiovascular system: research challenges and opportunities. J Am Coll Cardiol 45:1916–1924

Mäkelä SM, Jauhiainen M, Ala-Korpela M, Metso J, Lehto TM, Savolainen MJ, Hannuksela ML (2008) HDL_2 of heavy alcohol drinkers enhances cholesterol efflux from raw macrophages via phospholipid-rich HDL_{2b} particles. Alcohol Clin Exp Res 32:991–1000

Mandrekar P, Catalano D, White B, Szabo G (2006) Moderate alcohol intake in humans attenuates monocyte inflammatory responses: inhibition of nuclear regulatory factor kappa B and induction of interleukin 10. Alcohol Clin Exp Res 30:135–139

Mandrekar P, Jeliazkova V, Catalano D, Szabo G (2007) Acute alcohol exposure exerts anti-inflammatory effects by inhibiting IκB kinase activity and p65 phosphorylation in human monocytes. J Immunol 178:7686–7693

Mantzoros CS, Magkos F, Brinkoetter M, Sienkiewicz E, Dardeno TA, Kim SY, Hamnvik OP, Koniaris A (2011) Leptin in human physiology and pathophysiology. Am J Physiol Endocrinol Metab 301:E567–E584

Marumo M, Wakabayashi I (2010) Diverse effects of ethanol on Ca^{2+} entry and subsequent aggregation of platelets. Alcohol 44:343–350

Mazzaglia G, Britton AR, Altmann DR, Chenet L (2001) Exploring the relationship between alcohol consumption and non-fatal or fatal stroke: a systematic review. Addiction 96:1743–1756

McCarty MF (2001) Does regular ethanol consumption promote insulin sensitivity and leanness by stimulating AMP-activated protein kinase? Med Hypotheses 57:405–407

McFadden CB, Brensinger CM, Berlin JA, Townsend RR (2005) Systematic review of the effect of daily alcohol intake on blood pressure. Am J Hypertens 18:276–286

Meade TW, Imeson J, Stirling Y (1987) Effects of changes in smoking and other characteristics on clotting factors and the risk of ischaemic heart disease. Lancet 2:986–988

Modun D, Music I, Vukovic J, Brizic I, Katalinic V, Obad A, Palada I, Dujic Z, Boban M (2008) The increase in human plasma antioxidant capacity after red wine consumption is due to both plasma urate and wine polyphenols. Atherosclerosis 197:250–256

Mukamal KJ, Jadhav PP, D'Agostino RB, Massaro JM, Mittleman MA, Lipinska I, Sutherland PA, Matheney T, Levy D, Wilson PW, Ellison RC, Silbershatz H, Muller JE, Tofler GH (2001) Alcohol consumption and hemostatic factors: analysis of the Framingham Offspring cohort. Circulation 104:1367–1373

Mukamal KJ, Massaro JM, Ault KA, Mittleman MA, Sutherland PA, Lipinska I, Levy D, D'Agostino RB, Tofler GH (2005) Alcohol consumption and platelet activation and aggregation among women and men: the Framingham Offspring Study. Alcohol Clin Exp Res 29:1906–1912

Ndiaye M, Chataigneau T, Andriantsitohaina R, Stoclet JC, Schini-Kerth VB (2003) Red wine polyphenols cause endothelium-dependent EDHF-mediated relaxations in porcine coronary arteries via a redox-sensitive mechanism. Biochem Biophys Res Commun 310:371–377

Ndiaye M, Chataigneau M, Lobysheva I, Chataigneau T, Schini-Kerth VB (2005) Red wine polyphenol-induced, endothelium-dependent NO-mediated relaxation is due to the redox-sensitive PI3-kinase/Akt-dependent phosphorylation of endothelial NO-synthase in the isolated porcine coronary artery. FASEB J 19:455–457

Nelson S, Kolls JK (2002) Alcohol, host defence and society. Nat Rev Immunol 2:205–209

Nguyen A, Gemmell CH, Yeo EL, Packham MA, Rand ML (1998) Ethanol inhibits thrombin-induced secretion of the contents of human platelet dense and α-granules and lysosomes. Thromb Haemost 80:662–667

Osier M, Pakstis AJ, Kidd JR, Lee JF, Yin SJ, Ko HC, Edenberg HJ, Lu RB, Kidd KK (1999) Linkage disequilibrium at the ADH2 and ADH3 loci and risk of alcoholism. Am J Hum Genet 64:1147–1157

Pace-Asciak CR, Rounova O, Hahn SE, Diamandis EP, Goldberg DM (1996) Wines and grape juices as modulators of platelet aggregation in healthy human subjects. Clin Chim Acta 246:163–182

Pai JK, Hankinson SE, Thadhani R, Rifai N, Pischon T, Rimm EB (2006) Moderate alcohol consumption and lower levels of inflammatory markers in US men and women. Atherosclerosis 186:113–120

Paulson QX, Hong J, Holcomb VB, Nunez NP (2010) Effects of body weight and alcohol consumption on insulin sensitivity. Nutr J 9:14

Pignatelli P, Pulcinelli FM, Celestini A, Lenti L, Ghiselli A, Gazzaniga PP, Violi F (2000) The flavonoids quercetin and catechin synergistically inhibit platelet function by antagonizing the intracellular production of hydrogen peroxide. Am J Clin Nutr 72:1150–1155

Potter JF, Bannan LT, Saunders JB, Ingram MC, Beevers DG (1983) Blood pressure and pressor mechanisms during alcohol withdrawal. J Hypertens Suppl 1:97–99

Puddey IB, Beilin LJ (2006) Alcohol is bad for blood pressure. Clin Exp Pharmacol Physiol 33:847–852

Rehm J, Room R, Graham K, Monteiro M, Gmel G, Sempos CT (2003) The relationship of average volume of alcohol consumption and patterns of drinking to burden of disease: an overview. Addiction 98:1209–1228

Reynolds K, Lewis B, Nolen JD, Kinney GL, Sathya B, He J (2003) Alcohol consumption and risk of stroke: a meta-analysis. JAMA 289:579–588

Ridker PM, Vaughan DE, Stampfer MJ, Glynn RJ, Hennekens CH (1994) Association of moderate alcohol consumption and plasma concentration of endogenous tissue-type plasminogen activator. JAMA 272:929–933

Rimm EB, Williams P, Fosher K, Criqui M, Stampfer MJ (1999) Moderate alcohol intake and lower risk of coronary heart disease: meta-analysis of effects on lipids and haemostatic factors. BMJ 319:1523–1528

Rodrigo R, Miranda A, Vergara L (2011) Modulation of endogenous antioxidant system by wine polyphenols in human disease. Clin Chim Acta 412:410–424

Rosenkranz S, Knirel D, Dietrich H, Flesch M, Erdmann E, Böhm M (2002) Inhibition of the PDGF receptor by red wine flavonoids provides a molecular explanation for the "French paradox". FASEB J 16:1958–1960

Rubin R (1999) Effect of ethanol on platelet function. Alcohol Clin Exp Res 23:1114–1118

Samokhvalov AV, Irving HM, Rehm J (2010) Alcohol consumption as a risk factor for atrial fibrillation: a systematic review and meta-analysis. Eur J Cardiovasc Prev Rehabil 17:706–712

Savolainen MJ, Hannuksela M, Seppänen S, Kervinen K, Kesäniemi YA (1990) Increased high-density lipoprotein cholesterol concentration in alcoholics is related to low cholesteryl ester transfer protein activity. Eur J Clin Invest 20:593–599

Seitz HK, Meier P (2007) The role of acetaldehyde in upper digestive tract cancer in alcoholics. Transl Res 149:293–297

Sherman D, Davé V, Hsu LC, Peters TJ, Yoshida A (1993) Diverse polymorphism within a short coding region of the human aldehyde dehydrogenase-5 (ALDH5) gene. Hum Genet 92:477–480

Smirnoff N (2005) Antioxidants and reactive oxygen species in plants. Blackwell Publishing, Oxford

Soleas GJ, Diamandis EP, Goldberg DM (1997) Wine as a biological fluid: history, production, and role in disease prevention. J Clin Lab Anal 11:287–313

Stubbs CD, Rubin R (1992) Effect of ethanol on platelet phospholipase A_2. Lipids 27:255–260

Sumi H, Hamada H, Tsushima H, Mihara H (1988) Urokinase-like plasminogen activator increased in plasma after alcohol drinking. Alcohol Alcohol 23:33–43

Tabengwa EM, Wheeler CG, Yancey DA, Grenett HE, Booyse FM (2002) Alcohol-induced upregulation of fibrinolytic activity and plasminogen activators in human monocytes. Alcohol Clin Exp Res 26:1121–1127

Tan Y, Li X, Prabhu SD, Brittian KR, Chen Q, Yin X, McClain CJ, Zhou Z, Cai L (2012) Angiotensin II plays a critical role in alcohol-induced cardiac nitrative damage, cell death, remodeling, and cardiomyopathy in a protein kinase C/nicotinamide adenine dinucleotide phosphate oxidase-dependent manner. J Am Coll Cardiol 59:1477–1486

Tsuchihashi-Makaya M, Serizawa M, Yanai K, Katsuya T, Takeuchi F, Fujioka A, Yamori Y, Ogihara T, Kato N (2009) Gene-environmental interaction regarding alcohol-metabolizing enzymes in the Japanese general population. Hypertens Res 32:207–213
Urbano-Marquez A, Estruch R, Navarro-Lopez F, Grau JM, Mont L, Rubin E (1989) The effects of alcoholism on skeletal and cardiac muscle. N Engl J Med 320:409–415
Ursini F, Zamburlini A, Cazzolato G, Maiorino M, Bon GB, Sevanian A (1998) Postprandial plasma lipid hydroperoxides: a possible link between diet and atherosclerosis. Free Radic Biol Med 25:250–252
Van de Wiel A (2012) The effect of alcohol on postprandial and fasting triglycerides. Int J Vasc Med 2012:862504
van de Wiel A, de Lange DW (2008) Cardiovascular risk is more related to drinking pattern than to the type of alcoholic drinks. Neth J Med 66:467–473
van de Wiel A, van Golde PM, Kraaijenhagen RJ, von dem Borne PA, Bouma BN, Hart HC (2001) Acute inhibitory effect of alcohol on fibrinolysis. Eur J Clin Invest 31:164–170
van Golde PM, Hart HC, Kraaijenhagen RJ, Bouma BN, van de Wiel A (2002) Regular alcohol intake and fibrinolysis. Neth J Med 60:285–288
Vary TC, Lynch CJ, Lang CH (2001) Effects of chronic alcohol consumption on regulation of myocardial protein synthesis. Am J Physiol Heart Circ Physiol 281:H1242–H1251
Volpato S, Pahor M, Ferrucci L, Simonsick EM, Guralnik JM, Kritchevsky SB, Fellin R, Harris TB (2004) Relationship of alcohol intake with inflammatory markers and plasminogen activator inhibitor-1 in well-functioning older adults: the Health, Aging, and Body Composition study. Circulation 109:607–612
Wakabayashi I (2005) Sensitivity of circulatory response to alcohol influences the relationship between alcohol consumption and blood pressure in Orientals. Blood Press 14:238–244
Wakabayashi I (2010) History of antihypertensive therapy influences the relationships of alcohol with blood pressure and pulse pressure in older men. Am J Hypertens 23:633–638
Wakabayashi I (2011) Combined impact of alcohol and tobacco: implications for cardiovascular disease. Curr Cardiovasc Risk Rep 5:501–509
Wakabayashi I, Groschner K (2009) Modification of the association between alcohol drinking and non-HDL cholesterol by gender. Clin Chim Acta 404:154–159
Wakabayashi I, Marumo M (2002) Ethanol inhibits store-operated Ca^{2+} entry of platelets. Pharmacol Toxicol 90:226–228
Wakabayashi I, Masuda H (2006) Influence of drinking alcohol on atherosclerotic risk in alcohol flushers and non-flushers of Oriental patients with type 2 diabetes mellitus. Alcohol Alcohol 41:672–677
Wakabayashi I, Negoro M (2002) Mechanism of inhibitory action of ethanol on inducible nitric oxide synthesis in macrophages. Naunyn Schmiedebergs Arch Pharmacol 366:299–306
Wang Z, Barker TH, Fuller GM (1999) Alcohol at moderate levels decreases fibrinogen expression in vivo and in vitro. Alcohol Clin Exp Res 23:1927–1932
Whitfield JB, O'Brien ME, Nightingale BN, Zhu G, Heath AC, Martin NG (2003) ADH genotype does not modify the effects of alcohol on high-density lipoprotein. Alcohol Clin Exp Res 27:509–514
Wright JW, Morseth SL, Abhold RH, Harding JW (1986) Elevations in plasma angiotensin II with prolonged ethanol treatment in rats. Pharmacol Biochem Behav 24:813–818
Yamada Y, Sun F, Tsuritani I, Honda R (2002) Genetic differences in ethanol metabolizing enzymes and blood pressure in Japanese alcohol consumers. J Hum Hypertens 16:479–486
Yamamoto T, Moriwaki Y, Takahashi S (2005) Effect of ethanol on metabolism of purine bases (hypoxanthine, xanthine, and uric acid). Clin Chim Acta 356:35–57
Ying CJ, Xu JW, Ikeda K, Takahashi K, Nara Y, Yamori Y (2003) Tea polyphenols regulate nicotinamide adenine dinucleotide phosphate oxidase subunit expression and ameliorate angiotensin II-induced hyperpermeability in endothelial cells. Hypertens Res 26:823–828
Zhang WS, Jiang CQ, Cheng KK, Adab P, Thomas GN, Liu B, Lam KB, Schooling CM, Lam TH (2009) Alcohol sensitivity, alcohol use and hypertension in an older Chinese population: the Guangzhou Biobank Cohort Study. Hypertens Res 32:741–747

Tobacco

8

Vicki Myers, Laura Rosen, and Yariv Gerber

Abstract

Overwhelming epidemiological and pathophysiological evidence has shown that smokers are at significantly increased risk of developing acute coronary syndromes, while those with established cardiovascular disease have lower odds of survival. This chapter discusses the impact of firsthand, secondhand, and thirdhand exposure to tobacco smoke, as well as smokeless tobacco, on the development and progression of cardiovascular disease. Citing numerous examples, we demonstrate the extent of the cardiovascular damage caused not only to smokers themselves but to those around them and the processes by which these changes happen. Smoking is the leading preventable cause of cardiovascular mortality worldwide, making this a crucial issue to address. Strategies to reduce the global burden of tobacco are examined, including smoking cessation and, more controversially, reduction, on both an individual and population level. Biological pathways are reviewed, including atherosclerosis, endothelial dysfunction, and inflammatory processes, and methodological issues in the literature are addressed.

Keywords

Smoking • Cardiovascular disease • Secondhand exposure • Thirdhand exposure • Smoking cessation • Smoking reduction • Mortality

V. Myers, MSc
Sackler Faculty of Medicine, Department of Epidemiology and Preventive Medicine, Tel Aviv University, Tel Aviv, Israel

L. Rosen, PhD
Sackler Faculty of Medicine, Department of Health Promotion, Tel Aviv University, Tel Aviv, Israel

Y. Gerber, PhD (✉)
Sackler Faculty of Medicine, Department of Epidemiology and Preventive Medicine, School of Public Health, Tel Aviv University, Ramat Aviv, Tel Aviv 69978, Israel
e-mail: yarivg@post.tau.ac.il

I. Wakabayashi, K. Groschner (eds.), *Interdisciplinary Concepts in Cardiovascular Health*, DOI 10.1007/978-3-7091-1334-9_8, © Springer-Verlag Wien 2013

8.1 Introduction

Smoking is the leading preventable cause of mortality, accounting for 18 % of all US deaths (Mokdad et al. 2004; USDHHS 2010). Overwhelming evidence has shown that smoking is a risk factor for premature mortality and a causal factor in the development of a multitude of terminal diseases, including coronary heart disease (CHD) (Doll et al. 2004; USDHHS 2004; Wilhelmsson et al. 1975). Despite these catastrophic consequences, according to the WHO, 1.3 billion people smoke worldwide, comprising approximately a third of the world population aged 15 and over (Thun and Luiza da Costa e Silva 2003). While smoking is most commonly discussed in the public arena in the context of cancer, the adverse effects of tobacco smoke on cardiovascular disease (CVD) make sobering reading, and far more smokers die of cardiovascular causes than of lung cancer. This chapter will examine the epidemiological evidence concerning the link between exposure to tobacco smoke and CVD and the pathways through which these changes are effected in both the short and long term. We will look at four types of exposure: firsthand or active smoking, secondhand or passive smoking, thirdhand or environmental exposure, and smokeless tobacco. We will then examine the effects of smoking cessation and reduction on the cardiovascular system and finally the challenges and limitations present in smoking research.

8.2 Effects of Tobacco on CVD: Firsthand, Secondhand, Thirdhand, and Smokeless

8.2.1 Effects of Firsthand Exposure to Tobacco Smoke on CVD

Tobacco smoke contains over 4,000 chemicals which adversely affect every system in the human body, including the cardiovascular system (USDHHS 1983). Of the five million annual smoking-attributable deaths, 35 % are estimated to be due to cardiovascular causes, compared to 17.5 % from lung cancer and 20 % from chronic obstructive pulmonary disease (Ezzati and Lopez 2003). Eleven percent of total global cardiovascular deaths were attributable to smoking in the year 2000 (Ezzati et al. 2005). Among these 1.6 million smoking-attributable CV deaths, 54 % were estimated to be due to ischemic heart disease and 25 % to cerebrovascular disease.

The link between smoking and CVD has been known for decades, as detailed in successive US Surgeon General's reports, based on comprehensive epidemiological, clinical, and experimental evidence. While the early Surgeon General's reports focused on respiratory diseases and lung cancer, already in 1967 a probable link between smoking and coronary artery disease mortality was reported (USDHHS 1967). As far back as 1962, the Framingham study reported excess mortality in heavy smokers, mostly due to cardiovascular causes (Doyle et al. 1962). Later reports presented evidence for the effects of smoking on cerebrovascular disease and atherosclerosis, and by 1979 sufficient evidence was available to categorically define smoking as a causal factor in coronary heart disease (CHD). The US Surgeon General's report of 1983 cited cigarette smoking as the most important known modifiable risk factor for coronary heart disease in the United States, specifying that smoking is responsible for approximately 30 % of all CHD deaths (USDHHS 1983). Since CHD is much more prevalent than cancer, this means that smoking-attributable deaths caused by CHD are far more numerous than those caused by cancer. According to the 1983 report, smokers have a 70 % greater risk of CHD mortality than nonsmokers. In addition to cardiac mortality, smokers have increased risk of suffering a coronary or cerebrovascular event. A dose–response curve is evident; however, in recent years this has been shown to be nonlinear with a sharp increase in cardiovascular risk at low levels of smoking (USDHHS 2010).

Risk charts illustrating the dangers of tobacco exposure show that smoking increases CV risk at all ages, for example, a 35-year-old smoker has a sevenfold chance of developing heart disease compared to a never smoker (Woloshin et al. 2008). While the risk of heart disease increases steeply at a young age, stroke risk increases for smokers later in life, with a fivefold increase in stroke risk at age 50.

8.2.1.1 Myocardial Infarction

Smokers have higher odds of suffering a myocardial infarction (MI) than their nonsmoking counterparts, a fact evidenced in numerous studies showing overrepresentation of smokers among MI patients. An Israeli cohort study of first MI survivors aged 65 or less showed 53 % to be smokers, while smoking prevalence in the general population was a more modest 35 % (Gerber et al. 2009). A large-scale case–control study including MI patients from over 50 countries reported current smokers as having a threefold risk of nonfatal MI compared to never smokers (Teo et al. 2006). Risk increased with the number of cigarettes smoked (Fig. 8.1). The authors

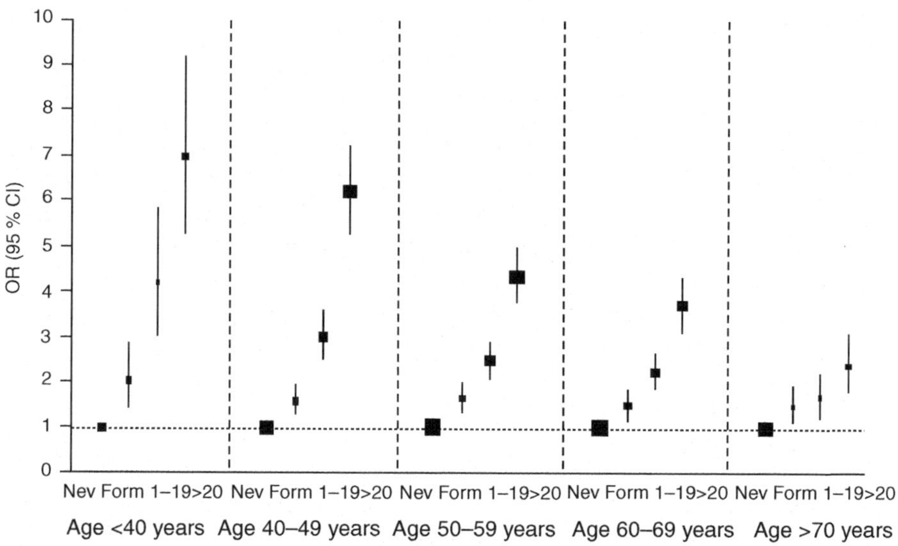

Fig. 8.1 Risk of AMI associated with number smoked, by age group. *P* for interaction <0.0001. *Nev* never smokers, *Form* former smokers. 1–19 and >20 refer to the number of cigarettes smoked daily (Reprinted from Teo et al. (2006), with permission from Elsevier, Licence no 2905480461542)

further estimated that each additional daily cigarette increased risk of MI by 5 %. This finding was replicated in a large cohort study, part of the WHO MONICA project, which reported increased risk of nonfatal MI in young smokers (Mähönen et al. 2004). Between the ages of 35 and 39, the risk of nonfatal MI was five times greater in smokers than in nonsmokers.

Furthermore, smokers who suffer MI are significantly younger than nonsmoking patients with the same condition. A retrospective analysis found that age of onset of first MI was directly related to the amount smoked, with heavy smokers suffering their first MI on average 8.7 years earlier than nonsmokers (Gottlieb et al. 1994). This effect has been suggested to be greater in women than men (Grundtvig et al. 2009). Figure 8.1 shows that while smoking intensity increases the risk of MI significantly, this effect is strongest at a younger age, for example, smoking more than 20 cigarettes per day is associated with a sevenfold increase in risk at age <40, while at age >70 the associated risk is approximately twofold (Teo et al. 2006).

Smoking not only increases the risk of developing CHD but also accelerates its development. Smokers who have already developed CHD have greater odds of mortality than their nonsmoking counterparts, for example, Grundtvig and colleagues (2011) reported 30 % greater post-MI mortality in smokers compared to nonsmokers. In a cohort of 1,521 first MI patients followed up for 13 years, persistent smokers had twice the odds of dying compared to both never smokers and those who had quit prior to MI, after controlling for clinical and socioeconomic characteristics and traditional risk factors (Gerber et al. 2009). Those who quit after MI also earned a survival benefit, reducing their risk substantially.

8.2.1.2 Stroke

Tobacco smoke is also an important cause of stroke. The Framingham study followed a cohort of over 4,000 people for 26 years and reported a significant increase in stroke risk in smokers compared to nonsmokers (Wolf et al. 1988). Additionally, a dose–response curve was evident, with heavy smokers at much greater risk compared to light smokers. Excess stroke risk reduced to nonsmoker levels 5 years after quitting smoking. A meta-analysis of 32 studies investigating the risk of stroke associated with cigarette smoke reported a relative risk of 1.5 (Shinton and Beevers 1989). Stroke risk in smokers is greater during middle age, with a risk of 2.9 under 55 years compared to 1.1 for smokers over 75, when stroke incidence is much higher in the general population. A recent review of studies conducted around the world put the excess risk of stroke in smokers at two to four times that of nonsmokers or those who had quit 10 years previously (Shah and Cole 2010).

8.2.1.3 Sudden Cardiac Death

The Framingham Heart Study reported an increase in sudden cardiac death for smokers compared to nonsmokers with a relative risk of 2.5 over 26 years of follow-up (Kannel et al. 1975). Other studies reported similar associations, for example, an odds ratio of 1.8 for sudden cardiac death in current smokers in the National Mortality Followback Survey (Escobedo and Caspersen 1997). Sudden cardiac

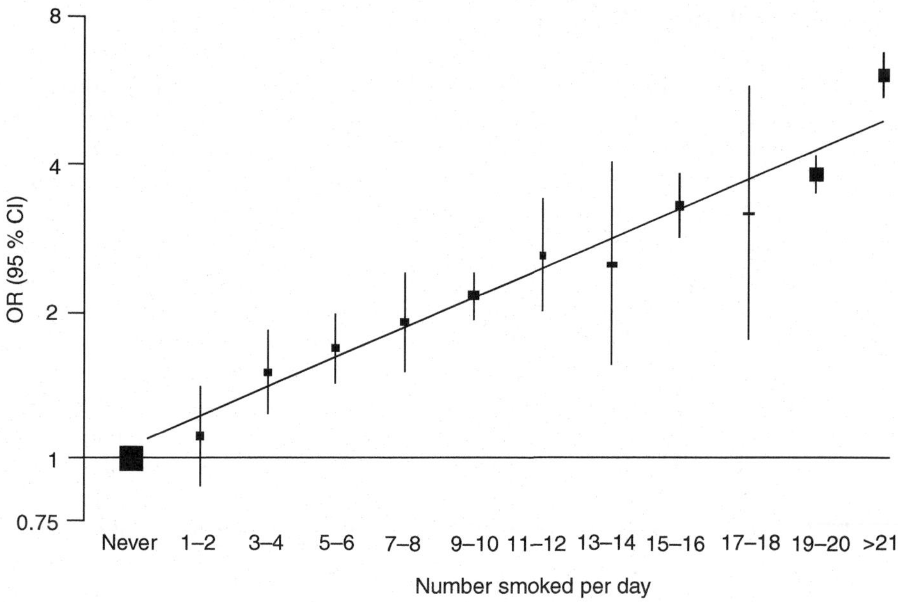

Fig. 8.2 Risk of AMI with increasing number of cigarettes smoked compared with never smokers (Reprinted from Teo et al. (2006), with permission from Elsevier, Licence no 2905480461542)

death may or may not be preceded by diagnosis of CHD, and it may be the first sign of underlying heart disease. Two-thirds of sudden cardiac deaths due to acute coronary thrombosis occur in cigarette smokers.

8.2.1.4 Dose–Response

It is generally accepted that a dose–response relationship exists between tobacco smoke and adverse cardiovascular outcomes (Fig. 8.2). Indeed, this is one of the features that allowed demonstration of a causal relationship. However, some studies do not support a linear relationship, and there is some evidence of a large effect from a small exposure. A meta-analysis of epidemiological studies found the risk of ischemic heart disease for smoking one cigarette per day to be approximately half that associated with a 20-a-day habit, but still 30 % more than unexposed nonsmokers (Law et al. 1997). Environmental exposure of individuals living with a smoker was similarly associated with a 30 % increased risk of heart disease, although the amount of smoke ingested is around 100 times less than that of an active smoker. This scale of effect from secondhand exposure was calculated as increasing the risk of CHD death in British nonsmoking men aged 60–69 from 5 to 6 % (Law et al. 1997). A prospective study conducted in Norway assessed the risks associated with smoking one to four cigarettes per day and reported that compared to never smokers, light smokers had a surprisingly high relative risk of 2.74 of ischemic heart disease mortality (Bjartveit and Tverdal 2005).

8.2.1.5 The Illusion of "Light" Cigarettes

In recent years, and due to the indisputable evidence concerning the adverse health effects of cigarettes, low tar or "light" brands have been increasingly marketed as less harmful than ordinary cigarettes. There is some debate however as to the true nature of this harm reduction strategy. A large prospective study found modest reductions in coronary heart disease mortality associated with a 15 mg reduction in tar (Tang et al. 1995). However, a review of research undertaken in this area concluded that only a minority of studies examining cardiovascular endpoints had evidence to support reduced risk associated with low tar cigarettes and that the reduction was at best modest, cited as approximately 10 % (Kabat 2003). Furthermore, in a case–control study of MI patients, no reduction in MI risk was seen in smokers of low-yield cigarettes (Negri et al. 1993). While reductions in cancer risk may be greater due to reduced tar content, cigarette components other than tar, such as nicotine and carbon monoxide, are largely responsible for the increased CV risk seen in smokers. It has additionally been proposed that laboratory smoking machines which test tar yields do not accurately portray the way that cigarettes are smoked and that smokers may compensate when smoking "light" cigarettes by smoking more intensively.

8.2.2 Effects of Secondhand Exposure to Tobacco Smoke on CVD

Secondhand exposure to tobacco smoke, or passive smoking, is known to have severe adverse health outcomes, via the same mechanisms as active smoking. A study of the burden of disease resulting from secondhand smoke in 192 countries estimated that 40 % of children and 34 % of adult nonsmokers are exposed to tobacco smoke worldwide, causing an estimated 379,000 annual deaths from ischemic heart disease (Oberg et al. 2011). Furthermore, almost three million disability-adjusted life years were attributed to CHD caused by passive smoking. Passive smoking is estimated to increase the risk of CHD by approximately 25–30 % (Barnoya and Glantz 2005; He et al. 1999), while the increased risk of an acute cardiovascular event associated with regular exposure to secondhand smoke (SHS) has been reported to be as high as 90 % (Panagiotakos et al. 2002). Even occasional exposure was associated with a 26 % increased risk in the same study. Indeed, acute effects of SHS have been demonstrated within hours of exposure including endothelial dysfunction, reduced heart rate variability, and increased arrhythmia (Kato et al. 2006; Chen et al. 2008). Further studies have shown impaired endothelial function and oxidative stress resulting from exposure to as little as 30 min of SHS (Kato et al. 2006; Raupach et al. 2006; Otsuka et al. 2001; Heiss et al. 2008). It has been estimated that passive smoking causes between 21,000 and 75,000 CHD deaths and up to 128,000 MIs per year in the USA alone, entailing a yearly health bill of $2–6 billion (Lightwood et al. 2009).

A systematic review of the literature reported that the adverse effects of passive smoking can already be seen in children, with exposed children – measured by serum cotinine – showing unfavorable blood lipid profiles and decreased vascular function, predisposing them to develop CVD (Metsios et al. 2010). Randomized controlled trials specifically found impaired endothelial function and aortic elasticity in children exposed to secondhand smoke (Kallio et al. 2007). Inflammatory markers have also been shown to be increased in passive smokers. A British study found that healthy adults with high levels of cotinine from secondhand exposure had higher risk of all-cause and cardiovascular deaths, with around half of the elevated risk explained by higher levels of C-reactive protein (Hamer et al. 2010); while a parallel study found elevated fibrinogen and homocysteine levels associated with passive exposure (Venn and Britton 2007).

Many studies assessing the risks of passive exposure to tobacco smoke involve nonsmokers who live with smokers and are therefore exposed on a daily basis. Research has shown significant increases in both total and cardiovascular death rates in this population (Svendsen et al. 1987; Helsing et al. 1988; Sandler et al. 1989; Garland et al. 1985). Indeed, a review by Wells suggested that "ischemic heart disease appears to be by far the major mortality risk from passive smoking" (Wells 1994). A review of the literature reported the effects of SHS to be almost as large as that of active smoking (Barnoya and Glantz 2005), somewhat contradicting the dose–response theory. While smokers receive about 100 times more smoke than do those affected by secondhand smoke, the risks do not increase in a comparative fashion, for example, relative risks of developing CHD according to a meta-analysis were 1.8 for smokers compared to 1.3 for nonsmokers who live with smokers (Law et al. 1997). A large-scale study of nonsmokers living with smokers followed up for 12 years found a death rate of 1.2 compared to non-exposed individuals, a rate comparative to that of ex-smokers and light smokers and about half the rate of heavy smokers (Sandler et al. 1989). The risk of cardiovascular death was even higher, reported as 1.3 in exposed men and 1.2 in exposed women (Helsing et al. 1988). Multiple investigations have shown a greater risk of SHS exposure at a younger age (Helsing et al. 1988; Sandler et al. 1989).

There are several limitations inherent to research on SHS. Exposed individuals (i.e., living with a smoker) are frequently compared to unexposed individuals (i.e., those not living with a smoker); however, this dichotomy is rather crude, since people may be exposed to SHS in other places such as their place of work. Later studies used biomarkers such as cotinine (metabolized nicotine in the blood) to assess exposure to smoke, with levels in a multicity study indicating that even the "non-exposed" control group had some level of SHS exposure (Whincup et al. 2004). Exposure to passive smoking was reported to be associated with between 68 and 86 % of the risk of light smoking, defined as smoking less than ten cigarettes per day.

8.2.3 Effects of Thirdhand Exposure to Tobacco Smoke on CVD

Recent evidence has accumulated concerning the dangers of thirdhand or environmental tobacco smoke, residual toxic compounds which remain on surfaces and in dust after a cigarette has been smoked. These are then reemitted into the air and react with other compounds, yielding secondary pollutants (Matt et al. 2011). An early study of this issue found household dust to contain modest amounts of nicotine (Hein et al. 1991). Thirdhand smoke is thought to particularly affect children living with smokers, where high levels of smoke pollution can be found on household materials. In a study of households with infants, levels of environmental smoke were five to seven times higher in smoking compared to nonsmoking households even when smokers actively tried to protect their children by smoking outdoors (Matt et al. 2004). Levels in households where smoking was permitted indoors were substantially higher. The health impact of thirdhand smoke remains controversial and requires further investigation.

8.2.4 Effects of Smokeless Tobacco on CVD

Around the world, tobacco habits vary, and in some countries, smokeless tobacco – which includes wet and dry snuff and chewing tobacco – is very common. Since the adverse effects of smoking have come to light and received so much negative press, tobacco companies are seeking new products to sell, including smokeless tobacco, which is marketed as a healthier option. The worldwide Interheart study examined the risk of MI in different types of tobacco users and found that tobacco chewers had more than double the odds of suffering an MI compared to non-tobacco users (Teo et al. 2006). The Atherosclerosis Risk in Communities study also found increased cardiovascular risk in users of smokeless tobacco (Yatsuya and Folsom 2010). The AHA released a statement in 2010 affirming that while smokeless tobacco is less harmful than cigarettes, long-term use is associated with increased risk of fatal MI and stroke, as well as reducing post-MI survival odds (Piano et al. 2010).

8.3 Changes in Smoking: Cessation and Reduction

8.3.1 Effects of Smoking Cessation on the Individual and the Community

Overwhelming evidence has accumulated demonstrating that quitting smoking significantly reduces risks of both cardiovascular mortality and morbidity (USDHHS 1990). Short-term effects of cessation include reduced blood pressure and heart rate (Minami et al. 1999), while in the long term, former smokers can expect to reduce their risk of developing and dying from CVD.

The good news is that even individuals who have smoked for many years can benefit and add years to their life expectancy by quitting smoking at almost any age (Fig. 8.3). After 1 year of abstinence, ex-smokers reduce their excess risk of CHD mortality by half, and after 15 years, the risk returns to that of never smokers (USDHHS 1990). Similar improvements are seen for stroke risk. The Nurses'

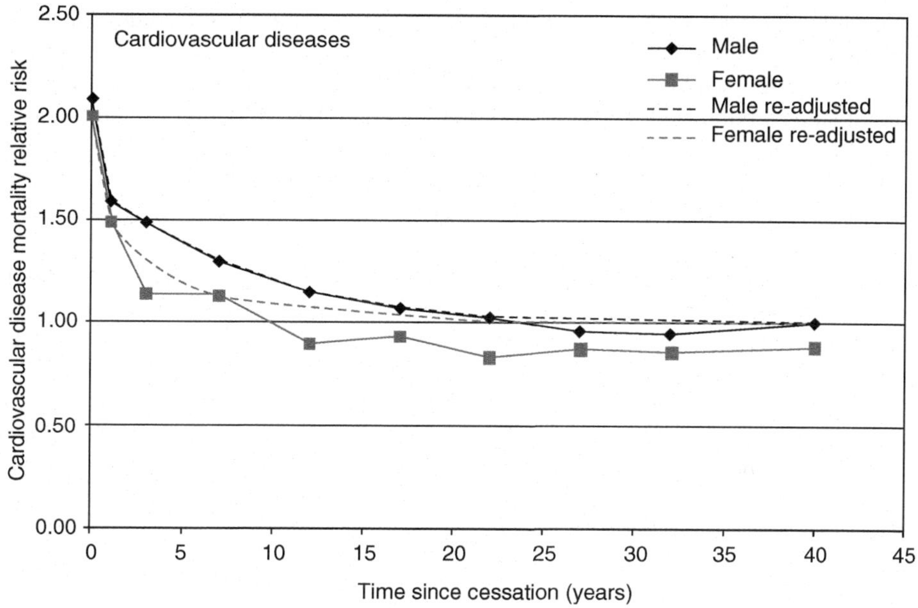

Fig. 8.3 Relative risk of CVD among ex-smokers. Zero years represent current smokers. The estimates at 40 years represent cessation of more than 35 years in the subjects. Data are from ACS CPS-II 1998 follow-up (Reproduced from WHO (2004), permission ID: 101036)

Health Study followed 100,000 women for 20 years and reported that compared to continuous smokers, those who quit smoking reduced their risk of death from CHD and cerebrovascular disease by 60 % (Kenfield et al. 2010). This survival benefit increased with time elapsed from cessation. Furthermore, survival benefit is seen even at older ages and in those with established CVD, with reduced risk of reinfarction and mortality, for example, healthy adults aged 60–64 reduced their mortality risk by 10 % after quitting, and post-MI patients with an average age of 54 who quit smoking after MI reduced their mortality risk by 40 % compared to persistent smokers (Gerber et al. 2009). A meta-analysis of 12 cohort studies investigating smoking cessation after MI reported a mortality odds ratio of 0.54 associated with quitting and a relative risk reduction of between 15 and 61 % for mortality (Wilson et al. 2000).

While the health effects of tobacco smoke are largely cumulative, reports show evidence of acute effects too. Reduced admissions for acute coronary events have been consistently reported following implementation of smoking bans (Sargent et al. 2004; Sims et al. 2010; Callinan et al. 2010). In 2011, 55 countries had implemented some kind of smoke-free legislation, as part of the WHO's tobacco control convention framework. Countrywide or statewide studies have been conducted where smoking bans have been enforced, in order to detect differences in mortality and hospitalizations associated with changes in smoking behavior and exposure.

Several meta-analyses have reported significant and seemingly immediate effects of smoking legislation consistently across various countries and cultures, with effect sizes ranging from 10 to 20 % reduction in acute cardiac events. Glantz reviewed eight studies published between 2004 and 2008 in different countries and found a reduction of 19 % in hospital admissions for acute MI associated with smoking legislation (Glantz 2008). Reductions in secondhand smoke exposure resulting from smoking legislation were even higher, averaging between 64 and 84 %. The benefits of smoke-free legislation seem to accrue with time; for instance, after 12 months, the rate of acute cardiac hospitalization had reduced by 15 % and after 3 years by 36 % (Lightwood and Glantz 2009). Another meta-analysis that reviewed 17 studies reported a pooled estimate of a 10 % reduction in acute coronary syndromes following smoking bans, improving by 6 % each year (Mackay et al. 2010). A third review of 11 international studies reported an overall decrease of 17 % in MI risk, emphasizing greater risk in younger people (Meyers et al. 2009). While some problems are inherent in these types of studies, including limited exposure assessment and short-term follow-up, results consistently show reduced cardiac risk following the introduction of legislation protecting people from exposure to tobacco smoke.

8.3.2 Smoking Reduction

Smoking cessation is the best way to reduce cardiovascular risk for smokers. However, since quit rates are low and many smokers are either unable or unwilling to quit, several studies have investigated whether a reduction in smoking intensity is associated with significantly reduced cardiovascular risk. While this issue remains controversial – indeed, the latest US Surgeon General's report does not advocate smoking reduction for CVD risk reduction (USDHHS 2010) – there is some evidence to support this theory. Several studies have demonstrated short-term benefits of reduction with regard to cardiovascular risk factors such as lipids and blood pressure (Pisinger and Godtfredsen 2007). A reduction of daily cigarette intake in MI survivors was related to decreased mortality risk; each reduction of five cigarettes per day was associated with an 18 % decline in mortality risk (Gerber et al. 2009). Furthermore, in a cohort of working men followed up for 40 years, a reduction in smoking intensity was associated with increased odds of reaching age 80 and a 15 % reduction in mortality, with the largest benefit for cardiovascular survival (Gerber et al. 2012). Despite this evidence, several large-scale studies found no significant survival benefit, or reduced risk of MI, due to smoking reduction (Tverdal and Bjartveit 2006; Godtfredsen et al. 2003). It has been hypothesized that a reduction in the number of cigarettes may not always equate to a reduction in the intake of nicotine or other components, due to compensatory smoking, whereby smokers smoke each cigarette more intensely and for longer (Godtfredsen et al. 2006). This issue requires further investigation before the benefits of smoking reduction can be dismissed.

8.4 Mechanisms

Tobacco smoke has both acute and chronic effects, being associated with vascular inflammation and autonomic dysfunction (Wells 1994). The components of tobacco smoke which increase the risk of developing CVD are oxidizing chemicals, nicotine, carbon monoxide, and particulate matter which together produce a chronic inflammatory state (USDHHS 2010). Short-term effects include decreased platelet sensitivity leading to increased platelet aggregation and decreased oxygen supply to the heart (Wells 1994). Platelet aggregation and blood viscosity affect the blood lipid profile in the long term, increasing triglycerides and lowering high-density lipoprotein cholesterol (Blache et al. 1992), all factors which contribute to the development of CHD. Of the many toxic components of tobacco smoke, nicotine seems to be largely responsible for the change in platelets and increased heart rate and blood pressure (Davis et al. 1985), while the reduction in the blood's ability to supply oxygen is due to a buildup of carbon monoxide (Wells 1994). Indeed, men who lived with smokers were found to have raised levels of expired carbon monoxide compared to controls (Svendsen et al. 1987). Longer-term effects of smoking begin from damage to the arterial endothelium and involve a buildup of plaque leading to atherosclerosis, increasing the risk of thrombosis, often detected via carotid artery wall thickness, and an imbalance of blood lipids.

The key pathways through which tobacco smoke causes and exacerbates CVD are atherosclerosis and endothelial injury (USDHHS 2004). Atheroma, the underlying pathological process preceding most coronary and cerebrovascular events, develops over years or decades. Atherosclerosis underlies CHD and involves a hardening of the arteries due to fat deposition and thickening of arterial walls, which cause a narrowing of the lumen and consequently reduced blood flow. A thrombus or clot can then form and break off, causing an infarct or stroke. The toxic components of tobacco smoke affect these processes. The 2004 Surgeon General's report found sufficient evidence to infer a causal relationship between smoking and subclinical atherosclerosis, the early development of atherosclerosis before clinical symptoms manifest (USDHHS 2004). The effects of smoking can be seen in the carotid and popliteal arteries, with greater presence of atherosclerosis. While earlier studies examined the correlation between smoking and cardiac endpoints such as MI or cardiac death, more recent research has investigated earlier effects which begin the cascade of events and go further in explaining the pathogenic pathways. Carotid intima medial thickness is often used as a marker of atherosclerosis since it is a subclinical sign yet is related to both CHD and stroke. Coronary calcium can also be measured to assess the presence of atherosclerosis. The Atherosclerosis Risk in Communities study followed up healthy participants for 3 years and found that smoking was strongly related to carotid atherosclerosis after controlling for age, with smokers showing greater intima medial thickness and incidence of CHD events (Sharrett et al. 1999). Studies have shown that atherosclerosis is already present at a young age, for example, an autopsy study of young people aged 15–34 showed atherosclerotic lesions in the carotid artery of the majority, increasing with age (Strong et al. 1999). Lesions have been shown to progress more rapidly in smokers compared to nonsmokers.

8.4.1 Endothelial Dysfunction

The endothelium, the upper layer of the arterial bed in contact with blood flow, controls vasodilation and constriction. Endothelial damage caused by tobacco smoke can impair the vessels' ability to dilate, at the same time increasing inflammation and cell proliferation, which contribute to the development and progression of atherosclerosis (Barnoya and Glantz 2005). Evidence from umbilical arteries and from the uterine arteries of pregnant smokers showed changes in the endothelium including intracellular holes and widening of intercellular junctions. Further studies in smokers showed elevated levels of circulating endothelial cells after smoking two cigarettes (Davis et al. 1985). Markers of endothelial damage such as von Willebrand factor have been shown to be increased minutes after smoking (Blann et al. 1998). Oxidizing chemicals and nicotine are thought to be responsible for endothelial dysfunction (USDHHS 2010). Once the endothelium is damaged, platelets rapidly stick to it and to each other, thereby increasing the risk of thrombosis (Meade 1994).

8.4.2 Inflammation

Exposure to tobacco smoke causes elevations in inflammatory markers. Cigarette smoking produces a chronic inflammatory state that contributes to the atherogenic disease processes. C-reactive protein is elevated in smokers, even years after quitting (Tracy et al. 1997). The same inflammatory marker has been show to predict CHD events. Fibrinogen, which is elevated in CVD, is strongly linked to smoking, with smokers displaying much higher levels than nonsmokers (Kannel et al. 1987; Miller et al. 1998). The level of fibrinogen increases with the amount smoked. Fibrinogen levels increase on starting smoking and decrease on cessation but remain somewhat elevated in ex-smokers for 5–10 years (Meade 1994; Rothwell et al. 1991).

8.4.3 Lipids

Widespread evidence, including meta-analyses, has demonstrated that smoking is associated with adverse lipid profiles (Craig et al. 1989), particularly high concentrations of low-density lipoprotein cholesterol (LDL) and reduced high-density lipoprotein cholesterol (HDL). Furthermore, HDL levels – which are essential in preventing atherosclerosis – were shown to decrease after starting smoking and to increase after quitting (Fortmann et al. 1986). Cigarette smoking produces an atherogenic lipid profile, primarily due to an increase in triglycerides and a decrease in HDL (USDHHS 2010). SHS has been shown to have the same effect, causing reduced HDL following both chronic (Mizoue et al. 1999) and acute exposure (Moffatt et al. 2004).

8.4.4 Cortisol/HPA Axis

Tobacco smoke causes an increase in cortisol secretion (Direk et al. 2011) and blunted cortisol response to stress (Rohleder and Kirschbaum 2006). Cortisol, produced by the hypothalamic–pituitary–adrenal (HPA) axis, is related to stress and to numerous chronic illnesses. Inhibition of cortisol response may be responsible for the increase in atherosclerosis seen in smokers. Chronic exposure to nicotine increases the activity of the HPA axis which, being involved in inflammatory pathways, may lead to increased inflammation and thereby contribute to CV morbidity (Rohleder and Kirschbaum 2006).

8.4.5 Blood Pressure

Blood pressure increases almost immediately during smoking, with the elevation largely caused by nicotine (Omvik 1996). Numerous studies have demonstrated short-term effects of tobacco, for example, Barutcu showed increased blood pressure and heart rate after smoking a single cigarette (Barutcu et al. 2004); and a study of the short-term effects of smoking cessation found blood pressure to be decreased during 1 week of abstinence compared to usual smoking behavior (Salonen et al. 1981). While epidemiological studies have not demonstrated chronically elevated blood pressure in smokers or have reported only weak associations (Bowman et al. 2007), results in an animal model demonstrated significant hypertension resulting from prolonged exposure to cigarette smoke (Talukder et al. 2011).

8.4.6 Platelet Aggregation

Studies of environmental exposure to tobacco smoke have shown a relationship with platelet aggregation, seen immediately after exposure and even at low doses (Law et al. 1997). Twenty minutes of exposure to SHS was associated with an increase in platelet aggregation related to a 34 % increased relative risk of CHD. SHS has been shown to activate blood platelets thereby increasing the risk of thrombosis (Barnoya and Glantz 2005). Further evidence of this causal pathway is the increased fibrinogen detected in passive smokers (Iso et al. 1996), since fibrinogen is a mediator of platelet activation, as well as an inflammatory marker.

8.4.7 Thrombosis

Arterial plaque is present in many people; however, smokers are more likely to have thromboses on their arterial walls (Spagnoli et al. 1994). Once the endothelium is damaged, circulating plasma coagulation factors are exposed to prothrombotic arterial and plaque tissue, and the interface between blood components and vessel walls may be disturbed. Tobacco smoke increases plasma concentration of clotting factors such as β-thromboglobulin and platelet factor (Davis et al. 1986).

8.4.8 Oxygen Demand

Tobacco smoke causes a rise in catecholamines such as adrenaline which lead to increased heart rate (USDHHS 2010). Nicotine causes blood pressure to increase in the short term, although this has not been demonstrated chronically. These factors combined increase the oxygen demand of the heart. In contrast, the effect of tobacco smoke, not least of carbon monoxide, is to reduce the blood oxygen supply. This oxidative stress puts a strain on the heart over time and eventually leads to ischemia (Ambrose and Barua 2004).

8.5 Limitations of Smoking Research

Due to the ethical obstacles of randomly assigning subjects to smoking, research in this field is largely based on observational studies, including case–control and cohort studies. Despite the limitations inherent in these types of research, including inability to control for all confounding factors, the sheer number of studies and weight of evidence have shown beyond a doubt that smokers have greatly increased risk of both developing and dying from CVD. Furthermore, the relationship between smoking and CVD fulfills many of the Bradford Hill criteria necessary to establish a causal relationship, including the consistently demonstrated strength of the relationship in numerous settings, countries, age groups, and ethnicities; a clear dose–response association; temporality, demonstrated by prospective investigations; and biological plausibility (Hill 1965).

Smoking status and intensity are most often self-reported, allowing for the introduction of bias and misclassification error. Although biomarkers of smoking are available, most commonly cotinine, these are not always verified. While some studies found that patients are usually truthful about their smoking status in a clinical context (Attebring et al. 2001), a systematic review reported that self-report often underestimates smoking prevalence (Gorber et al. 2009). Many studies of smoking cessation only assess 3–6 month outcomes – in light of the high recidivism rate, it seems presumptuous to assume that someone who has quit for 3 months will remain abstinent. Long-term follow-ups are therefore preferable. Studies of second- or thirdhand exposure encounter further problems associated with exposure assessment, for example, measurement of household exposure misses exposure at the workplace, entertainment venues, or other public areas.

Measurement accuracy has also been questioned with regard to smoking-attributable mortality estimates, which often fail to consider inaccuracies in death certificates, variation in diagnostic standards, and confounding factors (Lee 1996). Malarcher et al. compared estimates using data from the American Cancer Society's Cancer Prevention Study II and from the National Mortality Followback Survey and found a discrepancy of 19 % between smoking-attributable mortality for the most common smoking-related diseases including CHD and cerebrovascular disease (Malarcher et al. 2000). However, a report by Thun et al. (2000), which added multivariable adjustment for socioeconomic factors, alcohol consumption, and dietary factors to the surgeon general's estimates of smoking-attributable mortality, found the overall mortality estimate to be reduced by just 1 % (Thun et al. 2000). While certain limitations are involved in the estimation of the overall health burden of tobacco, published figures give an indication of the severity and scale of smoking-related diseases.

8.6 Summary and Perspectives

Exposure to tobacco, in all its forms, is undoubtedly a precursor to development and progression of CVD. Both smokers themselves and those in their immediate environment are continually exposed to a toxic cocktail which acts on the cardiovascular system in multiple ways, puts a strain on the heart, boosts atherosclerosis, and leads to MI and stroke at a relatively young age. Nationwide and worldwide research has shown time and again that smokers are at significantly increased risk of developing acute coronary syndromes and that those with already established CHD have lower odds of survival. Smoking cessation, particularly by the age of 30, greatly reduces cardiovascular risk in the long term, and smoking reduction has shown promising evidence although these benefits require additional investigation. As smoking rates decrease in developed countries, they continue to rise in many developing countries. Tobacco companies seek to fill the void left by falling rates of smoking in developed countries by promoting smoking in developing countries, marketing to women in traditional societies where smoking is regarded as a primarily male activity, developing new tobacco products for existing markets with promises of reduced harm, and finding ways to introduce the wide range of tobacco products found around the world into new markets. While forms of tobacco other than cigarettes are not well studied, the little research that exists shows that real risk is involved. Despite limitations in smoking research, the accumulation of epidemiological and pathophysiological evidence highlights increased cardiovascular trends in smokers and biologically plausible pathways for these processes. While there are numerous risk factors for CVD, all prevalent in modern western lifestyle, smoking is the most significant, preventable contributing factor, shaving years off smokers' lives.

References

Ambrose JA, Barua RS (2004) The pathophysiology of cigarette smoking and cardiovascular disease: an update. J Am Coll Cardiol 43(10):1731–1737

Attebring M, Herlitz J, Berndt AK, Karlsson T, Hjalmarson A (2001) Are patients truthful about their smoking habits? A validation of self-reports about smoking cessation with biochemical markers of smoking activity amongst patients with ischemic heart disease. J Int Med 249(2):145–151

Barnoya J, Glantz SA (2005) Cardiovascular effects of secondhand smoke: nearly as large as smoking. Circulation 111(20):2684–2698

Barutcu I, Esen AM, Degirmenci B, Acar M, Kaya D, Turkmen M, Melek M, Onrat E, Esen OB, Kirma C (2004) Acute cigarette smoking-induced hemodynamic alterations in the common carotid artery – a transcranial Doppler study. Circ J 68(12):1127–1131

Bjartveit K, Tverdal A (2005) Health consequences of smoking 1-4 cigarettes per day. Tob Control 14(5):315–320

Blache D, Bouthillier D, Davignon J (1992) Acute influence of smoking on platelet behaviour, endothelium and plasma lipids and normalization by aspirin. Atherosclerosis 93(3):179–188

Blann AD, Kirkpatrick U, Devine C, Naser S, McCollum CN (1998) The influence of acute smoking on leucocytes, platelets and the endothelium. Atherosclerosis 141(1):133–139

Bowman TS, Gaziano JM, Buring JE, Sesso HD (2007) A prospective study of cigarette smoking and risk of incident hypertension in women. J Am Coll Cardiol 50(21):2085–2092

Callinan JE, Clarke A, Doherty K, Kelleher C (2010) Legislative smoking bans for reducing secondhand smoke exposure, smoking prevalence and tobacco consumption. Cochrane Database Syst Rev (4):CD005992

Chen CY, Chow D, Chiamvimonvat N, Glatter KA, Li N, He Y, Pinkerton KE, Bonham AC (2008) Short-term secondhand smoke exposure decreases heart rate variability and increases arrhythmia susceptibility in mice. Am J Physiol Heart Circ Physiol 295(2):H632–H639

Craig WY, Palomaki GE, Haddow JE (1989) Cigarette smoking and serum lipid and lipoprotein concentrations: an analysis of published data. BMJ 298(6676):784–788

Davis JW, Shelton L, Eigenberg DA, Hignite CE, Watanabe IS (1985) Effects of tobacco and non-tobacco cigarette smoking on endothelium and platelets. Clin Pharmacol Ther 37(5):529–533

Davis JW, Hartman CR, Shelton L, Ruttinger HA (1986) Smoking-induced changes in endothelium and platelets are not affected by hydroxyethylrutosides. Br J Exp Pathol 67(5):765–771

Direk N, Newson RS, Hofman A, Kirschbaum C, Tiemeier H (2011) Short and long-term effects of smoking on cortisol in older adults. Int J Psychophysiol 80(2):157–160

Doll R, Peto R, Boreham J, Sutherland I (2004) Mortality in relation to smoking: 50 years' observations on male British doctors. BMJ 328(7455):1519

Doyle JT, Dawber TR, Kannel WB, Heslin AS, Kahn HA (1962) Cigarette smoking and coronary heart disease. Combined experience of the Albany and Framingham studies. N Engl J Med 26:796–801

Escobedo LG, Caspersen CJ (1997) Risk factors for sudden coronary death in the United States. Epidemiology 8(2):175–180

Ezzati M, Lopez AD (2003) Estimates of global mortality attributable to smoking in 2000. Lancet 362(9387):847–852

Ezzati M, Henley SJ, Thun MJ, Lopez AD (2005) Role of smoking in global and regional cardiovascular mortality. Circulation 112(4):489–497

Fortmann SP, Haskell WL, Williams PT (1986) Changes in plasma high density lipoprotein cholesterol after changes in cigarette use. Am J Epidemiol 124(4):706–710

Garland C, Barrett-Connor E, Suarez L, Criqui MH, Wingard DL (1985) Effects of passive smoking on ischemic heart disease mortality of nonsmokers. A prospective study. Am J Epidemiol 121(5):645–650

Gerber Y, Rosen LJ, Goldbourt U, Benyamini Y, Drory Y, Israel Study Group on First Acute Myocardial Infarction (2009) Smoking status and long-term survival after first acute myocardial infarction. J Am Coll Cardiol 54(25):2382–2387

Gerber Y, Myers V, Goldbourt U (2012) Smoking reduction at midlife and lifetime mortality risk in men: a prospective cohort study. Am J Epidemiol 175(10):1006–1012

Glantz SA (2008) Meta-analysis of the effects of smokefree laws on acute myocardial infarction: an update. Prev Med 47(4):452–453

Godtfredsen NS, Osler M, Vestbo J, Andersen I, Prescott E (2003) Smoking reduction, smoking cessation, and incidence of fatal and non-fatal myocardial infarction in Denmark 1976-1998: a pooled cohort study. J Epidemiol Community Health 57(6):412–416

Godtfredsen NS, Prescott E, Vestbo J, Osler M (2006) Smoking reduction and biomarkers in two longitudinal studies. Addiction 101(10):1516–1522

Gorber SC, Schofield-Hurwitz S, Hardt J, Levasseur G, Tremblay M (2009) The accuracy of self-reported smoking: a systematic review of the relationship between self-reported and cotinine-assessed smoking status. Nicotine Tob Res 11:12–24

Gottlieb S, Fallavollita J, McDermott M, Brown M, Eberly S, Moss AJ (1994) Cigarette smoking and the age at onset of a first non-fatal myocardial infarction. Coron Artery Dis 5(8): 687–694

Grundtvig M, Hagen TP, German M, Reikvam A (2009) Sex-based differences in premature first myocardial infarction caused by smoking: twice as many years lost by women as by men. Eur J Cardiovasc Prev Rehabil 16(2):174–179

Grundtvig M, Hagen TP, Amrud ES, Reikvam A (2011) Mortality after myocardial infarction: impact of gender and smoking status. Eur J Epidemiol 26(5):385–393

Hamer M, Stamatakis E, Kivimaki M, Lowe GD, Batty GD (2010) Objectively measured second-hand smoke exposure and risk of cardiovascular disease: what is the mediating role of inflammatory and hemostatic factors? J Am Coll Cardiol 56(1):18–23

He J, Vupputuri S, Allen K, Prerost MR, Hughes J, Whelton PK (1999) Passive smoking and the risk of coronary heart disease: a meta-analysis of epidemiologic studies. N Engl J Med 340(12):920–926

Hein HO, Suadicani P, Skov P, Gyntelberg F (1991) Indoor dust exposure: an unnoticed aspect of involuntary smoking. Arch Environ Health 46(2):98–101

Heiss C, Amabile N, Lee AC, Real WM, Schick SF, Lao D, Wong ML, Jahn S, Angeli FS, Minasi P, Springer ML, Hammond SK, Glantz SA, Grossman W, Balmes JR, Yeghiazarians Y (2008) Brief secondhand smoke exposure depresses endothelial progenitor cells activity and endothelial function: sustained vascular injury and blunted nitric oxide production. J Am Coll Cardiol 51(18):1760–1771

Helsing KJ, Sandler DP, Comstock GW, Chee E (1988) Heart disease mortality in nonsmokers living with smokers. Am J Epidemiol 127(5):915–922

Hill AB (1965) The environment and disease: association or causation? Proc R Soc Med 58:295–300

Iso H, Shimamoto T, Sato S, Koike K, Iida M, Komachi Y (1996) Passive smoking and plasma fibrinogen concentrations. Am J Epidemiol 144(12):1151–1154

Kabat GC (2003) Fifty years' experience of reduced-tar cigarettes: what do we know about their health effects? Inhal Toxicol 15(11):1059–1102

Kallio K, Jokinen E, Raitakari OT, Hämäläinen M, Siltala M, Volanen I, Kaitosaari T, Viikari J, Rönnemaa T, Simell O (2007) Tobacco smoke exposure is associated with attenuated endothelial function in 11-year-old healthy children. Circulation 115(25):3205–3212

Kannel WB, Doyle JT, McNamara PM, Quickenton P, Gordon T (1975) Precursors of sudden coronary death. Factors related to the incidence of sudden death. Circulation 51(4):606–613

Kannel WB, D'Agostino RB, Belanger AJ (1987) Fibrinogen, cigarette smoking, and risk of cardiovascular disease: insights from the Framingham Study. Am Heart J 113(4):1006–1010

Kato T, Inoue T, Morooka T, Yoshimoto N, Node K (2006) Short-term passive smoking causes endothelial dysfunction via oxidative stress in nonsmokers. Can J Physiol Pharmacol 84(5):523–529

Kenfield SA, Wei EK, Rosner BA, Glynn RJ, Stampfer MJ, Colditz GA (2010) Burden of smoking on cause-specific mortality: application to the Nurses' Health Study. Tob Control 19(3):248–254

Law MR, Morris JK, Wald NJ (1997) Environmental tobacco smoke exposure and ischaemic heart disease: an evaluation of the evidence. BMJ 315(7114):973–980

Lee PN (1996) Mortality from tobacco in developed countries: are indirect estimates reliable? Regul Toxicol Pharmacol 24(1 Pt 1):60–68

Lightwood JM, Glantz SA (2009) Declines in acute myocardial infarction after smoke-free laws and individual risk attributable to secondhand smoke. Circulation 120(14):1373–1379

Lightwood JM, Coxson PG, Bibbins-Domingo K, Williams LW, Goldman L (2009) Coronary heart disease attributable to passive smoking: CHD Policy Model. Am J Prev Med 36(1):13–20

Mackay DF, Irfan MO, Haw S, Pell JP (2010) Meta-analysis of the effect of comprehensive smoke-free legislation on acute coronary events. Heart 96(19):1525–1530

Mähönen MS, McElduff P, Dobson AJ, Kuulasmaa KA, Evans AE, WHO MONICA Project (2004) Current smoking and the risk of non-fatal myocardial infarction in the WHO MONICA Project populations. Tob Control 13(3):244–250

Malarcher AM, Schulman J, Epstein LA, Thun MJ, Mowery P, Pierce B, Escobedo L, Giovino GA (2000) Methodological issues in estimating smoking-attributable mortality in the United States. Am J Epidemiol 152(6):573–584

Matt GE, Quintana PJ, Hovell MF, Bernert JT, Song S, Novianti N, Juarez T, Floro J, Gehrman C, Garcia M, Larson S (2004) Households contaminated by environmental tobacco smoke: sources of infant exposures. Tob Control 13(1):29–37

Matt GE, Quintana PJ, Destaillats H, Gundel LA, Sleiman M, Singer BC, Jacob P, Benowitz N, Winickoff JP, Rehan V, Talbot P, Schick S, Samet J, Wang Y, Hang B, Martins-Green M, Pankow JF, Hovell MF (2011) Thirdhand tobacco smoke: emerging evidence and arguments for a multidisciplinary research agenda. Environ Health Perspect 119(9):1218–1226

Meade T (1994) Atheroma and thrombosis in cardiovascular disease: separate or complementary? In: Marmot M, Elliott P (eds) Coronary heart disease epidemiology: from aetiology to public health. Oxford University Press, New York

Metsios GS, Flouris AD, Angioi M, Koutedakis Y (2010) Passive smoking and the development of cardiovascular disease in children: a systematic review. Cardiol Res Pract 2011:pii587650

Meyers DG, Neuberger JS, He J (2009) Cardiovascular effect of bans on smoking in public places: a systematic review and meta-analysis. J Am Coll Cardiol 54(14):1249–1255

Miller GJ, Bauer KA, Cooper JA, Rosenberg RD (1998) Activation of the coagulant pathway in cigarette smokers. Thromb Haemost 79(3):549–553

Minami J, Ishimitsu T, Matsuoka H (1999) Effects of smoking cessation on blood pressure and heart rate variability in habitual smokers. Hypertension 33(1 Pt 2):586–590

Mizoue T, Ueda R, Hino Y, Yoshimura T (1999) Workplace exposure to environmental tobacco smoke and high density lipoprotein cholesterol among nonsmokers. Am J Epidemiol 150(10):1068–1072

Moffatt RJ, Chelland SA, Pecott DL, Stamford BA (2004) Acute exposure to environmental tobacco smoke reduces HDL-C and HDL2-C. Prev Med 38(5):637–641

Mokdad AH, Marks JS, Stroup DF, Gerberding JL (2004) Actual causes of death in the United States, 2000. JAMA 291(10):1238–1245

Negri E, Franzosi MG, La Vecchia C, Santoro L, Nobili A, Tognoni G (1993) Tar yield of cigarettes and risk of acute myocardial infarction. GISSI-EFRIM Investigators. BMJ 306(6892):1567–1570

Oberg M, Jaakkola MS, Woodward A, Peruga A, Prüss-Ustün A (2011) Worldwide burden of disease from exposure to second-hand smoke: a retrospective analysis of data from 192 countries. Lancet 377(9760):139–146

Omvik P (1996) How smoking affects blood pressure. Blood Press 5(2):71–77

Otsuka R, Watanabe H, Hirata K, Tokai K, Muro T, Yoshiyama M, Takeuchi K, Yoshikawa J (2001) Acute effects of passive smoking on the coronary circulation in healthy young adults. JAMA 286(4):436–441

Panagiotakos DB, Chrysohoou C, Pitsavos C, Papaioannou I, Skoumas J, Stefanadis C, Toutouzas P (2002) The association between secondhand smoke and the risk of developing acute coronary syndromes, among non-smokers, under the presence of several cardiovascular risk factors: the CARDIO2000 case-control study. BMC Public Health 2:9

Piano MR, Benowitz NL, Fitzgerald GA, Corbridge S, Heath J, Hahn E, Pechacek TF, Howard G, American Heart Association Council on Cardiovascular Nursing (2010) Impact of smokeless tobacco products on cardiovascular disease: implications for policy, prevention, and treatment: a policy statement from the American Heart Association. Circulation 122(15):1520–1544

Pisinger C, Godtfredsen NS (2007) Is there a health benefit of reduced tobacco consumption? A systematic review. Nicotine Tob Res 9(6):631–646

Raupach T, Schäfer K, Konstantinides S, Andreas S (2006) Secondhand smoke as an acute threat for the cardiovascular system: a change in paradigm. Eur Heart J 27(4):386–392

Rohleder N, Kirschbaum C (2006) The hypothalamic-pituitary-adrenal (HPA) axis in habitual smokers. Int J Psychophysiol 59(3):236–243

Rothwell M, Rampling MW, Cholerton S, Severs PS (1991) Haemorheological changes in the very short term after abstention from tobacco by cigarette smokers. Br J Haematol 79(3):500–503

Salonen JT, Puska P, Kottke TE (1981) Smoking, blood pressure and serum cholesterol as risk factors of acute myocardial infarction and death among men in Eastern Finland. Eur Heart J 2(5):365–373

Sandler DP, Comstock GW, Helsing KJ, Shore DL (1989) Deaths from all causes in non-smokers who lived with smokers. Am J Public Health 79(2):163–167

Sargent RP, Shepard RM, Glantz SA (2004) Reduced incidence of admissions for myocardial infarction associated with public smoking ban: before and after study. BMJ 328(7446):977–980

Shah RS, Cole JW (2010) Smoking and stroke: the more you smoke the more you stroke. Expert Rev Cardiovasc Ther 8(7):917–932

Sharrett AR, Sorlie PD, Chambless LE, Folsom AR, Hutchinson RG, Heiss G, Szklo M (1999) Relative importance of various risk factors for asymptomatic carotid atherosclerosis versus coronary heart disease incidence: the Atherosclerosis Risk in Communities Study. Am J Epidemiol 149(9):843–852

Shinton R, Beevers G (1989) Meta-analysis of relation between cigarette smoking and stroke. BMJ 298(6676):789–794

Sims M, Maxwell R, Bauld L, Gilmore A (2010) Short term impact of smoke-free legislation in England: retrospective analysis of hospital admissions for myocardial infarction. BMJ 340:c2161

Spagnoli LG, Mauriello A, Palmieri G, Santeusanio G, Amante A, Taurino M (1994) Relationships between risk factors and morphological patterns of human carotid atherosclerotic plaques. A multivariate discriminant analysis. Atherosclerosis 108(1):39–60

Strong JP, Malcom GT, McMahan CA, Tracy RE, Newman WP 3rd, Herderick EE, Cornhill JF (1999) Prevalence and extent of atherosclerosis in adolescents and young adults: implications for prevention from the Pathobiological Determinants of Atherosclerosis in Youth Study. JAMA 281(8):727–735

Svendsen KH, Kuller LH, Martin MJ, Ockene JK (1987) Effects of passive smoking in the Multiple Risk Factor Intervention Trial. Am J Epidemiol 126(5):783–795

Talukder MA, Johnson WM, Varadharaj S, Lian J, Kearns PN, El-Mahdy MA, Liu X, Zweier JL (2011) Chronic cigarette smoking causes hypertension, increased oxidative stress, impaired NO bioavailability, endothelial dysfunction, and cardiac remodeling in mice. Am J Physiol Heart Circ Physiol 300(1):H388–H396

Tang JL, Morris JK, Wald NJ, Hole D, Shipley M, Tunstall-Pedoe H (1995) Mortality in relation to tar yield of cigarettes: a prospective study of four cohorts. BMJ 311(7019):1530–1533

Teo KK, Ounpuu S, Hawken S, Pandey MR, Valentin V, Hunt D, Diaz R, Rashed W, Freeman R, Jiang L, Zhang X, Yusuf S, INTERHEART Study Investigators (2006) Tobacco use and risk of myocardial infarction in 52 countries in the INTERHEART study: a case-control study. Lancet 368(9536):647–658

Thun MJ, Luiza da Costa e Silva V (2003) Introduction and overview of global tobacco surveillance. In: Shafey O, Dolwick S, Guindon GE (eds) Tobacco control country profiles, 2nd edn. American Cancer Society, Atlanta

Thun MJ, Apicella LF, Henley SJ (2000) Smoking vs other risk factors as the cause of smoking-attributable deaths: confounding in the courtroom. JAMA 284(6):706–712

Tracy RP, Psaty BM, Macy E, Bovill EG, Cushman M, Cornell ES, Kuller LH (1997) Lifetime smoking exposure affects the association of C-reactive protein with cardiovascular disease risk factors and subclinical disease in healthy elderly subjects. Arterioscler Thromb Vasc Biol 17(10):2167–2176

Tverdal A, Bjartveit K (2006) Health consequences of reduced daily cigarette consumption. Tob Control 15(6):472–480

USDHHS (1967) The health consequences of smoking: a public health service review. US Department of Health, Education and Welfare, Washington DC

USDHHS (1983) The health consequences of smoking: cardiovascular disease. A report of the surgeon general. US Department of Health and Human Services, Public Health Service, Office on Smoking and Health, Rockville

USDHHS (1990) The health benefits of smoking cessation: a report of the surgeon general. U.S. Department of Health and Human Services, Public Health Service, Center for Disease Control, Rockville

USDHHS (2004) The health consequences of smoking: a report of the surgeon general. US Department of Health and Human services, Center for Disease Control and Prevention, Office on Smoking and Health, Atlanta

USDHHS (2010) How tobacco smoke causes disease: the biology and behavioral basis for smoking-attributable disease: a report of the surgeon general. U.S. Department of Health and Human Services, Centers for Disease Control and Prevention, National Center for Chronic Disease Prevention and Health Promotion, Office on Smoking and Health 2010, Atlanta

Venn A, Britton J (2007) Exposure to secondhand smoke and biomarkers of cardiovascular disease risk in never-smoking adults. Circulation 115(8):990–995

Wells AJ (1994) Passive smoking as a cause of heart disease. J Am Coll Cardiol 24(2):546–554

Whincup PH, Gilg JA, Emberson JR, Jarvis MJ, Feyerabend C, Bryant A, Walker M, Cook DG (2004) Passive smoking and risk of coronary heart disease and stroke: prospective study with cotinine measurement. BMJ 329(7459):200–205

WHO (2004) Smoking and oral tobacco use. In: Comparative quantification of health risks: global and regional burden of disease attributable to selected major risk factors, vol 1. World Health Organization, Geneva, Chapter 11

Wilhelmsson C, Vedin JA, Elmfeldt D, Tibblin G, Wilhelmsen L (1975) Smoking and myocardial infarction. Lancet 1(7904):415–420

Wilson K, Gibson N, Willan A, Cook D (2000) Effect of smoking cessation on mortality after myocardial infarction: meta analysis of cohort studies. Arch Intern Med 160:939–944

Wolf PA, D'Agostino RB, Kannel WB, Bonita R, Belanger AJ (1988) Cigarette smoking as a risk factor for stroke. The Framingham Study. JAMA 259(7):1025–1029

Woloshin S, Schwartz LM, Welch HG (2008) The risk of death by age, sex, and smoking status in the United States: putting health risks in context. J Natl Cancer Inst 100(12):845–853

Yatsuya H, Folsom AR (2010) Risk of incident cardiovascular disease among users of smokeless tobacco in the Atherosclerosis Risk in Communities (ARIC) Study. Am J Epidemiol 172(5):600–605

9 Socioeconomic Aspects of Cardiovascular Health

Vicki Myers and Yariv Gerber

Abstract

Extensive epidemiological evidence has demonstrated the socioeconomic gradient in cardiovascular health, with the most disadvantaged individuals at greatest risk of developing and dying from cardiovascular disease (CVD). Large-scale studies, beginning in the 1960s, reported significant associations between coronary mortality and occupation grade, income, and education. Individuals with low socioeconomic status (SES) additionally have a poorer prognosis post-myocardial infarction. Explanatory mechanisms explored here include differential access to medical care, differences in risk factor profiles, health literacy, and psychological and environmental factors. The influence of neighborhood SES is examined, beyond individual socioeconomic profile, as well as SES trajectory throughout life, and the relative importance of childhood versus adult SES in the development of CVD. Finally methodological factors are considered, examining the limitations of SES research, measurement issues, and risk prediction. It is proposed that SES should be considered in cardiovascular risk assessment, and that policy changes have the potential to minimize socioeconomic inequalities in cardiovascular health.

Keywords

Socioeconomic status • Cardiovascular disease • Epidemiology • Mortality • Prognosis • Myocardial infarction • Risk prediction

---

V. Myers, MSc
Sackler Faculty of Medicine,
Department of Epidemiology and Preventive Medicine,
Tel Aviv University, Tel Aviv, Israel

Y. Gerber, PhD (✉)
Sackler Faculty of Medicine,
Department of Epidemiology and Preventive Medicine,
School of Public Health, Tel Aviv University,
Ramat Aviv, Tel Aviv 69978, Israel
e-mail: yarivg@post.tau.ac.il

I. Wakabayashi, K. Groschner (eds.), *Interdisciplinary Concepts in Cardiovascular Health*, 187
DOI 10.1007/978-3-7091-1334-9_9, © Springer-Verlag Wien 2013

9.1 Introduction

Decades of research have consistently shown that socioeconomic status (SES) is intimately intertwined with health, be it through behavioral, psychosocial, or pathophysiological pathways, via education and knowledge, financial resources, environmental characteristics or access to health services, healthy food, and sports facilities. Where we live and work; who we socialize with; what we eat, drink, and breathe; and how we spend our free time all impact on our cardiovascular health.

Cardiovascular disease (CVD) is a leading cause of death and disability worldwide; however, it does not strike indiscriminately. The social gradient in health means that those at the lower end of the socioeconomic scale are more likely to develop CVD, contrary to the stereotypical image of the wealthy but stressed executive who dies of a heart attack. Among those who already have the disease, individuals with low SES are more likely to die. Interestingly, the linear relationship between SES and health does not apply only to the extremes of deprivation and wealth, but can be seen within relatively similar western populations (Macintyre 1994). Many factors are at play in this relationship, including individual, community, and environmental factors, and we will here examine socioeconomic factors related to cardiovascular health, the evidence for these associations, and the mechanisms which underlie them.

SES is a combination of the economic resources at one's disposal and social position and typically comprises educational level, type of occupation, income, social class, and wealth. SES is a relative measure and has no meaning in isolation; it is always examined in the context of society, relative to one's neighbors. The construct of SES is based on the assumption that resources are distributed unequally in society.

9.2 SES and Cardiovascular Outcomes

9.2.1 SES and Development of CVD

One of the first large-scale investigations to provide evidence of the social gradient in health was the Whitehall study, a prospective cohort study which recruited 18,000 male employees of the UK civil service in 1967, and discovered a striking association between occupational grade and both all-cause and coronary mortality. Men in the lowest job grade were much more likely to die prematurely than those in the highest grade, with a coronary mortality rate 3.6 times higher (Rose and Marmot 1981). Furthermore, traditional risk factors including high blood pressure, smoking, and elevated cholesterol only explained a third of this association with mortality, inferring a direct relation between occupational grade and coronary health (van Rossum et al. 2000). The subsequent Whitehall II study, which investigated the cardiovascular health of a further 10,000 employees beginning in 1985, reported higher prevalence of risk factors, including smoking and obesity, and lower prevalence of healthy behaviors such as physical activity and healthy diet, in lower-grade employees (Marmot et al. 1991). The Whitehall cohort, while large, has several limitations, comprising a majority of male white-collar workers in stable employment, and thus not being generalizable to the population as a whole. Among its strengths, the cohort included a wide range of job grades, from messengers to executive managers. In parallel, the Black report released in 1980 published evidence of large differences in mortality and morbidity according to social class (Smith et al. 1990). Far from improving over time, evidence from both the UK and USA showed that the gap between rich and poor grew between the 1970s and 1980s (Smith et al. 1990; Pappas et al. 1993).

While the Whitehall study used occupation as a marker of SES, representing salary, education, and social status, other facets have been equally assessed to represent SES.

The Tromso Heart Study found educational level to be associated with heart healthy behaviors in a cohort of 12,000 Norwegians: participants with the highest education were less likely to smoke and be overweight, were more physically active, and had a healthier diet (Jacobsen and Thelle 1988). It has been argued that education is a more reliable measure of SES when considering the relationship with health outcomes since it is usually fixed in early adulthood and does not tend to change, in contrast to occupation and income. As opposed to recruiting a single measure of SES, many studies have assessed multiple measures in order to gain a fuller picture.

Other longitudinal studies conducted in the USA, such as the Alameda County Study (Beebe-Dimmer et al. 2004), the Evans County Study (Johnson et al. 1986), Georgia Heart Study, the Charleston Heart Study (Nietert et al. 2006), and the US National Longitudinal Mortality Study, all found similar trends between SES and cardiovascular mortality. A recent meta-analysis comprising 70 studies reported an overall increased risk of myocardial infarction (MI) in low SES groups, when assessed on income, education, or occupation (Manrique-Garcia et al. 2011). The lowest income group had a 71 % increased MI risk compared to the highest income groups, while the least educated group had an increased risk of 34 % relative to the most educated.

While the Whitehall study examined only men, the Alameda County Study recruited over 3,000 women and collected socioeconomic data at four different time-points over a 30-year period, including childhood and current SES (Beebe-Dimmer et al. 2004). Low household income (set in 1965 as less than $5,000) was associated with elevated CVD mortality, with a hazard ratio of 1.5 compared to high household income (greater than $10,000), as was low childhood SES. Gender has been shown to interact with SES in the relationship with CVD, for example, in the First National Health and Nutrition Examination Survey, low education was associated with greater risk of coronary heart disease (CHD) in women than in men (Thurston et al. 2005).

9.2.1.1 Job Strain/Control

In recent years, research has investigated additional aspects of employment which may influence cardiovascular risk, beyond the mere fact of being employed, for example, the amount of control in the workplace or effort versus reward. A Swedish study of 8,000 white-collar workers found that increased job control was associated with reduced CHD score (comprising self-reported high blood pressure, chest pains, and trouble breathing) (Karasek 1990). Increased job strain was also found to be associated with increased blood pressure in an Italian study from the WHO-MONICA project (Cesana et al. 2003). A 50 % excess risk of CHD was reported in employees with high effort and low reward jobs and an increased risk of 43 % in high versus low strain jobs in a meta-analysis including over 80,000 workers (Kivimäki et al. 2006). This line of investigation attempts to explain the stark differences in CV risk seen between different job grades, based on the assumption that lower status jobs usually involve more strain and less control.

9.2.2 SES and Progression of CVD

In addition to increasing the risk of developing CVD, low SES is related to poorer outcome in individuals with established CVD. Research in post-MI patients has shown a clear inverse relationship between socioeconomic status and mortality. Following are several examples presenting evidence from epidemiological studies.

The FINMONICA study recorded all MI events in three Finnish regions over a 10-year period and classified patients according to income (Salomaa et al. 2001). Low-income men had more than twice the rate of pre-hospital coronary death compared to high-income men, and in those surviving MI the 12-month mortality rate was significantly higher in low-income patients. Furthermore, case fatalities showed a graded relationship with both income and education.

In a cohort of 1,500 Israeli MI survivors aged 65 or less followed up for 13 years, comprising 98 % of all incident MIs in the study area during a given year, patients with both low income and low education had a higher risk of mortality compared to their better-off counterparts (Gerber et al. 2008). Low income was more of a risk factor in individuals with less education (Fig. 9.1).

Several other studies reported similar findings, with income inversely related to post-MI mortality (Salomaa et al. 2001; Alter et al. 2006; Rao et al. 2004). Alter et al., in a study of 3,400 Canadian MI patients, reported this relation to be substantially attenuated on multivariable adjustment for age and CV risk factors (Alter et al. 2006), while Rao found that the poorest decile had a much higher short-term mortality rate – within 1 year – than the rest of the population (Rao et al. 2004). This relation is explained since poorer individuals presented later to the hospital and consequently received poorer treatment.

9.2.2.1 Why Are Low SES Patients More Likely to Die After Suffering an MI?

Access to care: There is evidence that lower-income patients are likely to receive poorer medical care. A US study of over 10,000 patients with acute coronary syndromes reported low-income patients as presenting with more severe disease compared to high-income patients (Rao et al. 2003). Furthermore, lower-income patients were less likely to receive evidence-based treatment including cardiac

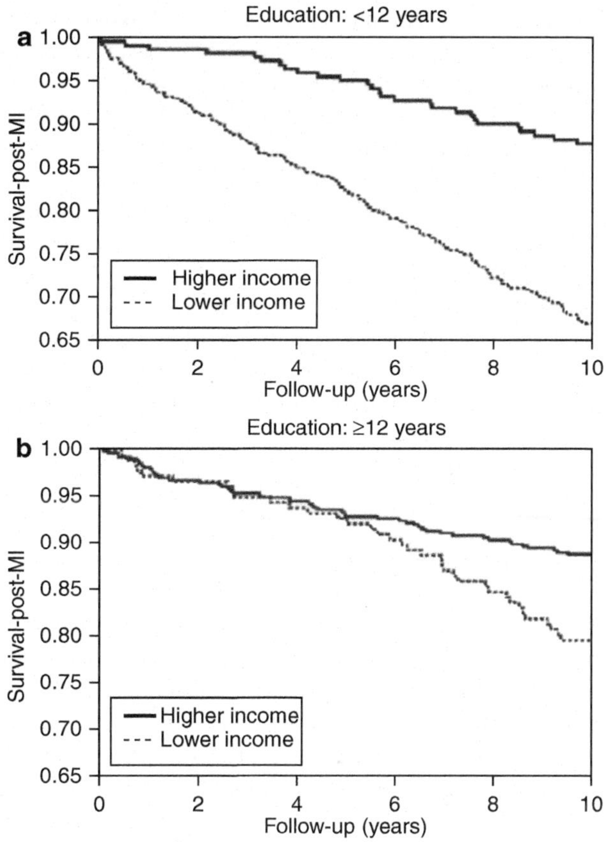

Fig. 9.1 Kaplan–Meier survival curves comparing lower-income patients with higher-income patients by education categories. (**a**) Less educated group [log-rank test, $P<0.001$; (**b**) More educated group [log-rank test, $P=0.003$] (Excerpt from Gerber et al. (2008), SAGE Publications, Inc.; all rights reserved)

catheterization, percutaneous coronary intervention, and prescription of aspirin or beta-blockers, although these differences were attenuated on multivariable adjustment. These trends may go some way to explaining the significantly higher 6-month mortality rate in the low-income group. Additional studies found discrepancies between treatments received by MI patients according to SES. A large-scale cohort study involving over 50,000 Canadian MI patients reported increased use of coronary angiography and reduced waiting times in the highest SES compared to lowest SES neighborhoods, based on census data (Alter et al. 1999). Furthermore, there was a strong inverse relationship between income and mortality 1 year post-MI, despite the universal healthcare provided in Canada. Similar results have been reported in numerous studies, presenting reduced use of invasive cardiac procedures in lower-income MI patients (Philbin et al. 2000; Rathore et al. 2000). Besides provision of treatment, access to medical facilities may differ according to SES. In the FINMONICA study, low-income males with MI were more likely to present with more than 4h delay compared to wealthier patients (Salomaa et al. 2001). Whether this delay is due to poorer access to appropriate care or to reduced help-seeking behavior remains open to debate.

Risk factors: An alternative or parallel explanation for the poorer survival odds of low SES MI patients is a difference in baseline risk factors (Ebrahim et al. 2004), which contribute to both the development and progression of CHD. While CHD mortality has declined over the years as has the prevalence of some primary CV risk factors such as smoking and physical inactivity, socioeconomic inequalities persist (Hotchkiss et al. 2011). Secondary risk factors such as diabetes and hypertension are on the increase. Less educated MI patients were more likely to have a history of diabetes mellitus and congestive heart failure in the Multicenter Investigation of the Limitation of Infarct Size (Tofler et al. 1993). Risk factors differed not only prior to MI but also during follow-up, with less educated patients less likely to stop smoking (never graduated 38 % vs. high school graduates 49 %). Patients who continued to smoke had increased mortality risk. This finding was replicated in an Israeli cohort study, with SES contributing to the likelihood of quitting smoking post-MI (Gerber et al. 2011a). Additionally, low neighborhood SES was associated with lower physical activity after MI (Gerber et al. 2011b), a factor strongly related to prognosis.

9.3 Mechanisms

The Black report identified four types of explanations for social inequalities in health. These are artifacts, or measurement errors in attributing social class, including the fact that lower or working classes are diminishing; social selection which proposes that health status determines socioeconomic status; behavioral, whereby unhealthy behaviors are more prevalent in lower social classes; and materialist, involving "hazards inherent in society," such as working in hazardous jobs or residing in heavily polluted areas (Smith et al. 1990). All these factors contribute to the socioeconomic gradient in cardiovascular health.

Risk factors: Much has been written about risk factors as the link between SES and cardiovascular outcomes. Evidence from the Framingham Heart Study – a long-term investigation which pioneered the concept of cardiac risk factors – has demonstrated that the primary risk factors for CVD are smoking, hypertension, high cholesterol (dyslipidemia), sedentary lifestyle, and diabetes, largely lifestyle-influenced factors alongside genetic predisposition (Mendis 2010). The Kuopio Ischemic Heart Disease Risk Factor Study investigated whether 23 biological, behavioral, psychological, and social risk factors could account for the association between income and CV mortality in men (Lynch et al. 1996). Adjustment for risk factors not only reduced but completely eliminated the association. Multivariable adjustment also attenuated the relation between SES and acute MI. Biological factors had the greatest effect in risk reduction. The question remains, why do low SES populations have a higher prevalence of CV risk factors, such as blood glucose, hypertension, and high cholesterol? There is direct evidence that SES affects behavior styles, coping styles, the endocrine system, the homeostasis system, and access to medical care (Kaplan and Keil 1993). While some evidence exists for psychological, physiological, and biochemical mediators of the relation between SES and disease, much remains open to speculation.

Hypertension, a risk factor for MI, stroke, and heart failure, has been frequently associated with SES (Cirera et al. 1998). This could be due to greater awareness of hypertension, the effects of diet and the importance of regular checkups, and better access to health services among more highly educated people or could be a by-product of a generally more stressful life associated with deprivation. Cumulative stress has an effect on the heart, increasing allostatic load, illustrated by delayed recovery of the cardiovascular system, specifically blood pressure and heart rate variability, after mental stress in low SES groups, in a sub-cohort of the Whitehall II study (Steptoe et al. 2002). This implies that certain characteristics of low SES – prolonged stress, dietary factors – may put a strain on the heart, making it more vulnerable to injury. Evidence has shown that acute stress can have adverse CV effects, for example, impairment of endothelial function or an increase in cytokine levels lasting for several hours (Steptoe et al. 2001). Cumulative stress is therefore likely to have an enduring effect on the CV system. Fibrinogen has also been demonstrated to be higher in lower socioeconomic groups, showing a significant association with four separate socioeconomic measures in the Kuopio Ischemic Heart Disease Risk Factor Study (Wilson et al. 1993).

Psychological factors: Certain psychological factors are associated with poorer outcomes in patients with established CHD and post-MI patients. Patients with depressive symptoms in the aftermath of MI are at significantly increased risk of mortality and re-infarction. Two meta-analyses of post-MI depression reported that patients diagnosed with depression within 3 months of MI had more than double the risk of all-cause and cardiac mortality than those without depression (van Melle et al. 2004; Meijer et al. 2011). Elevated rates of recurrent cardiac events were also detected. Depression is generally more prevalent among low SES backgrounds (Lorant et al. 2003), and low income has been associated with depression in CHD patients. A cohort study of post-MI patients found that those with depressive symptoms were less educated, had lower income, and were more likely to be unemployed than those without depressive symptoms (Myers et al. 2012). Furthermore, depression was associated with increased cardiac-related hospital admissions during 13 years of follow-up. In a British study of 300 patients with acute coronary syndromes, depression was also found to be more prevalent in lower SES individuals (Steptoe et al. 2011).

Health literacy: Various hypotheses have been suggested to determine why education is so strongly associated with health outcomes. The concept of health literacy posits that individuals with lower ability to read and comprehend medical information are likely to have poorer outcomes. This may be due to lack of awareness of the impact of lifestyle behaviors, nonadherence or incorrect adherence to medication, delayed presentation of symptoms, and poorer management of chronic disease due to poorer understanding of the condition. Scales have been devised to test health literacy, involving both reading and numeracy for health information, and studies have reported increased mortality in individuals with inadequate health literacy (Baker et al. 2007; Bostock and Steptoe 2012). A study of community-dwelling adults with heart failure found that patients with low health literacy were older, were less educated, and had more comorbidities than those who scored high (Morrow et al. 2006).

Lifestyle and environment: Behaviors associated with cardiovascular risk seem to be more prevalent in low SES individuals, whether defined by lower educational attainment or lower income, as evidenced in numerous studies. A study of socioeconomic differentials in CV and cancer mortality in Greece found not only a socioeconomic gradient in CV mortality but also that obesity, poor diet, and physical inactivity were more prevalent in the less educated participants (Naska et al. 2012). In fact, while smoking trends are decreasing in industrialized countries, this reduction is more evident in higher SES populations, while less educated sectors continue to smoke at high levels (Filion et al. 2012). Obesity is also strongly related to socioeconomic status (Wang and Beydoun 2007). Many cross-sectional analyses have found a connection between physical activity and both individual (Barnett et al. 2008) and neighborhood deprivation (Yen and Kaplan 1998; Lee et al. 2007). In order to establish a robust association, longitudinal cohort studies are required. Gerber et al., in a study of post-MI patients followed up for 10–13 years, reported neighborhood deprivation to be strongly associated with uptake of physical activity after MI (Gerber et al. 2011b). Some research has attempted to uncover which neighborhood features may influence exercise patterns and explain the discrepancy between high and low SES areas. Explanatory factors include both physical elements (such as lighting, street layout, and access to facilities) and social characteristics, particularly perceptions of others' behavior and perceived safety of the environment. An American study demonstrated that not only did deprived neighborhoods have fewer sports facilities including parks and gyms compared to high-SES neighborhoods, but they were also less likely to provide free sports facilities (Estabrooks et al. 2003). Further environmental factors, such as air pollution or poor living conditions, may also be involved in overall poorer health outcomes.

SES has also been associated with attendance at cardiac rehabilitation, a crucial component of post-MI recovery, but one for which uptake is low. A study of Danish MI survivors reported nonattendance to be associated with low income (Nielsen et al. 2008). A systematic review found nonattenders to be older and to have lower income or greater deprivation among other factors (Cooper et al. 2002).

9.4 Neighborhood SES: Location, Location, Location

Growing evidence suggests that our health may be influenced not only by our own SES but additionally by the socioeconomic characteristics of the neighborhood in which we live. Neighborhood SES may influence health through availability of health services and other resources, infrastructure, prevailing health attitudes and behaviors, social norms, environmental pollution, and stress (Pickett and Pearl 2001).

Epidemiological evidence has shown an increased risk of developing cardiovascular disease in more deprived areas (Diez-Roux et al. 1997, 2001; Sundquist et al. 2004). For example, in Sundquist et al., a random population sample followed up for incident CHD showed an increased risk associated with decreasing neighborhood income and education (Sundquist et al. 2004). By assessing the proportion of residents in each neighborhood with less than 10 years' education and the proportion in

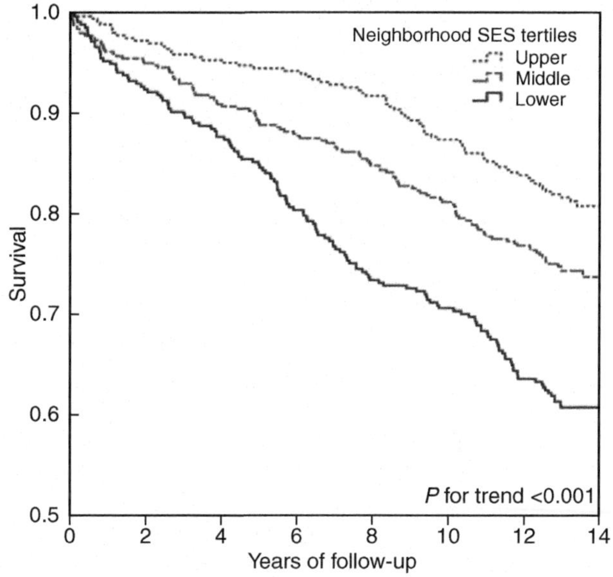

Fig. 9.2 Kaplan–Meier survival curves for neighborhood socioeconomic tertiles (based on 326 deaths occurring in 1,179 incident MI patients) (Reprinted from Gerber et al. (2010), with permission from Wolters Kluwer Health, license no.2874260929086 obtained March 22nd 2012)

the lowest national income quartile, a neighborhood SES score was assigned to each participant, enabling detection of this inverse association, which withstood multivariable adjustment. In addition to increased incidence of CHD, neighborhood deprivation has also been shown to be associated with increased case fatality. In a prospective study of almost four million Swedish men and women, CHD incidence was 1.9 times higher for women and 1.5 times higher for men in the most compared to the least deprived neighborhoods (Winkleby et al. 2007). Case fatality was similarly increased by around 1.6 times. This increased risk occurred regardless of individual SES.

Little data exists on the role of neighborhood SES after heart attack. The Israel Study of First Acute Myocardial Infarction assessed neighborhood SES by geocoding patients' residential addresses based on census data. The authors found neighborhood SES to be strongly related to survival in MI patients, with individuals from the most disadvantaged areas 47 % more likely to die than those in the best neighborhoods, even after controlling for clinical factors and individual SES characteristics (Gerber et al. 2010). There was a clear dose-response pattern between neighborhood SES and post-MI mortality (Fig. 9.2). The relationship with cardiac death was even stronger. Similar results were published from a US study of MI survivors, with a 30 % higher mortality rate in the most deprived neighborhoods compared to the wealthiest and a 47 % higher death rate for areas with the highest proportion of residents with less than high school education (Tonne et al. 2005).

Based on these findings, the Israeli study group went on to investigate the association between neighborhood SES and health behaviors which could potentially mediate the relationship with post-MI outcomes. Indeed, they reported that post-MI patients living in the most deprived areas were less likely to be physically active than their counterparts living in better-off areas (Gerber et al. 2011b) (Fig. 9.3).

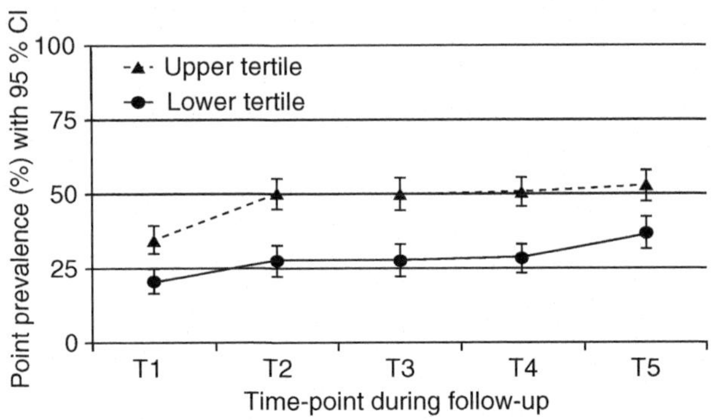

Fig. 9.3 Percentage of post-MI patients regularly engaged in leisure-time physical activity at different time-points by neighborhood SES group. *T1* baseline (pre-MI), *T2* 3–6 months, *T3* 1–2 years, *T4* 5 years, *T5* 10–13 years post-MI (Reprinted from Gerber et al. (2011b), with permission from Elsevier, license no.2874260218863 obtained March 22nd 2012)

Neighborhood SES is also likely to influence access to health services. A study of 50,000 post-MI patients found that not only were those in less deprived areas more likely to undergo angiography within 6 months than their less well-off counterparts, but they also experienced shorter waiting times as well as improved survival (Alter et al. 1999).

9.5 SES Trajectory: Change in SES Across the Life Span

Since SES is so strongly associated with cardiovascular development and progression, it stands to reason that by changing SES – not a trivial matter – cardiovascular risk may be altered. Several studies investigated the impact of social mobility on subsequent risk. The GAZEL French cohort study plotted socioeconomic trajectory by comparing father's occupational grade, own occupation in early adulthood, and midlife occupation. Premature mortality was associated both with persistently low SES/occupational grade (termed "lifelong socioeconomic disadvantage") and with downward mobility (moving from high to low-grade occupation). The strongest associations were for cancer and cardiovascular disease deaths (Melchior et al. 2006). The relationship was partially explained by tobacco and alcohol consumption, BMI, and diet. The authors concluded that while sustained socioeconomic disadvantage predicted premature mortality, occupational trajectory in adulthood played a greater part than socioeconomic circumstances in childhood. A study of Swedish women, also based on occupational class in childhood and adulthood, similarly found adult occupational status to be more strongly associated with CVD mortality than childhood status (Tiikkaja et al. 2009). Women whose occupational class went down (from nonmanual to manual) were twice as likely to die from CV cause compared to those who remained in nonmanual occupations, with a large percentage explained by educational level.

Barker theory lends support to the importance of childhood SES, proposing that early childhood factors influence the development of the heart, going as far back as pregnancy, with reports of low birth weight and small placental size being associated with development of CHD in adulthood (Barker et al. 2010). Childhood BMI measures were also related to the development of heart failure in adulthood. While research in this field is limited, these findings present the possibility of early intervention in childhood and even before birth to reduce levels of CHD in later life.

9.6 Methodological Issues

SES can be measured in a multitude of different ways, from single items to multidimensional indices or aggregate measures. While much earlier research into the relationship between SES and cardiovascular outcomes used single measures such as education or income, later studies noted the importance of multidimensional assessment. SES further operates on various levels including individual, household, and neighborhood levels. In addition to relying on a single SES measure, most health studies do not justify their choice of measure (Braveman et al. 2005). A critical analysis of standard SES measurement approaches proposed the inclusion of multiple SES indicators – including only those which are biologically plausible – the justification of the choice of factors and consideration of unmeasured factors (Braveman et al. 2005).

Education and income: Due to cultural taboos, income is often not directly measured, rather being self-reported as above or below average, thus being largely subjective and susceptible to bias. Education on the other hand is more readily available and people are less reticent about revealing this information, usually coded as years of formal schooling or qualifications achieved. So is it preferable to use one or both of these indices? While education and income are often correlated, it is recommended to include both if possible, since the correlation is not strong enough to risk collinearity, or to justify using one as a proxy for the other (Braveman et al. 2005). Indeed there are numerous examples of successful businesspeople with little in the way of formal education, and vice versa. Furthermore, income differs from wealth, or accumulation of economic resources. A low income may belie a large amount of wealth, thus distorting its effect on health. Further delving into the concept of education, three separate aspects have been recognized: quantity, credentials, and selectivity. However, quantity, or years of schooling, has been shown to have the largest effect on health (Ross and Mirowsky 1999). A workplace study including over 5,000 men aged between 35 and 64 years found both social class and education to be associated with blood pressure and mortality. Occupational social class was a better discriminator of socioeconomic differences in mortality than was education (Davey Smith et al. 1998).

Occupation: In order to determine SES, occupation has traditionally been classified according to skill level and responsibility, for example, manual versus nonmanual or administrative versus managerial, or by job grade as in the Whitehall study. Many SES indices, including Hollingshead's four-factor index, include lists of all possible occupations ranked into social categories, from architects and doctors in the top rank to cleaners and farm laborers in the bottom category. These classifications are subjective and have been widely criticized, being based either on public perception of their prestige or on the educational requirements required to gain access to them (Liberatos et al. 1988). Investigations of other aspects of work, such as job demand and control, attempt to classify occupation in a more meaningful way (Karasek 1990; Cesana et al. 2003).

Composite index: Hollingshead began examining social status in the 1940s and decided that occupation and years of schooling were the key ingredients in the SES equation (Hollingshead 1975). In 1975 she came up with the "four-factor index" comprising education, occupation, sex, and marital status. While criticisms have been directed at some indices since most have not been validated and may not be generalizable to different populations (Braveman et al. 2005), a comparison of different scales found high agreement between the Hollingshead index and two other SES scales (Cirino et al. 2002).

Census data/aggregate measures: Medical records do not usually include measures of SES; therefore, health studies which wish to consider this aspect must rely on indirect means such as census data. However, some doubt has been cast on the validity of these methods. Low correlations were reported between direct individual data and indirect census data of the same patients (Greenwald et al. 1994), and indirect measures may involve substantial error. Furthermore, associations of health outcomes with aggregate SES measures have been shown to be weaker than individual measures (Geronimus and Bound 1998). Other researchers strongly defend the use of aggregate measures, in order to incorporate macrolevel data, stating that group-level variables may be important in explaining the social gradient in health (Diez-Roux 1998). For example, mean neighborhood income is likely to provide a wealth of information about resources and facilities in the area, factors which affect health regardless of individual SES.

9.6.1 Causal Direction

Due to the nature of the socioeconomic field, little data exists to confirm the causal nature of the relationship between SES and CV outcomes. Since we cannot manipulate SES, we can only conduct observational studies, comparing groups based on their naturally occurring socioeconomic characteristics. Without empirical evidence, we cannot make any definitive claims about causal direction. However, some measures of SES do more than others to overcome this limitation, such as education level, which, since it is usually fixed in late childhood or early adulthood, precedes health outcomes, compared to income or occupation which are far more likely to change over the life course. A recent study, using data from the Whitehall II cohort, attempted to shed light on this conundrum, investigating two conflicting theories: the health-related selection hypothesis, which posits that health predicts social mobility, and the social causation hypothesis, which suggests that SES influences health. The report found that poorer childhood health was related to lower occupational grade in later life, but that health in adulthood, represented by cardiometabolic factors (e.g., high blood pressure, obesity, glycemia), did not predict chances of promotion to a higher grade (Elovainio et al. 2011). On the contrary, occupational grade did predict subsequent measures including BMI and glucose levels. The authors concluded that childhood health problems predicted lower SES in adulthood and that in adulthood, SES was associated with an increase in CV risk factors including adiposity and glucose metabolism.

9.6.2 Risk Prediction and Importance of Including SES

Since SES has such an impact on the risk of developing CVD, it stands to reason that it should be included, alongside standard risk factors, when calculating risk. Indeed, with regard to primary prevention, several studies reported that the Framingham score – typically used in estimation of CV risk – underestimates risk in individuals with the lowest SES and overestimates in the highest SES groups (Tunstall-Pedoe and Woodward 2006; Ramsay et al. 2011), a fact which could mislead treatment decisions. When SES was incorporated into a risk prediction model for post-MI patients, substantial gains were achieved in long-term mortality prediction (Molshatzki et al. 2011).

9.7 Summary and Perspectives

Epidemiological studies from diverse countries and eras have consistently demonstrated the social gradient in health, illustrating clear differences in cardiovascular risk for low and high socioeconomic groups, both for developing CVD in the general population and for worsening CVD in those with established disease. Pathophysiological investigations go some way to explaining the mechanisms underlying this relationship. SES significantly affects both the risk of developing CVD and mortality risk in established disease and should therefore be considered when assessing risk. It is clear that social inequalities in health need to be addressed and that many factors are at play in this relationship. Unfortunately, socioeconomic status is one CV risk factor that is difficult or impossible to alter; however, it should be considered in the overall CV risk profile. Furthermore, policy that addresses SES disparities in the population could minimize its consequences. While changing individual SES is beyond the scope of health services, several contributing factors can be improved, including access to healthcare and sports facilities, psychoeducation and improvement of health literacy, and improved awareness of other risk factors such as smoking, poor diet, and physical inactivity, with the aim of redressing the balance and improving outcomes for individuals from disadvantaged backgrounds.

References

Alter DA, Naylor CD, Austin P, Tu JV (1999) Effects of socioeconomic status on access to invasive cardiac procedures and on mortality after acute myocardial infarction. N Engl J Med 341(18):1359–1367

Alter DA, Chong A, Austin PC, Mustard C, Iron K, Williams JI, Morgan CD, Tu JV, Irvine J, Naylor CD, SESAMI Study Group (2006) Socioeconomic status and mortality after acute myocardial infarction. Ann Intern Med 144(2):82–93

Baker DW, Wolf MS, Feinglass J, Thompson JA, Gazmararian JA, Huang J (2007) Health literacy and mortality among elderly persons. Arch Intern Med 167(14):1503–1509

Barker DJ, Gelow J, Thornburg K, Osmond C, Kajantie E, Eriksson JG (2010) The early origins of chronic heart failure: impaired placental growth and initiation of insulin resistance in childhood. Eur J Heart Fail 12(8):819–825

Barnett TA, Gauvin L, Craig CL, Katzmarzyk PT (2008) Distinct trajectories of leisure time physical activity and predictors of trajectory class membership: a 22 year cohort study. Int J Behav Nutr Phys Act 5:57–64

Beebe-Dimmer J, Lynch JW, Turrell G, Lustgarten S, Raghunathan T, Kaplan GA (2004) Childhood and adult socioeconomic conditions and 31-year mortality risk in women. Am J Epidemiol 159(5):481–490

Bostock S, Steptoe A (2012) Association between low functional health literacy and mortality in older adults: longitudinal cohort study. BMJ 344:e1602

Braveman PA, Cubbin C, Egerter S, Chideya S, Marchi KS, Metzler M, Posner S (2005) Socioeconomic status in health research: one size does not fit all. JAMA 294(22):2879–2888

Cesana G, Sega R, Ferrario M, Chiodini P, Corrao G, Mancia G (2003) Job strain and blood pressure in employed men and women: a pooled analysis of four northern Italian population samples. Psychosom Med 65(4):558–563

Cirera L, Tormo MJ, Chirlaque MD, Navarro C (1998) Cardiovascular risk factors and educational attainment in Southern Spain: a study of a random sample of 3091 adults. Eur J Epidemiol 14(8):755–763

Cirino PT, Chin CE, Sevcik RA, Wolf M, Lovett M, Morris RD (2002) Measuring socioeconomic status: reliability and preliminary validity for different approaches. Assessment 9(2):145–155

Cooper AF, Jackson G, Weinman J, Horne R (2002) Factors associated with cardiac rehabilitation attendance: a systematic review of the literature. Clin Rehabil 16(5):541–552

Davey Smith G, Hart C, Hole D, MacKinnon P, Gillis C, Watt G, Blane D, Hawthorne V (1998) Education and occupational social class: which is the more important indicator of mortality risk? J Epidemiol Community Health 52(3):153–160

Diez-Roux AV (1998) Bringing context back into epidemiology: variables and fallacies in multilevel analysis. Am J Public Health 88(2):216–222

Diez-Roux AV, Nieto FJ, Muntaner C, Tyroler HA, Comstock GW, Shahar E, Cooper LS, Watson RL, Szklo M (1997) Neighborhood environments and coronary heart disease: a multilevel analysis. Am J Epidemiol 146(1):48–63

Diez-Roux AV, Merkin SS, Arnett D, Chambless L, Massing M, Nieto FJ, Sorlie P, Szklo M, Tyroler HA, Watson RL (2001) Neighborhood of residence and incidence of coronary heart disease. N Engl J Med 345(2):99–106

Ebrahim S, Montaner D, Lawlor DA (2004) Clustering of risk factors and social class in childhood and adulthood in British women's heart and health study: cross sectional analysis. BMJ 328(7444):861

Elovainio M, Ferrie JE, Singh-Manoux A, Shipley M, Batty GD, Head J, Hamer M, Jokela M, Virtanen M, Brunner E, Marmot MG, Kivimäki M (2011) Socioeconomic differences in cardiometabolic factors: social causation or health-related selection? Evidence from the Whitehall II Cohort Study, 1991-2004. Am J Epidemiol 174(7):779–789

Estabrooks PA, Lee RE, Gyurcsik NC (2003) Resources for physical activity participation: does availability and accessibility differ by neighborhood socioeconomic status? Ann Behav Med 25(2):100–104

Filion KB, Steffen LM, Duval S, Jacobs DR Jr, Blackburn H, Luepker RV (2012) Trends in smoking among adults from 1980 to 2009: the Minnesota heart survey. Am J Public Health 102(4):705–713

Gerber Y, Goldbourt U, Drory Y (2008) Interaction between income and education in predicting long-term survival after acute myocardial infarction. Eur J Cardiovasc Prev Rehabil 15(5):526–532

Gerber Y, Benyamini Y, Goldbourt U, Drory Y, Israel Study Group on First Acute Myocardial Infarction (2010) Neighbourhood socioeconomic context and long-term survival after myocardial infarction. Circulation 121:375–383

Gerber Y, Koren-Morag N, Myers V, Benyamini Y, Goldbourt U, Drory Y, Israel Study Group on First Acute Myocardial Infarction (2011a) Long-term predictors of smoking cessation in a cohort of myocardial infarction survivors: a longitudinal study. Eur J Cardiovasc Prev Rehabil 18(3):533–541

Gerber Y, Myers V, Goldbourt U, Benyamini Y, Drory Y, Israel Study Group on First Acute Myocardial Infarction (2011b) Neighborhood socioeconomic status and leisure-time physical activity after myocardial infarction: a longitudinal study. Am J Prev Med 41(3):266–273

Geronimus AT, Bound J (1998) Use of census-based aggregate variables to proxy for socioeconomic group: evidence from national samples. Am J Epidemiol 148(5):475–486

Greenwald HP, Polissar NL, Borgatta EF, McCorkle R (1994) Detecting survival effects of socioeconomic status: problems in the use of aggregate measures. J Clin Epidemiol 47(8):903–909

Hollingshead AB (1975) Four factor index of social status. Yale University, New Haven

Hotchkiss JW, Davies C, Gray L, Bromley C, Capewell S, Leyland AH (2011) Trends in adult cardiovascular disease risk factors and their socio-economic patterning in the Scottish population 1995-2008: cross-sectional surveys. BMJ Open 1(1):e000176

Jacobsen BK, Thelle DS (1988) Risk factors for coronary heart disease and level of education. The Tromso Heart Study. Am J Epidemiol 127(5):923–932

Johnson JL, Heineman EF, Heiss G, Hames CG, Tyroler HA (1986) Cardiovascular disease risk factors and mortality among black women and white women aged 40-64 years in Evans County, Georgia. Am J Epidemiol 123(2):209–220

Kaplan GA, Keil JE (1993) Socioeconomic factors and cardiovascular disease: a review of the literature. Circulation 88(4 Pt 1):1973–1998

Karasek R (1990) Lower health risk with increased job control among white collar workers. J Organ Behav 11:171–185

Kivimäki M, Virtanen M, Elovainio M, Kouvonen A, Väänänen A, Vahtera J (2006) Work stress in the etiology of coronary heart disease: a meta-analysis. Scand J Work Environ Health 32(6):431–442

Lee RE, Cubbin C, Winkleby M (2007) Contribution of neighbourhood socioeconomic status and physical activity among women. J Epidemiol Community Health 61:882–890

Liberatos P, Link BG, Kelsey JL (1988) The measurement of social class in epidemiology. Epidemiol Rev 10:87–121

Lorant V, Deliège D, Eaton W, Robert A, Philippot P, Ansseau M (2003) Socioeconomic inequalities in depression: a meta-analysis. Am J Epidemiol 157(2):98–112

Lynch JW, Kaplan GA, Cohen RD, Tuomilehto J, Salonen JT (1996) Do cardiovascular risk factors explain the relation between socioeconomic status, risk of all-cause mortality, cardiovascular mortality, and acute myocardial infarction? Am J Epidemiol 144(10):934–942

Macintyre S (1994) Understanding the social patterning of health: the role of the social sciences. J Public Health Med 16(1):53–59

Manrique-Garcia E, Sidorchuk A, Hallqvist J, Moradi T (2011) Socioeconomic position and incidence of acute myocardial infarction: a meta-analysis. J Epidemiol Community Health 65(4):301–309

Marmot MG, Smith GD, Stansfeld S, Patel C, North F, Head J, White I, Brunner E, Feeney A (1991) Health inequalities among British civil servants: the Whitehall II study. Lancet 337(8754):1387–1393

Meijer A, Conradi HJ, Bos EH, Thombs BD, van Melle JP, de Jonge P (2011) Prognostic association of depression following myocardial infarction with mortality and cardiovascular events: a meta-analysis of 25 years of research. Gen Hosp Psychiatry 33(3):203–216

Melchior M, Berkman LF, Kawachi I, Krieger N, Zins M, Bonenfant S, Goldberg M (2006) Lifelong socioeconomic trajectory and premature mortality (35-65 years) in France: findings from the GAZEL Cohort Study. J Epidemiol Community Health 60(11):937–944

Mendis S (2010) The contribution of the Framingham Heart Study to the prevention of cardiovascular disease: a global perspective. Prog Cardiovasc Dis 53(1):10–14

Molshatzki N, Drory Y, Myers V, Goldbourt U, Benyamini Y, Steinberg DM, Gerber Y (2011) Role of socioeconomic status measures in long-term mortality risk prediction after myocardial infarction. Med Care 49(7):673–678

Morrow D, Clark D, Tu W, Wu J, Weiner M, Steinley D, Murray MD (2006) Correlates of health literacy in patients with chronic heart failure. Gerontologist 46(5):669–676

Myers V, Gerber Y, Benyamini Y, Goldbourt U, Drory Y (2012) Post-myocardial infarction depression: increased hospital admissions and reduced adoption of secondary prevention measures – a longitudinal study. J Psychosom Res 72(1):5–10

Naska A, Katsoulis M, Trichopoulos D, Trichopoulou A (2012) The root causes of socioeconomic differentials in cancer and cardiovascular mortality in Greece. Eur J Cancer Prev 21(5):490–496

Nielsen KM, Faergeman O, Foldspang A, Larsen ML (2008) Cardiac rehabilitation: health characteristics and socio-economic status among those who do not attend. Eur J Public Health 18(5):479–483

Nietert PJ, Sutherland SE, Keil JE, Bachman DL (2006) Demographic and biologic influences on survival in whites and blacks: 40 years of follow-up in the Charleston Heart Study. Int J Equity Health 5:8

Pappas G, Queen S, Hadden W, Fisher G (1993) The increasing disparity in mortality between socioeconomic groups in the United States, 1960 and 1986. N Engl J Med 329(2):103–109

Philbin EF, McCullough PA, DiSalvo TG, Dec GW, Jenkins PL, Weaver WD (2000) Socioeconomic status is an important determinant of the use of invasive procedures after acute myocardial infarction in New York State. Circulation 102(19 Suppl 3):III107–III115

Pickett KE, Pearl M (2001) Multilevel analyses of neighbourhood socioeconomic context and health outcomes: a critical review. J Epidemiol Community Health 55(2):111–122

Ramsay SE, Morris RW, Whincup PH, Papacosta AO, Thomas MC, Wannamethee SG (2011) Prediction of coronary heart disease risk by Framingham and SCORE risk assessments varies by socioeconomic position: results from a study in British men. Eur J Cardiovasc Prev Rehabil 18(2):186–193

Rao SV, Kaul P, Newby LK, Lincoff AM, Hochman J, Harrington RA, Mark DB, Peterson ED (2003) Poverty, process of care, and outcome in acute coronary syndromes. J Am Coll Cardiol 41(11):1948–1954

Rao SV, Schulman KA, Curtis LH, Gersh BJ, Jollis JG (2004) Socioeconomic status and outcome following acute myocardial infarction in elderly patients. Arch Intern Med 164(10):1128–1133

Rathore SS, Berger AK, Weinfurt KP, Feinleib M, Oetgen WJ, Gersh BJ, Schulman KA (2000) Race, sex, poverty, and the medical treatment of acute myocardial infarction in the elderly. Circulation 102(6):642–648

Rose G, Marmot MG (1981) Social class and coronary heart disease. Br Heart J 45(1):13–19

Ross CE, Mirowsky J (1999) Refining the association between education and health: the effects of quantity, credential, and selectivity. Demography 36(4):445–460

Salomaa V, Miettinen H, Niemelä M, Ketonen M, Mähönen M, Immonen-Räihä P, Lehto S, Vuorenmaa T, Koskinen S, Palomäki P, Mustaniemi H, Kaarsalo E, Arstila M, Torppa J, Kuulasmaa K, Puska P, Pyörälä K, Tuomilehto J (2001) Relation of socioeconomic position to the case fatality, prognosis and treatment of myocardial infarction events: the FINMONICA MI Register Study. J Epidemiol Community Health 55:475–482

Smith GD, Bartley M, Blane D (1990) The Black report on socioeconomic inequalities in health 10 years on. BMJ 301(6748):373–377

Steptoe A, Willemsen G, Owen N, Flower L, Mohamed-Ali V (2001) Acute mental stress elicits delayed increases in circulating inflammatory cytokine levels. Clin Sci (Lond) 101(2):185–192

Steptoe A, Feldman PJ, Kunz S, Owen N, Willemsen G, Marmot M (2002) Stress responsivity and socioeconomic status: a mechanism for increased cardiovascular disease risk? Eur Heart J 23(22):1757–1763

Steptoe A, Molloy GJ, Messerly-Bürgy N, Wikman A, Randall G, Perkins-Porras L, Kaski JC (2011) Emotional triggering and low socio-economic status as determinants of depression following acute coronary syndrome. Psychol Med 41(9):1857–1866

Sundquist K, Winkleby M, Ahlen H, Johansson SE (2004) Neighborhood socioeconomic environment and incidence of coronary heart disease: a follow-up study of 25,319 women and men in Sweden. Am J Epidemiol 159(7):655–662

Thurston RC, Kubzansky LD, Kawachi I, Berkman LF (2005) Is the association between socioeconomic position and coronary heart disease stronger in women than in men? Am J Epidemiol 162(1):57–65

Tiikkaja S, Hemstrom O, Vagero D (2009) Intergenerational class mobility and cardiovascular mortality among Swedish women: a population-based register study. Soc Sci Med 68(4):733–739

Tofler GH, Muller JE, Stone PH, Davies G, Davis VG, Braunwald E (1993) Comparison of long-term outcome after acute myocardial infarction in patients never graduated from high school with that in more educated patients. Multicenter Investigation of the Limitation of Infarct Size (MILIS). Am J Cardiol 71(12):1031–1035

Tonne C, Schwartz J, Mittleman M, Melly S, Suh H, Goldberg R (2005) Long-term survival after acute myocardial infarction is lower in more deprived neighbourhoods. Circulation 111:3063–3070

Tunstall-Pedoe H, Woodward M (2006) By neglecting deprivation, cardiovascular risk scoring will exacerbate social gradients in disease. Heart 92(3):307–310

van Melle JP, de Jonge P, Spijkerman TA, Tijssen JG, Ormel J, van Veldhuisen DJ, van den Brink RH, van den Berg MP (2004) Prognostic association of depression following myocardial infarction with mortality and cardiovascular events: a meta-analysis. Psychosom Med 66(6):814–822

van Rossum CT, Shipley MJ, van de Mheen H, Grobbee DE, Marmot MG (2000) Employment grade differences in cause specific mortality. A 25 year follow up of civil servants from the first Whitehall study. J Epidemiol Community Health 54(3):178–184

Wang Y, Beydoun MA (2007) The obesity epidemic in the United States – gender, age, socioeconomic, racial/ethnic, and geographic characteristics: a systematic review and meta-regression analysis. Epidemiol Rev 29:6–28

Wilson TW, Kaplan GA, Kauhanen J, Cohen RD, Wu M, Salonen R, Salonen JT (1993) Association between plasma fibrinogen concentration and five socioeconomic indices in the Kuopio Ischemic Heart Disease Risk Factor Study. Am J Epidemiol 137(3):292–300

Winkleby M, Sundquist K, Cubbin C (2007) Inequities in CHD incidence and case fatality by neighborhood deprivation. Am J Prev Med 32(2):97–106

Yen IH, Kaplan GA (1998) Poverty area residence and changes in physical activity level: evidence from the Alameda County Study. Am J Public Health 88:1709–1712

Stress and Psychological Factors

10

Stefan Höfer, Nicole Pfaffenberger, and Martin Kopp

Abstract

This chapter explains how psychosocial factors like stress at work and/or home, depression, and anxiety, as well as personality characteristics contribute to the risk of developing coronary heart disease (CHD). Psychosocial factors seem to affect the heart through neuroendocrine (i.e., hypothalamic-pituitary-adrenal axis) and behavioral (i.e., sedentary lifestyle) pathways. In clinical practice, guidelines for the assessment of stress and psychological factors should be released. In case of elevated risk, multimodal, cognitive-behavioral intervention should be offered. New technologies (i.e., web applications) should be considered for the application of programs aimed at lifestyle change. Furthermore, the development of gender-specific approaches in cognitive-behavioral intervention guidelines should be an important aspect of future research.

Keywords

Stress • Anxiety • Depression • Cognitive behavioral therapy (CBT)

S. Höfer
Department of Medical Psychology,
Innsbruck Medical University, Innsbruck, Austria

N. Pfaffenberger
Department of Psychiatry and Psychotherapy,
Innsbruck Medical University, Innsbruck, Austria

M. Kopp (✉)
Department of Sport Science, University of Innsbruck,
Fürstenweg 185, 6020 Innsbruck, Austria
e-mail: martin.kopp@uibk.ac.at

I. Wakabayashi, K. Groschner (eds.), *Interdisciplinary Concepts in Cardiovascular Health*,
DOI 10.1007/978-3-7091-1334-9_10, © Springer-Verlag Wien 2013

10.1 Introduction

The linkage of psychosocial factors and coronary heart disease (CHD) has a long tradition. One of the earliest connections can be found in pre-Christian times, i.e., the Egyptians did not differ between heart and soul (von Fischer-Elfert 2005).

The INTERHEART study identified nine universal and modifiable risk factors contributing to the onset of myocardial infarction (Yusuf et al. 2004), one of which is described as psychosocial. This factor consists of two general aspects: stress and depression (Rosengren et al. 2004). Both contribute about 30 % of the risk for the onset of a myocardial infarction. Besides stress and depression, personality traits, anxiety, and social isolation have also been identified as independent risk factors (Rozanski et al. 1999).

Pathways by which psychosocial factors can be linked to the incidence and prognosis of CHD are both behavioral and biological (Hemingway et al. 2001; Kuper et al. 2002). Behavioral pathways focus on behavior impacting health, i.e., physical inactivity, smoking, poor diet, or barriers to adherence. This in turn can lead to the development of other cardiovascular risk factors, such as the metabolic syndrome. Biological pathways in relation to psychosocial risk factors are discussed on the basis of pathophysiological changes, i.e., autonomic disturbances, hormonal imbalance, inflammation, or endothelial dysfunction (Rozanski et al. 2005). According to the biopsychosocial model (Engel 1997), social pathways may also be considered (i.e., lack of resources, lack of social support) contributing to the onset or influencing the prognosis of CHD.

Psychosocial factors have been described not only as risk factors but also as protective factors. One of the more recent studies found that happiness was a protective factor preventing the onset of heart disease over a 10-year period in 1,700 healthy adults (Davidson et al. 2010).

10.2 Stress

While the term stress is widely used, in research as well as among the public, it is a poorly defined construct with many different meanings, making objective definition difficult (Rugulies et al. 2005). Referring to a meta-analytic study by Segerstrom and Miller (2004) which aimed to achieve a common definition, stressors were simply defined as circumstances that most people would appraise to be stressful. Taking into account that psychosocial stress has multiple etiologies including behavioral causes, acute events, or chronic stressors, Elliot and Eisdorfer (1982) adopted a more discerning categorization by distinguishing different stressors according to "duration" and "course."

Acute time-limited stressors involve laboratory challenges such as public speaking or mental arithmetic. Brief naturalistic stressors, such as academic examinations, involve a person confronting a real-life short-term challenge. Stressful event sequences focus on a special (life) event – such as the loss of a spouse or natural disaster. *Chronic stressors* force a person to restructure his or her identity or social roles. Another feature of chronic stressors is their stability. The person either does not know whether or when the challenge will end or cannot even be sure that it will ever end (e.g., suffering a traumatic injury that leads to physical disability or being a refugee forced out of one's native country by war). *Distant stressors* (e.g., having been sexually assaulted as a child, having been a prisoner of war) are traumatic experiences that occurred in the distant past yet having the potential to continue modifying immune system function because of their long-lasting cognitive and emotional sequelae (Elliot and Eisdorfer 1982).

Despite the absence of a common definition, several constructs within the broad conceptual framework of stress such as occupational stress, adverse life events, daily hassles, family relationships, and financial problems are increasingly regarded as being causally related to heart disease (Hemingway and Marmot 1999; Iso et al. 2002; Kivimäki et al. 2002; Stansfeld and Marmot 2002; Yusuf et al. 2004; French et al. 2005).

10.2.1 Stress at Work/Stress at Home

Most research activities have been conducted in the field of chronic stressors, particularly investigating the effect of occupational stress, while stress outside the workplace has received less attention to date.

With respect to job strain, two models serve as a basis for investigating different potentially stressful aspects of work. The Demand-Control-Support Model postulates that two keys of elements of work experience are level of demand and lack of control over how work is carried out and how skills are developed and utilized. Thus, occupational stress is caused by an imbalance between demands on a worker and the worker's ability to modify those demands. High strain or a stressful situation at work emerges when high demands such as time pressure are coupled with low job control such as low decision latitude (Karasek and Theorell 1990).

The Effort-Reward Imbalance Model by Siegrist (1996) focuses the imbalance with regard to effort and reward and postulates that stress responses arise when the effort resulting from the demands of the job and the personal commitment put into the work are not matched by rewards such as money, social esteem, job security, or career opportunities. In addition, the model includes a factor called work-related overcommitment, which is understood as an aspect of personality that makes some individuals more vulnerable to experience psychosocial stress reactions (Siegrist et al. 2004).

Both of these conceptualizations of occupational stress have been shown to be related to heart disease in cross-sectional and prospective studies (Steenland et al. 2000), in populations who were disease-free at baseline (Kivimäki et al. 2002; Stansfeld et al. 2002) and after controlling for other risk factors (Kivimäki et al. 2002; Rosengren et al. 2004).

Most of these studies focused on men with limited studies reporting on the relationship between work-related stress and heart disease in women. Women differ from men with heart disease in considerable ways that potentially interact with further psychosocial risk factors indicating that traditional models of job strain may be less important for women than for men (Low et al. 2010). The Stockholm Female Coronary Risk Study reported that stressful aspects of relationships almost triple the risk of recurrent cardiac events in women and demonstrated that the combination of occupational strain and marital stress is the strongest predictor for recurrence of the disease (Orth-Gomer and Leineweber 2005). Nevertheless, even though, for example, the hours spent caring for other family members are associated with heart disease (Lee et al. 2003a, b), the presence of social relationships and active involvement may also be a protective factor (Kim and McKenry 2002; Barefoot et al. 2005; Eaker et al. 2007). Whereas merely the presence of a partner seems to be protective for men, for women quality of partnership such as positive reciprocal social relationships seems to be more important (Gallo et al. 2003). Regarding the development and progression of heart disease, psychological stress resulting from interpersonal interactions seems to be a more substantial risk factor for women than for men, and furthermore, in women it seems to be more important than work-related stress (Low et al. 2010).

Acute stress or daily hassles can also be a trigger for acute onset of myocardial infarction as well as adverse disease progression. Studies looking into the frequency of myocardial infarction among viewers during major soccer games revealed that games with high emotional connotation are more likely to trigger myocardial infarction (Chi and Kloner 2003; Wilbert-Lampen et al. 2008). Furthermore, regarding lifestyle parameters, Twisk et al. (1999) found in a longitudinal study that a reduction in daily hassles over time increased the likelihood for daily physical activity and reduced smoking behavior, although not indicating that the presence of daily hassles leads to negative health behavior per se. In this context, Faulk and Batholomew (2012) focused on the moderating effect of exercise on cardiovascular reactivity following single fat feedings and a stress cue. Despite some limitations, results show a reduction in fat intake and cardiovascular reactivity after engaging in high-intensity exercise. Furthermore, data suggest a potential effect of postprandial high-intensity exercise to buffer the effect of fat consumption on cardiovascular reactivity. Adding data of reduced food craving being confronted with stress after single bouts of exercise (Taylor and Oliver 2009), the potential of physical activity in coping with acute stress and daily hassles seems to be an important area for future research.

10.2.2 Stressful Live Events

While less is known about the relationship between life events and the onset of heart disease, chronic stressors as well as acute stressful life events such as bereavement or retirement are associated with heart disease, especially with respect to acute coronary syndromes (Magni et al. 1983; Haldar et al. 2005). Rafanelli et al. (2005) investigated the presence of stressful life events in the year preceding heart disease and found that – with the exception of marriage – all other life events (socially desirable or undesirable, controlled versus uncontrolled events) were found more frequently in the group of patients compared to healthy controls. Furthermore, in line with the previous reports (Magni et al. 1983), the quality of life events was shown to be crucial for the perceived risk. Furthermore, it may be suggested that mood symptomatology acts as an intermediary factor in the relationship between acute stressful life events and risk of cardiac disease (Rafanelli et al. 2005).

10.2.3 Interventions to Reduce Stress

Psychological treatment strategies to reduce stress are usually based on cognitive behavioral therapy (CBT) approaches. Next to well-established programs for life-style changes in cardiovascular risk groups (Ornish 1990; Ornish et al. 1998), a recent clinical trial has shown a significant decrease in the risk of recurrent CHD and recurrent acute myocardial infarction in patients with CHD treated with CBT (Gulliksson et al. 2011). Intervention strategies used in this study were education, self-monitoring, skills training, cognitive restructuring, spiritual development, stress management, coping with stress, and reducing experience of daily stress, time urgency, and hostility, including brief relaxation and reflections of changed behavior. The program was performed in twenty 2h sessions over 1 year with five to nine patients per group, with separate groups for men and women (Gulliksson et al. 2011). This might be considered as a model for future psychological interventions in CHD patients and could be extended by technical innovations like smartphone apps (Bostock and Steptoe 2012).

10.3 Psychological Factors

Perceived stress is frequently moderated by psychological variables such as depression, anxiety, and personality factors, which have been shown to be related to the onset and prognosis of CHD.

10.3.1 Depression

Reviews and meta-analyses have examined the impact of depression on cardiovascular morbidity and mortality (Rugulies 2002; Wulsin and Singal 2003; Barth et al. 2004; Van der Kooy et al. 2007). Despite differences in samples, duration of follow-up, and methods of assessment of depression and depressive symptoms, these studies have demonstrated relatively consistent results.

Depression is at least three times more common in patients after acute myocardial infarction than in the general population. About 15–25 % of patients with MI meet the "Diagnostic and Statistical Manual of Mental Disorders" criteria for major depression, and an even higher portion (up to 50 %) shows elevated levels of depressive symptoms (Rugulies 2002; Wulsin and Singal 2003).

Postmyocardial infarction depression is associated with an increased risk for all-cause mortality [*OR* 2.25], cardiac mortality [2.71], and cardiac events [1.59]. This risk is stable over a period of 25 years (Meijer et al. 2011).

Depression also seems to be a risk factor for the development of CHD in healthy individuals. The Precursors Study (1,190 male medical students followed up for 40 years) showed that students with clinical depression at baseline were at twice the risk for subsequent CHD and MI (Ford et al. 1998). An increased risk for CHD was found not only in individuals with major depression but also in those with depressive symptoms. The risk between depression and cardiac risk occurs along a continuum of depression severity suggesting a dose-response relationship (Rugulies 2002).

Although depressed mood increases the risk of a wide range of CHD, diagnosed major depressive disorder is the most important risk factor for the onset of a myocardial infarction [*OR* 1.6] (Van der Kooy et al. 2007).

Psychological and psychopharmacological interventions (incl. SSRIs) may have a small, but clinically meaningful effect on depression outcomes in CHD patients. So far no beneficial evidence for psychological and psychopharmacological interventions on the reduction on mortality rates and cardiac events was found (Baumeister et al. 2011).

In summary, there is considerable evidence suggesting that depressive mood and clinically diagnosed depression are a risk factor for the development of CHD in healthy individuals and may lead to an increased risk of cardiac mortality in those with CHD. However, regardless whether depression affects cardiac outcomes directly or indirectly, depression is a clinical disorder by itself and must be treated accordingly to best practice.

10.3.2 Anxiety

In the 1970s some studies reported a positive association between anxiety and the onset of CHD while others did not (none of these studies were controlled for known cardiovascular risk factors). In the 1980s and 1990s, large prospective studies found a positive association between anxiety and the incidence of CHD, even when controlling for cardiovascular risk factors.

The most recent meta-analysis (including data from 1980 to 2009) by Roest et al. (2010) on the association of anxiety with the incidence of CHD combined data from 20 prospective studies, including approximately 250,000 initially healthy individuals having at baseline at least one self-report or interview-based assessment of anxiety symptoms or anxiety disorder. The mean follow-up period was years. End points included cardiac mortality or MI. Anxious people had an approximately 25 % greater risk of CHD and an almost 50 % higher risk of cardiac death than non-anxious ones. The most important finding of the meta-analysis was that anxiety was associated with the development of incident CHD in initially healthy persons (Roest et al. 2010). Since 2009 another large study including 49,000 Swedish people found that anxiety was a predictor for the incidence of CHD over a period of 37 years (Janszky et al. 2010); specifically, those with anxiety disorders were twice as likely to develop CHD or acute MI, even after adjustment for potential confounders.

According to findings from the Heart and Soul Study, CHD patients with generalized anxiety disorder were at higher risk for cardiovascular events (stroke, MI, and death) than patients with CHD only. Following adjustment for a variety of potentially confounding variables, including demographic characteristics, comorbid conditions (major depression), CHD severity, medication use, and behavioral and biological mediators, generalized anxiety disorder remained associated with a 74 % higher relative risk of cardiovascular events after an average follow-up of 5.6 years (Martens et al. 2010).

Psychological interventions, especially cognitive behavioral therapy, are well documented to have an impact on anxiety (Deacon and Abramowitz 2004) and are considered a reliable first-line therapy (Hofmann et al. 2012); the specific impact on cardiovascular outcomes is less well documented. While benzodiazepines are a pharmacological standard choice of treatment for free-floating anxiety (defined as persistent or recurrent worry or nervousness not meeting criteria for a formal anxiety disorder and common after an acute cardiac event), SSRIs (discussed above) are used as the treatment of choice for formal anxiety disorders (Sowden and Huffman 2009) with no sufficient evidence in impacting cardiovascular outcome per se.

Depression and anxiety are commonly treated with psychotropic drugs. Research has shown that obesity is a potential side effect of psychopharmacological treatment (Fleischhacker et al. 2008) and people with severe mental illness show a reduced life expectancy due to CHD (Colton and Manderscheid 2006). Therefore, research and interventions must also focus on health behavior in patients treated with psychotropic drugs (Kopp et al. 2011).

10.3.3 Personality

Over the years personality has often been discussed as a risk factor for developing heart disease. One of the earlier concepts was the type A pattern, which in 1960 was declared to be an independent risk factor (Friedman and Rosenman 1960). Type A refers to a number of personality trait characteristics, including rushed, ambitious, and competitive behavior, impatience, hostility, and intolerance. Early positive findings (Friedman and Rosenman 1971) were later displaced by the type D personality (Denollet et al. 1995). Type D is described as a combination of negative inhibition and social isolation, which can be assessed by the type D scale (Kupper et al. 2011). While findings from the research group of Denollet demonstrate that people showing type D characteristics have higher mortality rates, a first meta-analysis claims an overestimation of the prognostic relevance of type D personality (Grande et al. 2012).

10.3.4 Pathways Linking Stress, Depression, Anxiety, and CHD

There have been two types of mechanism proposed to explain the link between stress, anxiety, depression, and CHD – direct pathophysiological and indirect behavioral mechanisms – and are discussed in detail in Rozanski et al. (2005) (Table 10.1).

Table 10.1 Pathways linking stress, depression, anxiety, and CHD

Behavioral mechanisms	Pathophysiological mechanisms
Unhealthy lifestyle (unhealthy diet, smoking, drinking more alcohol, being sedentary)	Hypothalamic-pituitary-adrenal axis and autonomic nervous system dysfunction: increased sympathetic activity and/or reduced vagal activity, hypercortisolemia and elevated levels of corticotropin-releasing factor (increased free fatty acids), and reduced heart rate variability (a powerful predictor of sudden cardiac death)
Reduced chances of successful modifications of risk factors (such as smoking cessation)	Reduced level of serotonin leading to platelet dysfunction and hyperactivity and hypercoagulability
Decreased adherence to medications and increased risk of noncompliance with medical treatment regimens	Ω-3 fatty acid deficiency
Decreased participation in cardiac rehabilitation	Elevated homocysteine levels Endothelial dysfunction (measured by brachial artery flow-mediated vasodilation) Elevated levels of inflammatory biomarkers (C-reactive protein, interleukin-6, intercellular adhesion molecule-1)

10.4 Summary and Perspectives

Data published within the last decade have revealed important influences of stress and psychological factors on the development and course of CHD, ranging from experimental data demonstrating cardiovascular consequences of stress (Steptoe and Kivimaki 2012) to life stress as a noteworthy risk factor for myocardial infarction (Yusuf et al. 2004).

Despite the growing body of evidence, stress and psychological factors are not included in a recent review focusing on risk factors for cardiovascular incidence (Berry et al. 2012). This absence may be in part explained by the lack of a clear definition of stress and clear guidelines on how to assess and record psychological factors in the general population as well as in patients with CHD.

Therefore, international boards working on this important topic should join and publish guidelines for the assessment of stress and psychological factors (i.e., considering the inclusion of technical developments like remote online assessments). This might help to gather important data in routine care as well as to collect important information about these risk factors.

From a behavioral approach, research focusing on stress and psychological factors should include moderating variables such as diet and physical activity. Different long-term patterns and acute effects of these behaviors seem to be connected to cardiovascular reactivity which might be a key determinant of cardiovascular disease.

From a psychological point of view, it is important to note that providing information alone about healthy lifestyle or strategies to reduce stress does not lead to desired changes in behavior (Soureti et al. 2012). Health psychological intervention strategies based on sound theories like self-determination theory or health action process approach including individual goal setting and implementation strategies seem to be effective in changing behavior (Platter et al. 2012) and might help to prevent future cardiac incidents (Ornish 1990; Ornish et al. 1998). Limiting aspects are that CBT programs are time consuming and rarely available outside university cities. Research has shown important developments in web-based health-behavior change programs (Civljak et al. 2010), and trials bringing more knowledge about web-based CBT approaches in cardiovascular prevention and treatment of CHD might help to implement best practice approaches within routine care. Lieber et al. (2012), for example, report on encouraging results regarding the effects of a 12-week, web-based, and self-directed lifestyle intervention for women.

Research has shown considerable gender-specific differences indicating that women are more sensitive with regard to psychosocial risk factors. Focus on research and development of gender-specific prevention strategies and treatment options is becoming more important (Hallman et al. 2001).

Overall, psychosocial factors have been shown to be primary risk factors. A variety of intervention strategies exist to improve and foster a healthy lifestyle to tackle CHD as the lifestyle disease of the twenty-first century.

- The convincing evidence for the influence of stress and psychological factors on coronary heart disease should lead to general accepted guidelines for assessment and treatment.
- New technologies like web-based interventions should be assessed with regard to feasibility and success rates.
- Research may focus on the development of adapted programs for special risk groups (i.e., severe mental illness) and gender-specific approaches in cognitive-behavioral intervention.

References

Barefoot JC, Grønbaek M, Jensen G, Schnohr P, Prescott E (2005) Social network diversity and risks of ischemic heart disease and total mortality: findings from the Copenhagen City Heart Study. Am J Epidemiol 161(10):960–967

Barth J, Schumacher M, Herrmann-Lingen C (2004) Depression as a risk factor for mortality in patients with coronary heart disease: a meta-analysis. Psychosom Med 66(6):802–813

Baumeister H, Hutter N, Bengel J (2011) Psychological and pharmacological interventions for depression in patients with coronary artery disease. Cochrane Database Syst Rev (9):CD008012

Berry JD, Dyer A, Cai X, Garside DB, Ning H, Thomas A, Greenland P, Van Horn L, Tracy RP, Lloyd-Jones DM (2012) Lifetime risks of cardiovascular disease. N Engl J Med 366(4):321–329

Bostock S, Steptoe A (2012) Work stress? There's an app for that. A randomized controlled trial of a mindfulness meditation smartphone app. UK Society of Behavioural Medicine. 8th annual scientific meeting, Manchester, 10 Dec 2012

Chi JS, Kloner RA (2003) Stress and myocardial infarction. Heart 89(5):475–476

Civljak M, Sheikh A, Stead LF, Car J (2010) Internet-based interventions for smoking cessation. Cochrane Database Syst Rev (9):CD007078

Colton CW, Manderscheid RW (2006) Congruencies in increased mortality rates, years of potential life lost, and causes of death among public mental health clients in eight states. Prev Chronic Dis 3(2):A42

Davidson KW, Mostofsky E, Whang W (2010) Don't worry, be happy: positive affect and reduced 10-year incident coronary heart disease: the Canadian Nova Scotia Health Survey. Eur Heart J 31(9):1065–1070

Deacon BJ, Abramowitz JS (2004) Cognitive and behavioral treatments for anxiety disorders: a review of meta-analytic finding. J Clin Psychol 60(4):429–441

Denollet J, Sys SU, Brutsaert DL (1995) Personality and mortality after myocardial infarction. Psychosom Med 57(6):582–591

Eaker ED, Sullivan LM, Kelly-Hayes M, D'Agostino RB Sr, Benjamin EJ (2007) Marital status, marital strain, and risk of coronary heart disease or total mortality: the Framingham Offspring Study. Psychosom Med 69(6):509–513

Elliot P, Eisdorfer C (1982) Stress and human health: an analyses and implications of research. A study by the Institute of Medicine National Academy of Sciences. Springer, New York

Engel GL (1997) From biomedical to biopsychosocial. Being scientific in the human domain. Psychosomatics 38(6):521–528

Faulk KE, Bartholomew JC (2012) The moderating effect of physical activity on cardiovascular reactivity following single fat feedings. Psychophysiology 49(1):145–149

Fleischhacker WW, Cetkovich-Bakmas M, De Hert M, Hennekens CH, Lambert M, Leucht S, Maj M, McIntyre RS, Naber D, Newcomer JW, Olfson M, Osby U, Sartorius N, Lieberman JA (2008) Comorbid somatic illnesses in patients with severe mental disorders: clinical, policy, and research challenges. J Clin Psychiatry 69(4):514–519

Ford DE, Mead LA, Chang PP, Cooper-Patrick L, Wang NY, Klag MJ (1998) Depression is a risk factor for coronary artery disease in men: the precursors study. Arch Intern Med 158(13):1422–1426

French D, Maissi E, Marteau TM (2005) The purpose of attributing cause: beliefs about the causes of myocardial infarction. Soc Sci Med 60:1411–1421

Friedman M, Rosenman RH (1960) Overt behavior pattern in coronary disease. Detection of overt behavior pattern A in patients with coronary disease by a new psychophysiological procedure. JAMA 173:1320–1325

Friedman M, Rosenman RH (1971) Type A behavior pattern: its association with coronary heart disease. Ann Clin Res 3(6):300–312

Gallo LC, Troxel WM, Kuller LH, Sutton-Tyrrell K, Edmundowicz D, Matthews KA (2003) Marital status, marital quality, and atherosclerotic burden in postmenopausal women. Psychosom Med 65(6):952–962

Grande G, Romppel M, Barth J (2012) Association between type D personality and prognosis in patients with cardiovascular diseases: a systematic review and meta-analysis. Ann Behav Med 43(3):299–310

Gulliksson M, Burell G, Vessby B, Lundin L, Toss H, Svärdsudd K (2011) Randomized controlled trial of cognitive behavioral therapy vs standard treatment to prevent recurrent cardiovascular events in patients with coronary heart disease: Secondary Prevention in Uppsala Primary Health Care project (SUPRIM). Arch Int Med 171(2):134–140

Haldar A, Saha S, Mandal S, Haldar S, Mundle M, Mitra SP (2005) Life events as risk factors for myocardial infarction: a pilot case-control study in Kolkata, India. J Health Popul Nutr 23(2):131–136

Hallman T, Burell G, Setterlind S, Odén A, Lisspers J (2001) Psychosocial risk factors for coronary heart disease, their importance compared with other risk factors and gender differences in sensitivity. J Cardiovasc Risk 8(1):39–49

Hemingway H, Marmot M (1999) Evidence based cardiology: psychosocial factors in the aetiology and prognosis of coronary heart disease. Systematic review of prospective cohort studies. BMJ 318(7196):1460–1467

Hemingway H, Whitty CJ, Shipley M, Stansfeld MS, Brunner E, Fuhrer R, Marmot M (2001) Psychosocial risk factors for coronary disease in White, South Asian and Afro-Caribbean civil servants: the Whitehall II study. Ethn Dis 11(3):391–400

Hofmann SG, Asnaani A, Vonk IJ, Sawyer AT, Fang A (2012) The efficacy of cognitive behavioral therapy: a review of meta-analyses. Cogn Ther Res 36(5):427–440

Iso H, Date C, Yamamoto A, Toyoshima H, Tanabe N, Kikuchi S, Kondo T, Watanabe Y, Wada Y, Ishibashi T, Suzuki H, Koizumi A, Inaba Y, Tamakoshi A, Ohno Y (2002) Perceived mental stress and mortality from cardiovascular disease among Japanese men and women: the Japan Collaborative Cohort Study for Evaluation of Cancer Risk Sponsored by Monbusho (JACC Study). Circulation 106(10):1229–1236

Janszky I, Ahnve S, Lundberg I, Hemmingsson T (2010) Early-onset depression, anxiety, and risk of subsequent coronary heart disease: 37-year follow-up of 49,321 young Swedish men. J Am Coll Cardiol 56(1):31–37

Karasek R, Theorell T (1990) Healthy work: stress, productivity and the reconstruction of working life. Basic Books, New York

Kim H, McKenry P (2002) The relationship between marriage and psychosocial well-being. A longitudinal analysis. J Family Issues 23:885–911

Kivimäki M, Leino-Arjas P, Luukkonen R, Riihimäki H, Vahtera J, Kirjonen J (2002) Work stress and risk of cardiovascular mortality: prospective cohort study of industrial employees. BMJ 325(7369):857

Kopp M, Fleischhacker WW, Stürz K, Ruedl G, Kumnig M, Rumpold G (2011) Poor health behaviour and reduced quality of life of people treated with psychotropic drugs. Hum Psychopharmacol 26(2):161–167

Kuper H, Marmot M, Hemingway H (2002) Systematic review of prospective cohort studies of psychosocial factors in the etiology and prognosis of coronary heart disease. Semin Vasc Med 2(3):267–314

Kupper N, Pedersen SS, Höfer S, Saner H, Oldridge N, Denollet J (2013) Cross-cultural analysis of type D (distressed) personality in 6222 patients with ischemic heart disease: a study from the International HeartQoL Project. Int J Cardiol 166(2):327–333. http://dx.doi.org/10.1016/j.ijcard.2011.10.084

Lee S, Colditz G, Berkman L, Kawachi I (2003a) Caregiving to children and grandchildren and risk of coronary heart disease in women. Am J Public Health 93(11):1939–1944

Lee S, Colditz GA, Berkman LF, Kawachi I (2003b) Caregiving and risk of coronary heart disease in U.S. women: a prospective study. Am J Prev Med 24(2):113–119

Lieber SB, Redberg RF, Blumenthal RS, Gandhi A, Robb KJ, Mora S (2012) A national interactive web-based physical activity intervention in women, evaluation of the American Heart Association choose to move program 2006-2007. Am J Cardiol 109(12):1754–1760

Low CA, Thurston RC, Matthews KA (2010) Psychosocial factors in the development of heart disease in women: current research and future directions. Psychosom Med 72(9):842–854

Magni G, Corfini A, Berto F, Rizzardo R, Bombardelli S, Miraglia G (1983) Life events and myocardial infarction. Aust N Z J Med 13(3):257–260

Martens EJ, de Jonge P, Na B, Cohen BE, Lett H, Whooley MA (2010) Scared to death? Generalized anxiety disorder and cardiovascular events in patients with stable coronary heart disease: the Heart and Soul Study. Arch Gen Psychiatry 67(7):750–758

Meijer A, Conradi HJ, Bos EH, Thombs BD, van Melle JP, de Jonge P (2011) Prognostic association of depression following myocardial infarction with mortality and cardiovascular events: a meta-analysis of 25 years of research. Gen Hosp Psychiatry 33(3):203–216

Ornish D (1990) Dr. Dean Ornish's program for reversing heart disease. Ballantine Books, New York

Ornish D, Scherwitz LW, Billings JH, Brown SE, Gould KL, Merritt TA, Sparler S, Armstrong WT, Ports TA, Kirkeeide RL, Hogeboom C, Brand RJ (1998) Intensive lifestyle changes for reversal of coronary heart disease. JAMA 280(23):2001–2007

Orth-Gomer K, Leineweber C (2005) Multiple stressors and coronary disease in women. The Stockholm Female Coronary Risk Study. Biol Psychol 69(1):57–66

Platter M, Hölzl C, Hofer M, Huber A, Höfer S (2012) Impulse for heart-healthy diet: a short health psychological intervention at the cardiology ward. J Clin Exp Cardiol. 2012, S:5. http://dx.doi.org/10.4172/2155-9880.S5-007

Rafanelli C, Roncuzzi R, Milaneschi Y, Tomba E, Colistro MC, Pancaldi LG, Di Pasquale G (2005) Stressful life events, depression and demoralization as risk factors for acute coronary heart disease. Psychother Psychosom 74(3):179–184

Roest AM, Martens EJ, de Jonge P, Denollet J (2010) Anxiety and risk of incident coronary heart disease: a meta-analysis. J Am Coll Cardiol 56(1):38–46

Rosengren A, Hawken S, Ounpuu S, Sliwa K, Zubaid M, Almahmeed WA, Blackett KN, Sitthi-amorn C, Sato H, Yusuf S, INTERHEART Investigators (2004) Association of psychosocial risk factors with risk of acute myocardial infarction in 11119 cases and 13648 controls from 52 countries (the INTERHEART study): case-control study. Lancet 364(9438):953–962

Rozanski A, Blumenthal JA, Kaplan J (1999) Impact of psychological factors on the pathogenesis of cardiovascular disease and implications for therapy. Circulation 99(16):2192–2217

Rozanski A, Blumenthal JA, Davidson KW, Saab PG, Kubzansky L (2005) The epidemiology, pathophysiology, and management of psychosocial risk factors in cardiac practice: the emerging field of behavioral cardiology. J Am Coll Cardiol 45(5):637–651

Rugulies R (2002) Depression as a predictor for coronary heart disease. A review and meta-analysis. Am J Prev Med 23(1):51–61

Rugulies R, Aust B, Syme SL (2005) Epidemiology of health and illness: a socio-psycho-physiological perspective. In: Sutton S, Baum A, Johnston M (eds) The SAGE handbook of health psychology. Thousand Oaks, London

Segerstrom SC, Miller GE (2004) Psychological stress and the human immune system: a meta-analytic study of 30 years of inquiry. Psychol Bull 130(4):601–630

Siegrist J (1996) Adverse health effects of high-effort/low-reward conditions. J Occup Health Psychol 1(1):27–41

Siegrist J, Starke D, Chandola T, Godin I, Marmot M, Niedhammer I, Peter R (2004) The measurement of effort-reward imbalance at work: European comparisons. Soc Sci Med 58:1483–1499

Soureti A, Hurling R, van Mechelen W, Cobain M, ChinAPaw M (2012) Moderators of the mediated effect of intentions, planning, and saturated-fat intake in obese individuals. Health Psychol 31(3):371–379

Sowden GL, Huffman JC (2009) The impact of mental illness on cardiac outcomes: a review for the cardiologist. Int J Cardiol 132(1):30–37

Stansfeld AG, Marmot M (eds) (2002) Stress and the heart. Psychosocial pathways to coronary heart disease. BMJ Books, London

Stansfeld SA, Fuhrer R, Shipley MJ, Marmot MG (2002) Psychological distress as a risk factor for coronary heart disease in the Whitehall II Study. Int J Epidemiol 31(1):248–255

Steenland K, Fine L, Belkić K, Landsbergis P, Schnall P, Baker D, Theorell T, Siegrist J, Peter R, Karasek R, Marmot M, Brisson C, Tüchsen F (2000) Research findings linking workplace factors to CVD outcomes. Occup Med 15(1):7–68

Steptoe A, Kivimaki M (2012) Stress and cardiovascular disease. Nat Rev Cardiol 9(6):360–370

Taylor AH, Oliver AJ (2009) Acute effects of brisk walking on urges to eat chocolate, affect, and responses to a stressor and chocolate cue. An experimental study. Appetite 52(1):155–160

Twisk JW, Snel J, Kemper HC, van Mechelen W (1999) Changes in daily hassles and life events and the relationship with coronary heart disease risk factors: a 2-year longitudinal study in 27-29-year-old males and females. J Psychosom Res 46(3):229–240

Van der Kooy K, van Hout H, Marwijk H, Marten H, Stehouwer C, Beekman A (2007) Depression and the risk for cardiovascular diseases: systematic review and meta analysis. Int J Geriatr Psychiatry 22(7):613–626

von Fischer-Elfert HW (2005) Papyrus Ebers und die antike Heilkunde, The Harrassowitz Publishing House, Wiesbaden

Wilbert-Lampen U, Leistner D, Greven S, Pohl T, Sper S, Völker C, Güthlin D, Plasse A, Knez A, Küchenhoff H, Steinbeck G (2008) Cardiovascular events during world cup soccer. N Engl J Med 358(5):475–483

Wulsin LR, Singal BM (2003) Do depressive symptoms increase the risk for the onset of coronary disease? A systematic quantitative review. Psychosom Med 65(2):201–210

Yusuf S, Hawken S, Ounpuu S, Dans T, Avezum A, Lanas F, McQueen M, Budaj A, Pais P, Varigos J, Lisheng L, INTERHEART Study Investigators (2004) Effect of potentially modifiable risk factors associated with myocardial infarction in 52 countries (the INTERHEART study): case-control study. Lancet 364(9438):937–952

Index

I. Wakabayashi, K. Groschner (eds.), *Interdisciplinary Concepts in Cardiovascular Health*,
DOI 10.1007/978-3-7091-1334-9, © Springer-Verlag Wien 2013